WS. James

d 32.

E4L

AF791217

SEGMENTED NEGATIVE STRAND VIRUSES

Academic Press Rapid Manuscript Reproduction

Proceedings of a Symposium on the
Molecular Biology of Negative Strand Viruses
Sponsored by the University of Alabama in Birmingham
Held in Hilton Head, South Carolina
September 11–18, 1983

SEGMENTED NEGATIVE STRAND VIRUSES

Arenaviruses, Bunyaviruses, and Orthomyxoviruses

Edited by

Richard W. Compans
David H. L. Bishop
Department of Microbiology
University of Alabama in Birmingham
Birmingham, Alabama

1984

ACADEMIC PRESS, INC.
(Harcourt Brace Jovanovich, Publishers)
Orlando San Diego San Francisco New York London
Toronto Montreal Sydney Tokyo São Paulo

COPYRIGHT © 1984, BY ACADEMIC PRESS, INC.
ALL RIGHTS RESERVED.
NO PART OF THIS PUBLICATION MAY BE REPRODUCED OR TRANSMITTED IN ANY FORM OR BY ANY MEANS, ELECTRONIC OR MECHANICAL, INCLUDING PHOTOCOPY, RECORDING, OR ANY INFORMATION STORAGE AND RETRIEVAL SYSTEM, WITHOUT PERMISSION IN WRITING FROM THE PUBLISHER.

ACADEMIC PRESS, INC.
Orlando, Florida 32887

United Kingdom Edition published by
ACADEMIC PRESS, INC. (LONDON) LTD.
24/28 Oval Road, London NW1 7DX

Library of Congress Cataloging in Publication Data

Main entry under title:

Segmented negative strand viruses.

Proceedings of the 5th International Symposium on Negative Strand Viruses, held at Hilton Head, S.C., Sept. 11-17, 1983.
Includes index.
1. Arenaviruses--Congresses. 2. Bunyaviruses--Congresses. 3. Orthomyxoviruses--Congresses. I. Compans, Richard W. II. Bishop, David H. L. III. International Symposium on Negative Strand Viruses (5th : 1983 : Hilton Head, S.C.) [DNLM: 1. Arenaviridae--Congresses. 2. Bunyaviridae--Congresses. 3. Orthomyxoviridae--Congresses. QW 168 S454 1983]
QR395.S44 1984 576'.64 84-6276
ISBN 0-12-183501-4 (alk. paper)

PRINTED IN THE UNITED STATES OF AMERICA

84 85 86 87 9 8 7 6 5 4 3 2 1

Contents

Genome Structure, Transcription, and Genetics

Gene Expression, Protein Synthesis, and Protein Modification

Viral Proteins: Antigenic and Functional Analyses

Biology

Contributors

Numbers in parentheses indicate the pages on which the authors' contributions begin.

Gordon Abraham (37), *School of Science, Griffith University, Nathan, Queensland 4111, Australia*

Rafi Ahmed (333), *Department of Immunology, Scripps Clinic and Research Foundation, La Jolla, California 92037*

Katsutoshi Aihara[1] (387), *Department of Bacteriology, Nagasaki University School of Medicine, Nagasaki 852, Japan*

G. M. Air[2] (225), *John Curtin School of Medical Research, Canberra City, ACT 2601, Australia*

Hiroomi Akashi (3), *Department of Microbiology, University of Alabama in Birmingham, Birmingham, Alabama 35294*

Ramesh K. Akkina (85), *Department of Microbiology and Immunology, School of Medicine, University of California, Los Angeles, California 90024*

Lisa M. Allison (209), *Department of Medical Microbiology, London School of Hygiene and Tropical Medicine, London WC1E 7HT, United Kingdom*

Firelli V. Alonso-Caplen (179), *Department of Microbiology, University of Alabama in Birmingham, Birmingham, Alabama 35294*

David D. Auperin (51), *Department of Microbiology, University of Alabama in Birmingham, Birmingham, Alabama 35294*

Thomas Bächi (247), *Institute for Immunology and Virology, University of Zürich, 8028 Zürich, Switzerland*

Sukla Basak (361), *Department of Microbiology, University of Alabama in Birmingham, Birmingham, Alabama 35294*

William J. Bean (315), *Division of Virology and Molecular Biology, St. Jude Children's Research Hospital, Memphis, Tennessee 38101*

Serge Belloncik (3), *Virology Research Centre, Institut Armand-Frappier, Quebec H7N 4Z3, Canada*

David H. L. Bishop (3, 51), *Department of Microbiology, University of Alabama in Birmingham, Birmingham, Alabama 35294*

[1]Present address: The Chemo-Sero-Therapeutic Research Institute, Kumamoto 860, Japan.

[2]Present address: Laboratories of Special Cancer Research, University of Alabama in Birmingham, Birmingham, Alabama 35294.

Timothy J. Bos (125), *Department of Microbiology and Immunology, School of Medicine, University of California, Los Angeles, California 90024*

F. X. Bosch (273), *Institut für Virologie, Justus-Liebig-Universität Giessen, D-6300 Giessen, Federal Republic of Germany*

G. W. Both (233), *Division of Molecular Biology, CSIRO, North Ryde, New South Wales, Australia*

M. Bouloy (29), *Unité de Virologie Moléculaire, Institut Pasteur, 75724 Paris Cedex 15, France*

Janet Braam (73), *Molecular Biology and Virology Program, Graduate School of Medical Sciences, Memorial Sloan–Kettering Cancer Center, New York, New York 10021*

George G. Brownlee (253), *Sir William Dunn School of Pathology, University of Oxford, Oxford OX1 3RE, United Kingdom*

D. J. Bucher (373), *Department of Microbiology, Mount Sinai School of Medicine, New York, New York 10029*

Michael J. Buchmeier (193, 201, 209), *Department of Immunology, Scripps Clinic and Research Foundation, La Jolla, California 92037*

Alicia J. Buckler-White (395), *Laboratory of Infectious Diseases, National Institute of Allergy and Infectious Diseases, National Institutes of Health, Bethesda, Maryland 20205*

Deborah A. Buonagurio (147), *Mount Sinai School of Medicine of the City University of New York, New York, New York 10029*

A. S. Carver (355), *Department of Biological Sciences, University of Warwick, West Midlands CV4 7AL, United Kingdom*

Andrew J. Caton (253), *Sir William Dunn School of Pathology, University of Oxford, Oxford OX1 3RE, United Kingdom*

Thomas M. Chambers (85), *Department of Microbiology and Immunology, School of Medicine, University of California, Los Angeles, California 90024*

Robert M. Chanock (395), *Laboratory of Infectious Diseases, National Institute of Allergy and Infectious Diseases, National Institutes of Health, Bethesda, Maryland 20205*

Purnell W. Choppin (131), *The Rockefeller University, New York, New York 10021*

W. Clark (45), *MRC Virology Unit, Institute of Virology, Glasgow G11 5JR, Scotland*

J. C. S. Clegg (341), *Public Health Laboratory Service, Centre for Applied Microbiology and Research, Salisbury SR4 OJG, United Kingdom*

Mary Lou Clements (395), *Center for Vaccine Development, Department of Medicine, University of Maryland School of Medicine, Baltimore, Maryland 21201*

John P. M. Clerx[3] (321), *Institute for Medical Microbiology and Hygiene, Technical University, 8000 Munich 40, Federal Republic of Germany*

M. S. Collett (159), *Molecular Genetics, Inc., Minnetonka, Minnesota 55343*

[3]Present address: Department of Cellular and Developmental Biology, The University of Arizona, Tucson, Arizona 85721.

P. M. Colman (225), *John Curtin School of Medical Research, Canberra City, ACT 2601, Australia, and CSIRO Division of Protein Chemistry, Melbourne, Victoria 3052, Australia*

Richard W. Compans (179, 361), *Department of Microbiology, University of Alabama in Birmingham, Birmingham, Alabama 35294*

Nancy J. Cox (101, 253), *Influenza Branch 7-106, Centers for Disease Control, Atlanta, Georgia 30333*

Janusz Dabrowski (289), *Max-Planck-Institut für Medizinische Forschung, Heidelberg, Federal Republic of Germany*

Joel M. Dalrymple (21, 117, 159), *U.S. Army Medical Research Institute of Infectious Diseases, Fort Detrick, Frederick, Maryland 21701*

Roelf Datema[4] (187), *Institut für Virologie, Justus-Liebig-Universität Giessen, D-6300 Giessen, Federal Republic of Germany*

Alan R. Davis[5] (125), *Department of Microbiology and Immunology, School of Medicine, University of California, Los Angeles, California 90024*

Dan C. DeBorde (147, 379), *School of Public Health, University of Michigan, Ann Arbor, Michigan 48109*

N. J. Dimmock (355), *Department of Biological Sciences, University of Warwick, West Midlands CV4 7AL, United Kingdom*

Armen M. Donabedian (379), *Department of Epidemiology, School of Public Health, University of Michigan, Ann Arbor, Michigan 48109*

R. R. Dourmashkin (373), *Department of Microbiology, Mount Sinai School of Medicine, New York, New York 10029*

L. H. Elliott (201), *Special Pathogens Branch, Division of Viral Disease, Centers for Disease Control, Atlanta, Georgia 30333*

Richard M. Elliott (13, 45, 349), *MRC Virology Unit, Institute of Virology, Glasgow G11 5JR, Scotland*

Yuki Eshita (3), *Department of Microbiology, University of Alabama in Birmingham, Birmingham, Alabama 35294*

John T. Finch (65), *MRC Laboratory of Molecular Biology, Cambridge CB2 2PH, United Kingdom*

Frederick Fuller[6] (3), *Department of Microbiology, University of Alabama in Birmingham, Birmingham, Alabama 35294*

S. Gerbaud (29), *Unité de Virologie Moléculaire, Institut Pasteur, 75724 Paris, Cedex 15, France*

Walter Gerhard (233, 247), *The Wistar Institute, Philadelphia, Pennsylvania 19104*

Rudolf Geyer (289), *Biochemisches Institut, Justus-Liebig-Universität Giessen, D-6300 Giessen, Federal Republic of Germany*

M. Girard (29), *Unité de Virologie Moléculaire, Institut Pasteur, 75724 Paris Cedex 15, France*

[4]Present address: Department of Antiviral Chemotherapy, Astra Läkemedel AB, Södertälje, Sweden.

[5]Present address: Wyeth Laboratories, Radnor, Pennsylvania 19108.

[6]Present address: Department of Microbiology, Pathology, and Immunology, North Carolina State University, Raleigh, North Carolina 27606.

J. P. Gonzalez[7] (201), *Office de la Recherche Scientifique et Technique Outre Mer, Paris, France*

J. A. Greenberg (373), *Department of Microbiology, Mount Sinai School of Medicine, New York, New York 10029*

Anastasia Gregoriades[8] (299), *The Public Health Research Institute of the City of New York, New York, New York 10016*

Lynne F. Haber (239), *Department of Biological Chemistry, School of Medicine, University of California, Los Angeles, California 90024*

Otto Haller[9] (131), *The Rockefeller University, New York, New York 10021*

Joanna Hansen (139), *Molecular Viral Biology Section, Laboratory of Infectious Diseases, National Institute of Allergy and Infectious Diseases, National Institutes of Health, Bethesda, Maryland 20205*

Martinez J. Hewlett (17), *Department of Cellular and Developmental Biology, University of Arizona, Tucson, Arizona 85721*

V. S. Hinshaw (267), *Division of Virology and Molecular Biology, St. Jude Children's Research Hospital, Memphis, Tennessee 38101*

Robert S. Hodges (109), *Department of Biochemistry and Medical Research Council Group in Protein Structure and Function, University of Alberta, Edmonton, Alberta T6G 2H7, Canada*

Yasuhiro Hosaka (365), *Department of Preventive Medicine, Research Institute for Microbial Diseases, Osaka University, Osaka 565, Japan*

Colin R. Howard (209), *Department of Medical Microbiology, London School of Hygiene and Tropical Medicine, London WC1E 7HT, United Kingdom*

Takeshi Ihara (3), *Department of Microbiology, University of Alabama in Birmingham, Birmingham, Alabama 35294*

Lori D. Ishizawa (217), *Department of Microbiology, California State University, Long Beach, California 90840*

Philip A. Jennings (65), *Laboratory of Molecular Biology, Medical Research Council Centre, Cambridge, United Kingdom*

Guozhong Jing (109), *Department of Medicine, University of Alberta, Edmonton, Alberta T6G 2G3, Canada*

Walter Keil (289), *Institut für Virologie, Justus-Liebig-Universität Giessen, D-6300 Giessen, Federal Republic of Germany*

Alan P. Kendal (101, 253, 259), *Influenza Branch 7-106, Centers for Disease Control, Atlanta, Georgia 30303*

E. D. Kilbourne (233), *Department of Microbiology, Mount Sinai School of Medicine, New York, New York 10029*

M. P. Kiley (201), *Special Pathogens Branch, Division of Viral Diseases, Centers for Disease Control, Atlanta, Georgia 30333*

[7]Present address: Special Pathogens Branch, Division of Viral Diseases, Centers for Disease Control, Atlanta, Georgia 30333.

[8]Present address: New York College of Podiatric Medicine, New York, New York 10035.

[9]Present address: Institute for Immunology and Virology, University of Zürich, Zürich, Switzerland.

Laura Kingsford (217), *Department of Microbiology, California State University, Long Beach, California 90840*

Hans-Dieter Klenk (93, 289), *Institut für Virologie, Justus-Liebig-Universität Giessen, D-6300 Giessen, Federal Republic of Germany*

Robert M. Krug (73), *Molecular Biology and Virology Program, Memorial Sloan–Kettering Cancer Center, New York, New York 10021*

Mark Krystal (147), *Mount Sinai School of Medicine of the City University of New York, New York, New York 10029*

Ching-Juh Lai (139, 167), *Molecular Viral Biology Section, Laboratory of Infectious Diseases, National Institute of Allergy and Infectious Diseases, National Institutes of Health, Bethesda, Maryland 20205*

Robert A. Lamb[10] (131), *The Rockefeller University, New York, New York 10021*

W. G. Laver (225), *John Curtin School of Medical Research, Canberra City, ACT 2601, Australia*

Ming-Fan Law (167), *Laboratory of Pathology, National Cancer Institute, Bethesda, Maryland 20205*

Janice F. Lees (13, 45), *MRC Virology Unit, Institute of Virology, Glasgow G11 5JR, Scotland*

Wai-Choi Leung (109), *Department of Medicine, University of Alberta, Edmonton, Alberta T6G 2G3, Canada*

Hanna Lewicki (209), *Department of Immunology, Scripps Clinic and Research Foundation, La Jolla, California 92037*

Bor-Chian Lin (167), *Molecular Viral Biology Section, Laboratory of Infectious Diseases, National Institute of Allergy and Infectious Diseases, National Institutes of Health, Bethesda, Maryland 20205*

G. Lloyd (341), *Public Health Laboratory Service, Centre for Applied Microbiology and Research, Salisbury SR4 0JG, United Kingdom*

William T. London (395), *National Institute of Neurological and Communicative Disorders and Stroke, National Institutes of Health, Bethesda, Maryland 20205*

H. F. Maassab (147, 379), *School of Public Health, University of Michigan, Ann Arbor, Michigan 48109*

B. W. J. Mahy (173), *Division of Virology, Laboratories Block, Addenbrooke's Hospital, Cambridge CB2 2QQ, United Kingdom*

Lewis Markoff (139), *Molecular Viral Biology Section, Laboratory of Infectious Diseases, National Institute of Allergy and Infectious Diseases, National Institutes of Health, Bethesda, Maryland 20205*

Yumiko Matsuoka (3, 179), *Department of Microbiology, University of Alabama in Birmingham, Birmingham, Alabama 35294*

[10]Present address: Department of Biochemistry, Molecular, and Cell Biology, Northwestern University, Evanston, Illinois 60201.

John W. McCauley (281), *Division of Virology, Department of Pathology, University of Cambridge, Laboratories Block, Addenbrooke's Hospital, Cambridge CB2 2QQ, United Kingdom*

J. B. McCormick (201), *Special Pathogens Branch, Division of Viral Diseases, Centers for Disease Control, Atlanta, Georgia 30333*

Jerry R. McGhee (307), *Department of Microbiology, University of Alabama in Birmingham, Birmingham, Alabama 35294*

Nancy L. McQueen (125), *Department of Microbiology and Immunology, School of Medicine, University of California, Los Angeles, California 90024*

Herbert Meier-Ewert (321), *Institute for Medical Microbiology and Hygiene, Technical University, 8000 Munich 40, Federal Republic of Germany*

Suzanne M. Michalek (307), *University of Alabama in Birmingham, Birmingham, Alabama 35294*

S. W. Mitchell (201), *Special Pathogens Branch, Division of Viral Diseases, Centers for Disease Control, Atlanta, Georgia 30333*

Tsutomu Miyamoto (387), *Department of Bacteriology, Nagasaki University School of Medicine, Nagasaki 852, Japan*

Kam Mong (109), *Department of Medicine, University of Alberta, Edmonton, Alberta T6G 2G3, Canada*

Karin Müller (93), *Institut für Virologie, Justus-Liebig-Universität Giessen, D-6300 Giessen, Federal Republic of Germany*

Brian R. Murphy (307, 395), *Laboratory of Infectious Diseases, National Institute of Allergy and Infectious Diseases, National Institutes of Health, Bethesda, Maryland 20205*

C. W. Naeve (267), *Division of Virology and Molecular Biology, St. Jude Children's Research Hospital, Memphis, Tennessee 38101*

Susumu Nakada (147), *Mount Sinai School of Medicine of the City University of New York, New York, New York 10029*

K. Nakajima (259), *The Institute of Public Health, Tokyo 108, Japan*

S. Nakajima (259), *The Institute of Public Health, Tokyo 108, Japan*

Debi P. Nayak (85, 125), *Department of Microbiology and Immunology, School of Medicine, University of California, Los Angeles, California 90024*

Heiner Niemann (289), *Institut für Virologie, Justus-Liebig-Universität Giessen, D-6300 Giessen, Federal Republic of Germany*

Michael B. A. Oldstone (59, 333, 361), *Department of Immunology, Scripps Clinic and Research Foundation, La Jolla, California 92037*

Peter Palese (147), *Mount Sinai School of Medicine, New York, New York 10029*

N. Pardigon (29), *Unité de Virologie Moléculaire, Institut Pasteur, 75724 Paris Cedex 15, France*

M. D. Parker (21, 159), *Department of Viral Biology, Virology Division, U.S. Army Medical Research Institute of Infectious Diseases, Fort Detrick, Frederick, Maryland 21701*

Asit K. Pattnaik (37), *School of Science, Griffith University, Nathan, Queensland 4111, Australia*

James C. Paulson (239), *Department of Biological Chemistry, School of Medicine, University of California, Los Angeles, California 90024*

C. R. Penn (173), *Division of Virology, Laboratories Block, Addenbrooke's Hospital, Cambridge CB2 2QQ, United Kingdom*

Susan M. Peters (379), *Department of Epidemiology, School of Public Health, University of Michigan, Ann Arbor, Michigan 48109*

Craig R. Pringle[11] (13, 45, 349), *MRC Virology Unit, Institute of Virology, Glasgow G11 5JR, Scotland*

Thomas J. Pritchett (239), *Department of Biological Chemistry, School of Medicine, University of California, Los Angeles, California 90024*

A. F. Purchio (159), *Molecular Genetics, Inc., Minnetonka, Minnesota 55343*

Arlene Ramsingh[12] (109), *Department of Medicine, University of Alberta, Edmonton, Alberta T6G 2G3, Canada*

F. Lucy Raymond (253), *Sir William Dunn School of Pathology, University of Oxford, Oxford OX1 3RE, United Kingdom*

James S. Robertson[13] (65), *Division of Virology, University of Cambridge, Addenbrooke's Hospital, Cambridge CB2 2QQ, United Kingdom*

Gary N. Rogers (239), *Department of Biological Chemistry, School of Medicine, University of California, Los Angeles, California 90024*

Victor Romanowski (51), *Department of Microbiology, University of Alabama in Birmingham, Birmingham, Alabama 35294*

Pedro A. Romero (187), *Institut für Virologie, Justus-Liebig-Universität Giessen, D-6300 Giessen, Federal Republic of Germany*

R. Rott (273), *Institut für Virologie, Justus-Liebig-Universität Giessen, D-6300 Giessen, Federal Republic of Germany*

Erling Rud (3), *Virology Research Centre, Institut Armand-Frappier, Quebec H7N 4Z3, Canada*

Aimo Salmi[14] (333), *Department of Immunology, Scripps Clinic and Research Foundation, La Jolla, California 92037*

Mark Salter (209), *Department of Medical Microbiology, London School of Hygiene and Tropical Medicine, London WC1E 7HT, United Kingdom*

Connie S. Schmaljohn (117, 159), *Department of Viral Biology, U.S. Army Medical Research Institute of Infectious Diseases, Fort Detrick, Frederick, Maryland 21701*

Christoph Scholtissek (93), *Institut für Virologie, Justus-Liebig-Universität Giessen, D-6300 Giessen, Federal Republic of Germany*

[11]Present address: Department of Biological Sciences, University of Warwick, Coventry CV4 7AL, United Kingdom.

[12]Present address: Department of Human Genetics, Yale University School of Medicine, New Haven, Connecticut 06510.

[13]Present address: National Institute for Biological Standards and Control, Hampstead, London, United Kingdom.

[14]Present address: Department of Virology, University of Turko, Turko, Finland.

Ralph T. Schwarz (93, 187), *Institut für Virologie, Justus-Liebig-Universität Giessen, D-6300 Giessen, Federal Republic of Germany*

Kenji Sekikawa (139), *Molecular Viral Biology Section, Laboratory of Infectious Diseases, National Institute of Allergy and Infectious Diseases, National Institutes of Health, Bethesda, Maryland 20205*

Michael W. Shaw (131), *The Rockefeller University, New York, New York 10021*

Kazufumi Shimizu[15] (387), *Department of Bacteriology, Nagasaki University School of Medicine, Nagasaki 852, Japan*

J. F. Smith (21), *U.S. Army Medical Research Institute of Infectious Diseases, Fort Detrick, Frederick, Maryland 21701*

Peter J. Southern (59), *Department of Immunology, Scripps Clinic and Research Foundation, La Jolla, California 92037*

Melanie K. Spriggs (17), *Department of Cellular and Developmental Biology, University of Arizona, Tucson, Arizona 85721*

Ashok K. Taneja (109), *Department of Biochemistry and Medical Research Council Group in Protein Structure and Function, University of Alberta, Edmonton, Alberta T6G 2H7, Canada*

H. P. Taylor (355), *Department of Biological Sciences, University of Warwick, West Midlands CV4 7AL, United Kingdom*

Shu-fang Tian (395), *Laboratory of Infectious Diseases, National Institute of Allergy and Infectious Diseases, National Institutes of Health, Bethesda, Maryland 20205*

Ismo Ulmanen[16] (73), *Molecular Biology and Virology Program, Graduate School of Medical Sciences, Memorial Sloan–Kettering Cancer Center, New York, New York 10021*

Kathleen L. van Wyke (307), *Laboratory of Infectious Diseases, National Institute of Allergy and Infectious Diseases, National Institutes of Health, Bethesda, Maryland 20205*

J. N. Varghese (225), *CSIRO Division of Protein Chemistry, Parkville, Victoria 3052, Australia*

P. Vialat (29), *Unité de Virologie Moléculaire, Institut Pasteur, 75724 Paris Cedex 15, France*

Xiao-fan Wang (239), *Department of Biological Chemistry, School of Medicine, University of California, Los Angeles, California 90024*

C. W. Ward (225), *CSIRO Division of Protein Chemistry, Melbourne, Victoria 3052, Australia*

Gillian E. Watret (349), *MRC Virology Unit, Institute of Virology, Glasgow G11 5JR, Scotland*

R. G. Webster (225, 267), *Division of Virology and Molecular Biology, St. Jude Children's Research Hospital, Memphis, Tennessee 38101*

[15]Present address: Department of Bacteriology, Nihon University School of Medicine, Tokyo 173, Japan.

[16]Present address: Recombinant DNA Laboratory, University of Helsinki, Helsinki, Finland.

Greg Winter (65), *Laboratory of Molecular Biology, Medical Research Council Centre, Cambridge, United Kingdom*

Jonathan W. Yewdell (247, 307), *The Wistar Institute of Anatomy and Biology, Philadelphia, Pennsylvania 19104, and Imperial College of Science and Technology, London, United Kingdom*

Preface

Negative strand viruses include several families of large, enveloped RNA viruses. They are responsible for many important common or exotic, human and animal diseases, including influenza, mumps, measles, rabies, canine distemper, some human encephalitis, and hemorrhagic fevers. Four previous symposia have been held on negative strand viruses. Three were organized by Drs. R. D. Barry and B. W. J. Mahy and were held in Cambridge, England, in 1969, 1973, and 1977. A fourth symposium, which we organized, was held in St. Thomas, Virgin Islands, in 1980.

The investigation of negative strand viruses at the molecular level has continued at an accelerated pace, and the current knowledge in this field was reviewed at a symposium held September 11–17, 1983, at Mariner's Inn, Hilton Head, South Carolina. The papers presented at this 1983 symposium are being organized into two volumes. This volume concerns the three virus families with segmented RNA genomes: arenaviruses, bunyaviruses, and orthomyxoviruses. A second volume contains papers on two virus families with nonsegmented genomes: paramyxoviruses and rhabdoviruses.

The three families of segmented-genome negative strand viruses, while sharing some features of their replication strategy, exhibit significant differences in gene organization and biochemical and biological properties. The orthomyxoviruses have been studied in greatest detail: complete nucleotide sequences have been determined for several viruses, and extensive information has been obtained on their replication process and on the three-dimensional structure and antigenic properties of their surface glycoproteins. Recent results on these subjects are reported in this volume, as are such new findings as the discovery of a second polypeptide encoded by the neuraminidase gene of influenza B viruses and the transcription and translation of RNA species contained in defective-interfering virus particles. Studies of the structure and replication of arenaviruses and bunyaviruses are not as advanced; however, important findings concerning gene organization and coding and the antigenic properties of viral proteins are presented for the first time in this volume.

We believe that this book will provide a unique source of information on recent studies of negative strand viruses, including research on their molecular biology, genetics, viral immunology and vaccine development, and biological properties.

Besides being of interest to investigators in these areas, the subjects covered will also be of interest to microbiologists, molecular and cell biologists, and investigators of infectious diseases.

We would like to acknowledge the sponsorship of the symposium by the University of Alabama in Birmingham and the financial support provided by the following organizations: Amicon; Biocell Laboratories; Elsevier/North Holland; Hoffmann–La Roche, Inc.; Imperial Chemical Industries; New Brunswick Scientific Company, Inc.; Pitman–Moore, Inc., Rheem Manufacturing Company, Inc.; Vangard International, Inc.; and Wheaton Instruments.

We also wish to thank all of the participants who contributed to the success of the meeting and the authors who have provided an excellent and timely series of manuscripts. Our special thanks are also due to the following participants who served as chairmen of the scientific sessions: Drs. G. M. Air, M. A. Bratt, J. M. Dalrymple, N. J. Dimmock, S. U. Emerson, A. Flamand, C. Y. Kang, E. D. Kilbourne, D. W. Kingsbury, H.-D. Klenk, D. Kolakofsky, W. G. Laver, R. A. Lazzarini, D. P. Nayak, M. B. A. Oldstone, P. Palese, M. Schubert, and G. W. Wertz.

Finally, we would like to thank Betty Jeffrey and Denice L. Montgomery, both of whom worked cheerfully and extraordinarily hard in organizing the symposium and in preparing these volumes for publication.

R. W. COMPANS
D. H. L. BISHOP

Contents of *Nonsegmented Negative Strand Viruses*

Genome Structure

Transcription and Replication

Gene Expression, Protein Synthesis, and Protein Modification

Viral Proteins: Antigenic and Functional Analyses

Biology

Defective Viruses and Virus Persistence

Genome Structure, Transcription, and Genetics

CODING ANALYSES OF BUNYAVIRUS RNA SPECIES[1]

David H.L. Bishop, Erling Rud, Serge Belloncik, Hiroomi Akashi, Frederick Fuller, Takeshi Ihara, Yumiko Matsuoka and Yuki Eshita

Department of Microbiology
University of Alabama in Birmingham
Birmingham, AL 35294
and
Virology Research Centre
Institut Armand-Frappier
Quebec,
CANADA H7N 4Z3

CODING ASSIGNMENTS OF BUNYAVIRUS VIRAL RNA SPECIES

The genome of bunyaviruses (family Bunyaviridae, genus Bunyavirus) consists of three single-stranded, negative-sense, RNA molecules, designated L, M and S (1,2). For snowshoe hare (SSH) virus, a member of the California encephalitis serogroup of bunyaviruses, it has been demonstrated by genetic, molecular and DNA sequencing studies that the S RNA ($3.3x10^5$ daltons) codes for the structural nucleocapsid polypeptide N ($26.8x10^3$ daltons) as well as a non-structural protein NS_S ($10.5x10^3$ daltons) that is found in virus infected cells (3-5). The function of the NS_S protein is unknown. The two gene products are read from overlapping open reading frames in the S viral complementary RNA species (4). The M RNA (estimated to be $2x10^6$ daltons) codes for the two viral glycoproteins (G1, $115x10^3$ daltons; G2, $35x10^3$ daltons) and a non-structural protein designated NS_M ($12x10^3$ daltons) (4,6,7). The function of NS_M is unknown, nor is it known if it is related to G1, or G2, or shares a common precursor. It is presumed that the SSH L RNA (estimated to $3x10^6$ daltons in size) codes for the viral transcriptase - replicase enzymes.

[1]This work was supported by NIH grant AI-15492 and contract DAMD-17-78-C-8017 from the U.S. Army Medical Research and Development Command.

Copyright © 1984 by Academic Press, Inc.
All rights of reproduction in any form reserved.
ISBN 0-12-183501-4

CONSERVED REGIONS OF BUNYAVIRUS S RNA SEQUENCES

As shown in Figure 1, the S RNA species and encoded gene products of SSH and LAC viruses are comparable in sequence. Although most of the 114 nucleotide differences do not predict amino acid changes there are a total of 22 N and 12 NS_S differences (8). Direct RNA sequencing of SSH and LAC RNA species has demonstrated that the 3' ends of the three viral RNA species (L, M and S) are conserved, at least for 10-15 nucleotides (9). In the non-coding viral complementary sequences of LAC and SSH S RNA species the 5' ends are conserved for some 40 nucleotides (8). However after this region there is little sequence conservation before the coding region that begins with residues 80-82 (8). By contrast, the last 150 nucleotides of the viral complementary RNA are highly conserved (i.e., at the non-coding 3' end of the sequence, Figure 1). As expected for the region of the genome that codes for the two gene products there are fewer nucleotide differences between the two sequences than in the region that codes only for N (Figure 1).

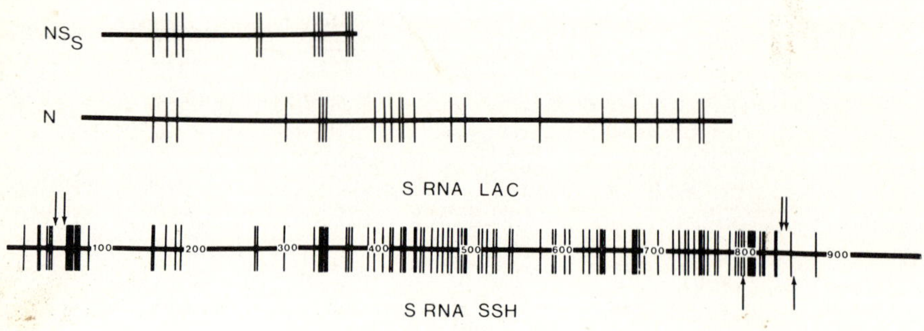

FIGURE 1. Comparison of the S RNA viral complementary sequences of SSH and LAC viruses (5,8). The positions of nucleotide substitutions are indicated by vertical bars on the lower line, nucleotide additions by arrows. The region of the genome coding for N and NS_S proteins are depicted by the upper 2 horizontal lines; amino acid substitutions in gene products are indicated by vertical bars.

In order to investigate further the question of which regions of the SSH S RNA species are conserved, an alternate virus isolate obtained from Quebec Province (Canada) was plaque cloned and the S RNA sequenced. Shown in Figure 2 is the comparison of the S viral complementary RNA sequence of prototype SSH virus and that of the Canadian isolate. Seven nucleotide differences were found (prototype to Canadian isolate: residue #382 A to G; #412 A to G; #451 C to U; #472 U to C; #541 G to A; #571 A to G; #637 C to U, see reference 5). None of these changes causes an amino acid change. All are located in the region of the genome coding only for N protein.

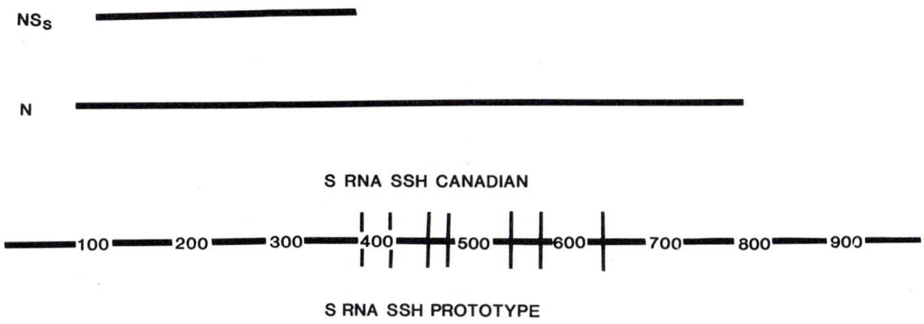

FIGURE 2. Comparison of the S RNA viral complementary sequences of prototype SSH virus (5) and an alternate isolate obtained from Canada (see legend to Figure 1 for details).

In the viral sense S RNA species five of the nucleotide differences between prototype SSH virus and the Canadian isolate predict changes in the ribonuclease T1 derived oligonucleotides that are 10, or more, nucleotides in length. Each of these changes was identified in an oligonucleotide fingerprint comparison of the S RNA species of the two viruses (Figures 3,4).

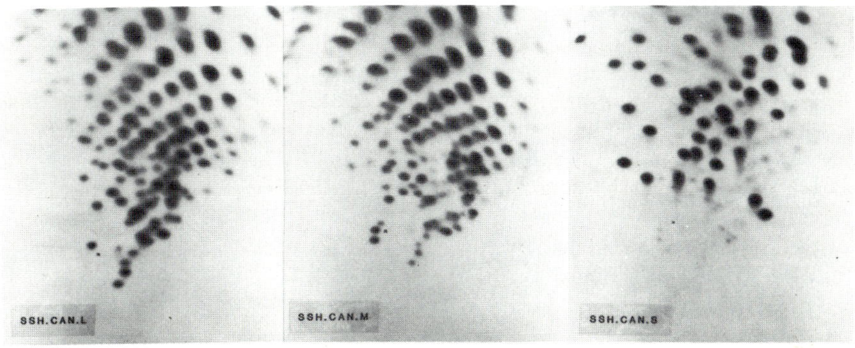

FIGURE 3. Oligonucleotide fingerprints of the L, M and S RNA species of a Canadian SSH virus isolate.

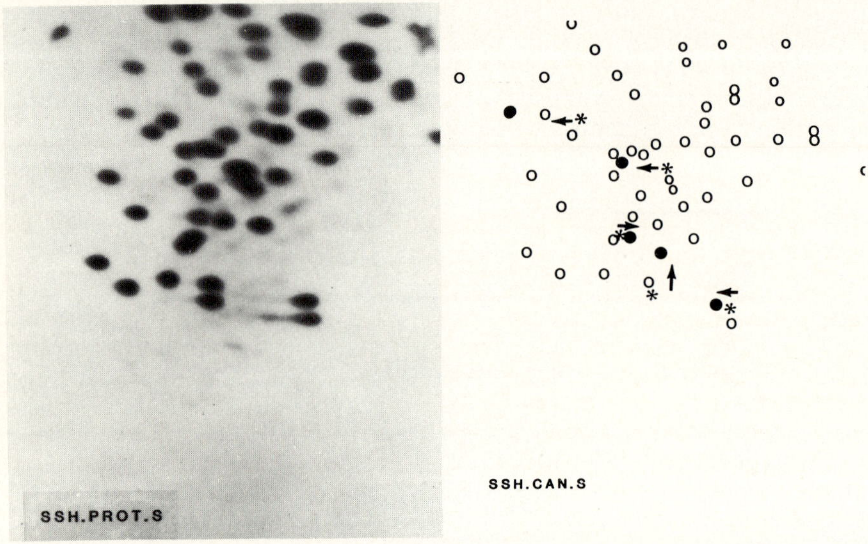

FIGURE 4. Oligonucleotide fingerprint of the S RNA of prototype SSH virus (left) and a schematic comparison (right) with that of the Canadian isolate. Stars represent oligonucleotides missing from the Canadian S RNA fingerprint; filled circles are new oligonucleotides (Fig. 3) that by size and position (arrows) could be related to oligonucleotides missing from the prototype virus S RNA pattern.

The results of these studies indicate that for SSH and LAC viruses, nucleotide substitutions in the S RNA coding region that do not cause an amino acid change are more prevelant than those that do cause a change, particularly by comparison to sequences that code for gene products read in overlapping reading frames. The distribution of nucleotide substitutions in the non-coding region of the S RNA species is remarkable not just for the conserved (and complementary) end 10-15 nucleotides (which must be involved in replicase/transcriptase enzyme recognition) but also for the unilateral conservation of an extended region of the 3' end of the viral complementary sequence (5' end of the viral RNA sequence). Whether this sequence is important for packaging and other morphogenetic processes is not known.

COMPARISON OF AINO BUNYAVIRUS S RNA GENE PRODUCTS TO THOSE OF SSH AND LAC BUNYAVIRUSES

Correlation of the predicted gene products of the S RNA of SSH virus to proteins recovered from virus (N), or from virus infected cells (NS_S), has been made by tryptic peptide composition and sequence analyses (4). However the fact that proteins have been identified in infected cell extracts and that S mRNA is translated <u>in vitro</u> to yield N and NS_S proteins does not prove that these proteins have roles in the infection process. N protein is evidently an important structural element for the virus. However the role of the NS_S protein is not known. Its occurrence may be the fortuitous result of the existence of an open reading frame in the viral complementary sequence that is initiated by a methionine codon.

In order to determine if both gene products are found in other bunyaviruses, the 3' end sequence of the S RNA of Aino virus, a member of the Simbu serogroup of bunyaviruses, was determined (10), a complementary oligonucleotide primer synthesized and used to make cDNA. After cloning, the viral DNA inserts were sequenced. The results obtained (Figure 5) indicate that the first S coded gene product begins with a methionine codon at viral complementary RNA residues #35-37 and terminates with a UAA anticodon at residues #734-736. A second gene product begins with a methionine codon at viral complementary RNA residues #60-62 and terminates with a UAG anticodon at residues #333-345.

```
M F L N G I S L R L T R R S G M W H L L L
M A N Q F I F Q D V P Q R N L A T F N P E V G Y V A F I A
AGUAGUGUACUCCACUAUAGAACAAAGCUUUUAAAUGGCAAACCAAUUUAUUUUCCAAGAUGUUCCUCAACGGAAUCUCGCUACGUUUAACCCGGAGGUCGGGUAUGUGGCAUUUAUUGC
10 20 30 40 50 60 70 80 90 100 110 120

N M G P N S I S I P L D S S S S I R R R P R W Y S V R R H N Q V L I L H L V A S
K H G A G L N F D T V R F F F L N Q K K A K M V L S K T A Q P S V D L T F G G I
UAAACAUGGGGCCCAACUCAAUUUCGAUACCGUUAGAUUCUUCUUCCUCAAUCAGAAGAAGGCCAAGAUGGUACUCAGUAAGACGGCACAACCAAGUGUUGAUCUUACAUUUGGUGGCAU
130 140 150 160 170 180 190 200 210 220 230 240

S L H W L I T I F P N T Q Q I L C Q T L P S L S I V S Q D I
K F T L V N N H F P Q Y T A N P V P D T A L T L H R L S G Y L A K W V A D Q C K
CAAGUUUACACUGGUUAAUAACCAUUUUCCCCAAUACACAGCAAAUCCUGUGCCAGACACUGCCCUCACUCUCCAUCGUCUCUCAGGAUAUCUAGCAAAAUGGGUUGCAGACCAAUGUAA
250 260 270 280 290 300 310 320 330 340 350 360

T N Q I K L A E A M E K I V M P L A E V K G C T W T E G L T M Y L G F A P G A E
AACGAAUCAGAUUAAAUUGGCUGAGGCCAUGGAAAAAAUUGUCAUGCCACUGGCUGAAGUGAAAGGUUGCACCUGGACUGAAGGGCUGACUAUGUAUCUGGGAUUUGCACCAGGUGCUGA
370 380 390 400 410 420 430 440 450 460 470 480

M F L E T F E F Y P L V I D M H R V L K D G M D V N F M R K V L R Q R Y G T L T
AAUGUUUUUAGAAACAUUUGAAUUCUACCCUUUGGUUAUUGAUAUGCACAGAGUGCUGAAGGAUGGAAUGGAUGUCAACUUCAUGAGGAAGGUCCUUCGCCAGCGCUACGGCACAUUGAC
490 500 510 520 530 540 550 560 570 580 590 600

A E Q W M T Q K I D A V R A A F N A V G Q L S W A K S G F S P A A R A F L A Q F
UGCAGAACAGUGGAUGACUCAAAAAAUAGAUGCUGUCCGUGCAGCCUUCAAUGCUGUUGGGCAGCUAAGCUGGGCUAAAUCAGGAUUCUCACCAGCUGCUAGAGCCUUCCUUGCCCAAUU
610 620 630 640 650 660 670 680 690 700 710 720

G I N I
UGGCAUAAACAUCUAAAUUCCUGU
730 740
```

FIGURE 5. The viral complementary sequences of the S RNA of Aino virus and its predicted gene products. Although the sequence is incomplete (the 3' end was not represented in the clone due to a strong stop sequence that curtailed cDNA synthesis by reverse transcriptase) the two open reading frames predict two Aino S coded gene products.

Shown in Figures 6 and 7 are best fit comparisons of the presumptive Aino N (26,249 daltons) and NS_S (10,501 daltons) protein sequences and those of SSH virus. Some 45% of the N protein amino acids of the two viruses were in similar positions, 33% of the NS_S protein amino acids. Comparisons of SSH and LAC N proteins gave values of 91% (N) and 87% (NS_S). These data indicate that not only are N and NS_S gene products predicted for the S RNA of Aino virus but also that both share conserved sequences with SSH (and LAC) viruses. Apparently the N proteins of these bunyaviruses are more conserved than their NS_S proteins.

```
  1          11         21         31         41         51
  MSL--VFYDV ASTGANGFDP DAGYMAFCVK YAESVNLAAV RIFFLNAAKA KAALSRKPER
  *     * **        * *   ** **  *      *   * * ****  ** *  **
  MANQFIFQDV PQRNLATFNP EVGYVAFIAK HGAQLNFDTV RFFFLNQKKA KMVLSKTAQP
  1          11         21         31         41         51
 61          71         81         91         101        111
  KANPKFGEWQ VEVVNNHFPG NRNNPINSDD LTIHRLSGYL ARWVLEQYKE NEDESRRELI
       **       ******     **      ** ******* * **  * *  *
  SVDLTFGGIK FTLVNNHFPQ YTANPVPDTA LTLHRLSGYL AKWVADQCKT NQIKLAEAME
 61          71         81         91         101        111
121          131        141        151        161        171
  KTTIINPIAE SNGVRWDSGA EIYLSFFPGT EMFLETFKFY PLTIGIYRVK QGMMDPQYLK
  *  *  * **   *  *  *    ** * **  ******* ** ** *   **     **
  K--IVMPLAE VKGCTWTEGL TMYLGFAPGA EMFLETFEFY PLVIDMHRVL KDGMDVNFMR
121          131        141        151        161        171
181          191        201        211        221        231
  KALRGRYGSL TADKWMSQKV TAIAKSLKEV EQLKWGRGGL SDTARSFLQK FGIRLP
  * ****** * **  ** **   *       *  ** *   *  *  ** **   ***
  KVLRGRYGTL TAEQWMTQKI DAVRAAFNAV GQLSWAKSGF SPAARAFLAQ FGINI*
181          191        201        211        221        231
```

FIGURE 6. Best fit comparison of the primary amino acid sequence of SSH N protein (upper line) and the presumptive Aino N protein.

```
  1          11         21         31         41         51
  MMSHQQVQMD LILMQGIWHS VLNMQNQSIL LQLGSSSSMP QRPRLLSRVS QRGRQILNLE
   *         *    * **   ***   **    *  ***    ***  *         ** *
  -MFLNGISLR LTRRSGMWHL LLNMGPNSIS IPLDPSSSIR RRPRWYSVRR HNQVLILHLV
  1          11         21         31         41         51
 61          71         81         91         101        111
  SGRWRLSIII FLETGTIQLT ATILPSTDCQ DI
         * * *  *  *         *   * **
  ASSLHWLITI FPNTQQILCQ TLPSLSIVSQ DI
 61          71         81         91         101        111
```

FIGURE 7. Best fit comparison of the primary amino acid sequence of SSH NS_S protein (upper line) and the presumptive Aino NS_S protein.

ANALYSES OF S MESSENGER AND OTHER S PLUS SENSE RNA SPECIES OF SSH BUNYAVIRUS

In order to determine the 5' sequences of SSH S messenger and other plus sense RNA species, SSH infected cell RNA was resolved by sucrose gradient centrifugation (Figure 8, upper panel) and the fractions containing S mRNA species identified by in vitro translation using rabbit reticulocyte lysates (Figure 8, lower panel).

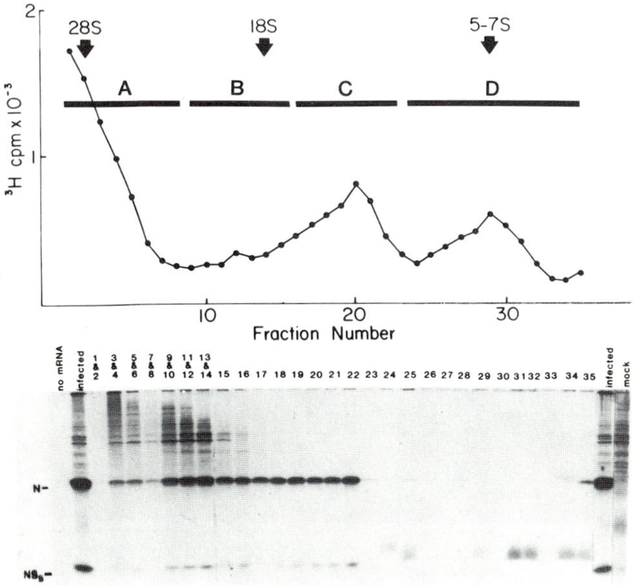

FIGURE 8. Sucrose gradient resolution of ^{3}H-adenosine labeled SSH infected cell RNA (upper panel) and in vitro translation by mock, or infected cell RNA, or no RNA, or RNA recovered from the gradient fractions (lower panel). The indicated pooled fractions (A-D) were used for cDNA transcription (see text).

Based on the amino acid incorporation data and the proteins synthesized, fractions of a similar gradient were pooled (A-D) and employed to template cDNA synthesis using a viral sense primer representing S RNA residues 121-149. As shown in the left panel of Figure 9, fraction C (which on the basis of the translation data contained the most S mRNA species) yielded the largest amount of two transcripts estimated to be approximately 150 and 160-170 nucleotides in length (Figure 9, right panel). Smaller transcripts came from RNA recovered from the top of the gradient (i.e., non-messenger RNA). After polydA tailing, backcopying and cloning, the sequences were determined for 3 clones derived from the 160–170 nucleotide transcripts and 2 of the 150 nucleotide transcripts (Figure 10).

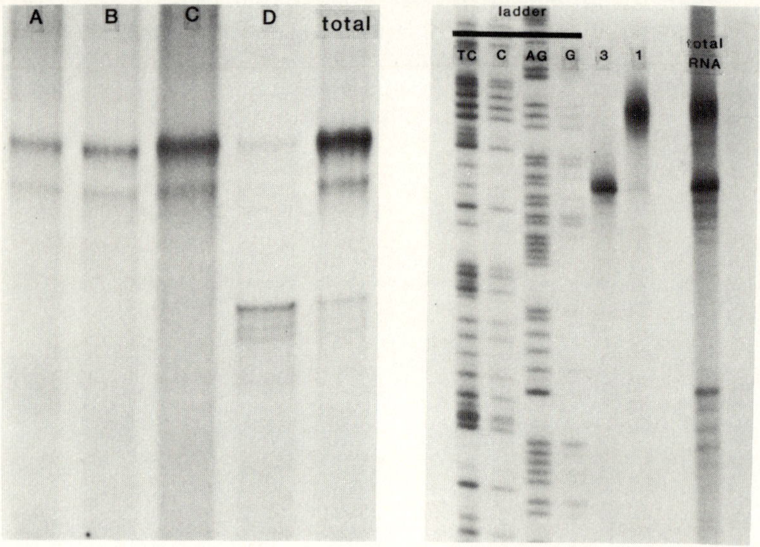

FIGURE 9. In the left panel gradient purified RNA (A–D, see Figure 8) was used with an S RNA viral sense primer (residues 121–149) to synthesize cDNA transcripts that were resolved by 7M urea, 6% polyacrylamide gel electrophoresis. In the right panel are shown the positions of the two largest transcripts (identified as 3 and 1) compared to a ladder and cDNA made from total infected cell RNA.

```
                                                                  1
Clone 1-6   TTTTTTTTTT   GGTCTCTTGCTGT                            AGTAGTGTAC  TCCACTCGAA  TACTTTGAAA
Clone 1-2   TTTTTTTTTT   GATCTTGGCGGCTG                            GTAGTGTAC  TCCACTCGAA  TACTTTGAAA
Clone 1-3   TTTTTTTTTT   GAGGTGACGGTTCC                            GTAGTGTAC  TCCACTCGAA  TACTTTGAAA

Clone 3-1   TTTTTTTTTT                                              TAGTGTAC  TCCACTCGAA  TACTTTGAAA
Clone 3-2   TTTTTTTTTT                                              TAGTGTAC  TCCACTCGAA  TACTTTGAAA

                                                                                          N
            31
Clone 1-6   ATACTTTGTT   GCTAATTGTT   TTTACCTAAG   GGATTGACTT   GGAAGTGTGA  TGTCCGATTT
Clone 1-2   ATACTTTGTT   GCTAATTGTT   TTTACCTAAG   GGATTGACTT   GGAAGTGTGA  TGTCCGATTT
Clone 1-3   ATACTTTGTT   GCTAATTGTT   TTTACCTAAG   GGATTGACTT   GGAAGTGTGA  TGTCCGATTT

Clone 3-1   ATACTTTGTT   GCTAATTGTT   TTTACCTAAG   GGATTGACTT   GGAAGTGTGA  TGTCCGATTT
Clone 3-2   ATACTTTGTT   GCTAATTGTT   TTTACCTAAG   GGATTGACTT   GGAAGTGTGA  TGTCCGATTT

                     NS_S
            91
Clone 1-6   GGTGTTTTAT   GATGTCGCAT   CAACAGGTGC   AAATGGATTT   GATCCTGATG  CAGGGTATA
Clone 1-2   GGTGTTTTAT   GATGTCGCAT   CAACAGGTGC   AAATGGATTT   GATCCTGATG  CAGGGTATA
Clone 1-3   GGTGTTTTAT   GATGTCGCAT   CAACAGGTGC   AAATGGATTT   GATCCTGATG  CAG    TATA

Clone 3-1   GGTGTTTTAT   GATGTCGCAT   CAACAGGTGC   AAATGGATTT   GATCCTGATG  CAGGGTATA
Clone 3-2   GGTGTTTTAT   GATGTCGCAT   CAACAGGTGC   AAATGGATTT   GATCCTGATG  C      TATA
                                                   -----------------------------------
                                                                 PRIMER
```

FIGURE 10. Sequences of the 5' ends of SSH S viral complementary RNA species as determined by backcopying plus sense RNA, cloning and sequencing. The initiating codons for N and NS_S are indicated.

The results of these analyses indicate that there are two types of bunyavirus S plus strand species in infected cells (11), one is presumably a replicative plus sense species (the 150 nucleotide transcripts). By contrast, the other has heterogeneous, non-viral, sequences at their 5' ends (the 160-170 transcripts). Such non-viral sequences are similar to the heterogeneous cellular primers used to initiate influenza mRNA species (12-15). Whether bunyaviruses use host cell derived primers to initiate mRNA synthesis is currently under investigation.

The results described here do not agree with the data presented by others (16) that are interpreted to indicate that bunyaviruses make leader transcripts and distinct S mRNA species for each S coded gene product. Those analyses did not purify mRNA size templates, rather they used unresolved infected cell RNA. Although direct mRNA 5' analyses are required to clarify the issue, it is of interest to note that the first AUG predicted for the Aino S viral complementary RNA is at residue 35, much nearer the end of the RNA than that of SSH, or LAC viruses.

REFERENCES

1. Bishop, D.H.L., and Shope, R. E. (1979). In "Comprehensive Virology" Volume 14, (H. Fraenkel-Conrat and R. R. Wagner eds.), p. 1. Plenum Press, New York.
2. Bishop, D.H.L., Calisher, C., Casals, J., Chumakov, M.P., Gaidamovich, S.Ya., Hannoun, C., Lvov, D.K., Marshall, I.D., Oker-Blom, N., Pettersson, R.F., Porterfield, J.S., Russell, P.K., Shope, R.E., and Westaway, E.G. (1980). Intervirol. 14, 125.
3. Gentsch, J., and Bishop, D.H.L. (1978). J. Virol. 28, 417.
4. Fuller, F., and Bishop, D.H.L. (1982). J. Virol. 41, 643.
5. Bishop, D.H.L., Gould, K.G., Akashi, H., and Clerx-van Haaster, C.M. (1982). Nucl. Acids Res. 10, 3703.
6. Gentsch, J.R., and Bishop, D.H.L. (1979). J. Virol. 30, 767.
7. Fuller, F., Bhown, A.S., and Bishop, D.H.L. (1983). J. Gen. Virol., in press.
8. Akashi, H., and Bishop, D.H.L. (1983). J. Virol. 45, 1155.
9. Clerx-van, Haaster, C.M., Akashi, H., Auperin, D.D., and Bishop, D.H.L. (1982). J. Virol. 41, 119.
10. Clerx-van Haaster, C.M., Clerx, J.P.M., Ushijima, H., Akashi, H., Fuller, F., and Bishop, D.H.L. (1982). J. Gen. Virol. 61, 289.
11. Bishop, D.H.L., Gay, M.E. and Matsuoko, Y. (1983). Nucl. Acids Research in press.
12. Caton, A.H., and Robertson, J.S. (1980). Nucl. Acids Res. 8, 2591.
13. Dhar, R., Chanock, R.M. and Lai, C-J. (1980). Cell 21, 495.
14. Beaton, A.R., and Krug, R.M. (1981). Nucl. Acids Res. 9, 4423.
15 Lamb, R.A., Lai, C-J., and Choppin, P.W. (1981). Proc. Natl. Acad. Sci. USA 78, 4170.
16. Patterson, J.L., Cabradilla, C., Holloway, B.P., Obijeski, J.F., and Kolakofsky, D. (1983). Cell 33, 791.

MOLECULAR CLONING OF THE BUNYAMWERA VIRUS GENOME

Janice F. Lees, Craig R. Pringle and Richard M. Elliott

MRC Virology Unit, Institute of Virology, Church Street, Glasgow, G11 5JR Scotland

The genome of bunyaviruses comprises three segments of negative-stranded RNA designated L (large), M (medium) and S (small) (1). In the Bunyavirus genus, the L RNA is believed to encode the L protein, the M segment encodes two virion glycoproteins G1 and G2 and at least one non-structural protein (NS_m) and the S RNA encodes the nucleocapsid protein, N, and a further non-structural protein NS_s (1, 2, 3). It has been shown by nucleotide sequence analysis of a cloned DNA copy of the S RNA segment that N and NS_s are encoded by overlapping reading frames (4), but little is known about the synthesis of bunyavirus mRNAs. With a view to studying the coding potential of bunyavirus genome segments and to examine transcription of the viral RNA, we undertook to produce complementary DNA (cDNA) copies of the individual gene segments for use as specific probes. Recombinant plasmids containing bunyavirus genetic information have also proved valuable in determining the genotypes of bunyavirus recombinants by a dot-hybridization procedure (see Pringle et al., this volume).

cDNA CLONING

The terminal nucleotides of the three genome segments of bunyaviruses are conserved and exhibit partial inverted complementarity; furthermore the conserved sequences appear to be genus specific (5). In the genus Bunyavirus, the conserved sequences are:

Copyright © 1984 by Academic Press, Inc.
All rights of reproduction in any form reserved.
ISBN 0-12-183501-4

3' UCAUCACAUG.....
and
5' AGUAGUCAC...... (5, 6).

We have exploited this fact in our cloning protocol which employs synthetic oligonucleotides to prime specifically both first and second strand cDNA synthesis (Elliott et al., in preparation). The single-stranded cDNA products of one experiment, resolved on an alkaline-agarose gel, are shown in Figure 1. Prominent bands corresponding to full-length cDNA copies of the L and M segments of Bunyamwera virus were clearly seen; a band corresponding to the S gene was less distinct on this gel. The sizes of the cDNAs, as estimated from their electrophoretic migration under denaturing conditions, are compared with estimates made from RNA (1) in Table 1. Also included is the actual length of the snowshoe hare virus S gene obtained by sequence analysis (7). The cDNA estimates are generally shorter than those made from RNA but are probably more accurate because of the difficulties in sizing single-stranded RNA directly.

Double-stranded cDNA was inserted into the plasmid pBR322, and recombinant plasmids identified by in situ hybridization using ^{32}P-labelled individual virion RNA species as gene specific probes.

CHARACTERIZATION OF RECOMBINANT PLASMIDS

Partial characterization of a number of

TABLE 1
COMPARISON OF THE LENGTHS OF BUNYAMWERA GENES ESTIMATED FROM ANALYSES OF RNA AND cDNA

Segment	RNA (ref 1)	cDNA	Sequence (ref 7)
L	8200-9400b	7000b	-
M	5450-6970b	4300b	-
S	850-1500b	900b	980b

recombinant plasmids is given in Table 1 of Pringle et al., this volume. We have not so far identified any plasmids with full length inserts, though by cross hybridizations some of the cloned inserts are non-overlapping (data not shown). It should be possible, therefore, to identify plasmids containing inserts covering all of the gene sequences from our cDNA library. One surprising feature of the library is the relative lack of S gene specific plasmids, and also that the S gene plasmids so far identified all have short inserts. This may indicate greater secondary structure in the S RNA which affects reverse transcription of this gene.

1
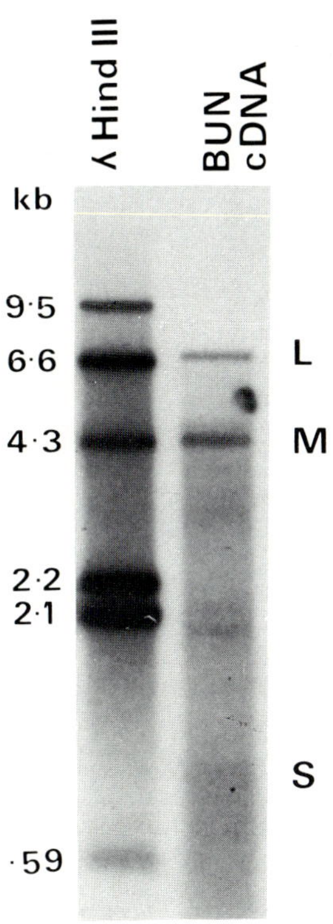

2
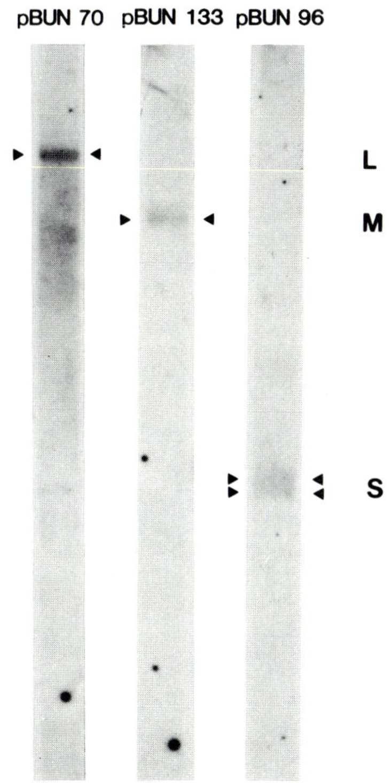

FIGURE 1. Alkaline agarose gel analysis of Bunyamwera virus first-strand cDNAs.

FIGURE 2. Northern blot analysis of Bunyamwera virus infected cell specific RNAs.

IDENTIFICATION OF INFECTED CELL SPECIFIC RNA SPECIES

Preliminary experiments have shown that Bunyamwera recombinant plasmids can be used as hybridization probes to identify virus specific RNA species in infected cells. Unfractionated RNA extracted from Bunyamwera virus infected BHK cells at 20 hours post infection was separated on a formaldehyde/agarose gel and transferred to a nitrocellulose filter. Strips cut from the filter were hybridized with plasmids labelled with ^{32}P by nick translation. Representative examples are shown in Figure 2.

The L gene probe (pBUN70) and M gene probe (pBUN133) hybridized only to single RNA species, whereas the S gene probe (pBUN96) detected two closely migrating RNAs. These results agree with observations of Fuller et al. (4) that the L gene and M gene mRNAs comigrate with virion RNAs, whereas the S gene mRNA migrates slightly faster than the S gene virion RNA species. Other recombinant plasmids will be used in an attempt to enumerate other RNA species in infected cells.

ACKNOWLEDGMENTS

We thank Carolyn Blance for technical assistance. J.F.L. was in receipt of a Medical Research Council post-graduate studentship.

REFERENCES

1. Bishop, D.H.L. and Shope, R.E. (1979). Comprehensive Virology **14**, 1-156.
2. Gentsch, J.R. and Bishop, D.H.L. (1979). J. Virol. **30**, 767-776.
3. Fuller, F. and Bishop, D.H.L. (1982). J. Virol. **41**, 634-648.
4. Fuller, F., Bhown, A.S. and Bishop, D.H.L. (1983). J. gen. Virol. **64**, 1705-1714.
5. Clerx-van Haaster, C.M., Clerx, J.P.M., Ushijima, H., Akashi, H., Fuller, F. and Bishop, D.H.L. (1982). J. gen. Virol. **61**, 289-292.
6. Obijeski, J.F., McCauley, J. and Skehel, J.J. (1980). Nucl. Acids Res. **8**, 2431-2438.
7. Bishop, D.H.L., Gould, K.G., Akushi, H. and Clerx-van Haaster, C.M. (1982). Nucl. Acids Res. 10, 3703-3713.

HYBRID SELECTION OF VIRAL mRNA USING cDNA CLONES OF INKOO VIRUS M GENE [1]

Melanie K. Spriggs and Martinez J. Hewlett

Department of Cellular and Developmental Biology
University of Arizona, Tucson, AZ 85721

The members of the family Bunyaviridae possess a single strand, tripartite RNA genome of negative sense. Little is known about the mRNA species of these viruses, aside from the fact that they do not possess a 3' poly A tail of sufficient length to adhere to an oligo[dT] column (Parker and Hewlett, unpublished observations). Thus it is difficult to obtain purified virus specific messages for a detailed study of their structural properties and coding capacities. In order to purify bunyavirus messages, we have attempted to clone Inkoo virus, a member of the California Encephalitis serogroup, and use these clones to hybrid select viral mRNA.

The cloning procedure we employed is one used by J.F. Obijeski (personal communication) in cloning the LaCrosse S genome segment. Nucleocapsid RNA, representing full length plus and minus strands, was purified away from cellular RNA and viral mRNA in a 20-40% CsCl gradient (1) (Fig. 1). Nucleocapsid bands were harvested, phenol extracted and denatured to give the complementary genome length templates used in first strand synthesis. Figure 1 contains an autoradiogram of a 1.5% acid urea gel on which ^{3}H-labeled nucleocapsid RNA (NC) has been electrophoresed (2). This system, which separates on the basis of composition as well as size, resolves the plus and minus genome length RNA of L,M, and S (3).

[1]This work was supported by NSF Grant PCM-8310815

Copyright © 1984 by Academic Press, Inc.
All rights of reproduction in any form reserved.
ISBN 0-12-183501-4

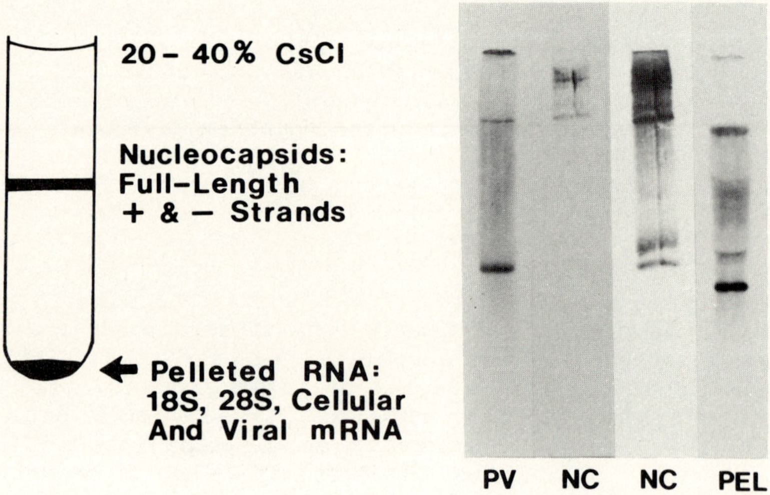

FIGURE 1. Schematic representation of the location of viral RNAs in a 20-40% CsCl gradient. The right panel is a composite autoradiogram of a 1.5% acid urea gel showing purified Inkoo virus (PV); Nucleocapsids (NC), full genome length plus and minus strand RNA and viral RNA found in the pellet (PEL).

At this point, nucleocapsid RNA was 3' polyadenylated, denatured and first strand synthesis carried out using an oligo[dT] primer. Alternatively, two artificial oligodeoxynucleotide primers representing the complement of the first 11 nucleotides of the 3' end of the two strands was used and subsequent extension with reverse transcriptase carried out.

At the completion of first strand synthesis the mixture was treated with base to remove RNA and the remaining DNA allowed to self-anneal. The large fragment of E. coli DNA polymerase 1 (Klenow) was used to extend any incomplete complementary strands (second strand synthesis). The resulting double strand cDNA was electrophoresed on a 4% polyacrylimide gel, stained with ethidium bromide and the proper size bands electroeluted. Using this method, we have obtained full length cDNA (as discerned by gel electrophoresis) of all three Inkoo virus genome segments. This cDNA was then C-tailed with terminal transferase, annealed to G-tailed pBR322 (opened at the PST1 site in the amp^R gene), and transformed into E. coli strain mm294 (Fig.2).

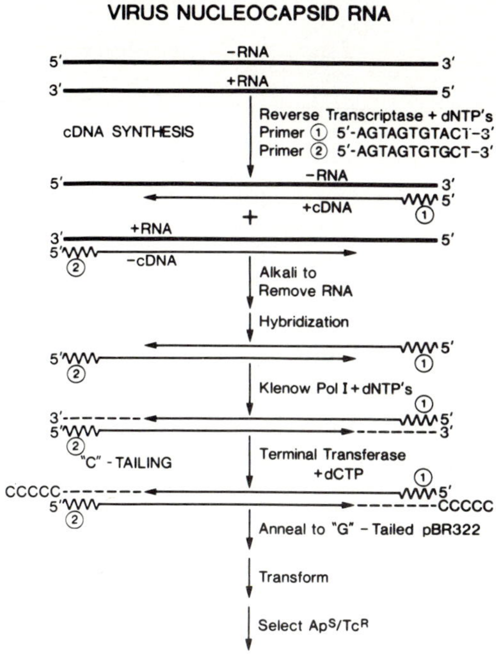

FIGURE 2. Methodology used in cloning Inkoo RNA genome segments. Full length nucleocapsid RNA is polyadenylated or annealed to synthesized primers and serves as the template for first strand synthesis.

Potential Inkoo clones were identified as virus specific by dot-blot analysis. Viral segment specificity was determined by nick translation of plasmids containing viral sequences and hybridization to purified virion RNA electrophoresed on a denaturing gel and transferred to nitrocellulose.

Messenger species corresponding to a desired segment were then obtained by hybrid selection (4). In brief, plasmid DNA was bound to a nitrocellulose filter, baked, and hybridized overnight with RNA obtained from the pellet of a 20-40% CsCl gradient containing virus infected cell lysates (Fig.1). Filters were washed and bound RNA eluted by freeze-thawing. Recovered RNA was then visualized by gel electrophoresis or used to program a cell-free reticulocyte translation reaction. Figure 3 is an autoradiogram of hybrid selected RNA. Each of our M segment specific clones, which together represent at least 60% of the genome, selected only one subgenomic RNA species while the LaCrosse S segment

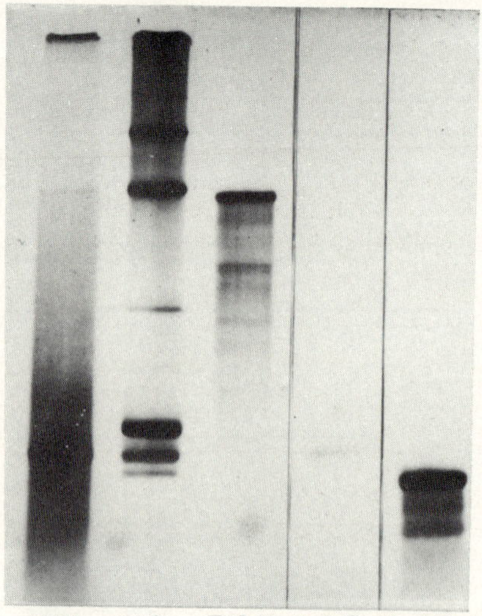

FIGURE 3. Autoradiogram of a composite gel. PV:purified Inkoo virus, NC: nucleocapsids (showing cross contamination of subgenomic S RNA species), B35: subgenomic RNA species hybrid selected using M segment specific cDNA clone, LAC (uninf): hybrid selection of ^{3}H-labaeled, actinomycin D treated uninfected cell pellet RNA with LaCrosse cDNA clone, LAC: hybrid selection of ^{3}H labaeled pelleted Inkoo RNA with LaCrosse cDNA clone.

clone selected three distinct size classes of pelleted Inkoo specific RNA. To date we have been unsuccessful in translating our M subgenomic RNA in vitro, but are able to translate immunoprecipitable nucleocapsid protein from material selected with an S clone. Separation of the three S subgenomic species and subsequent translation of each RNA size class is in progress.

REFERENCES

1. Leppert, M.L., Rittenhouse, L., Perrault,J., Summers,D.F. and Kolakofsky, D. (1979). Cell 18, 735-747.
2. Lehrach, H., Diamond, D., Worney, J.M., and Boedtker,H., (1977). Biochem. 16, 4743-4751.
3. Lindsey, Helen Ph.D. thesis, Emory University, 1983.
4. Ricciardi, R., Miller, J.S. and Roberts, B., (1979). P.N.A.S. 76: 4927-4931.

RIFT VALLEY FEVER VIRUS INTRACELLULAR RNA: A FUNCTIONAL ANALYSIS

M. D. Parker, J. F. Smith, and J. M. Dalrymple

U.S. Army Medical Research Institute
of Infectious Diseases
Fort Detrick
Frederick, Maryland 21701

Rift Valley fever (RVF) virus is a member of the genus Phlebovirus in the family Bunyaviridae. Bunyaviridae are composed of three, unique, negative stranded RNA genome segments associated with a major nucleocapsid protein (N) in the form of helical ribonucleoproteins. These nucleocapsids are surrounded by a lipid envelope containing two virus specific glycoproteins, G1 and G2. In addition, a high molecular weight protein is also located in the interior of the particle which may be responsible for the RNA polymerase activity found in preparations of purified virus (1).

In addition, nonstructural polypeptides have been identified in infected cells (2,3,4). Using snowshoe hare virus, it has been determined that the small (S) genome segment encodes the nucleocapsid protein and a nonstructural polypeptide, NSs. The M segment has been shown to direct the synthesis of the envelope glycoproteins and an additional nonstructural polypeptide, NSm (2,5). The large segment (L) presumably encodes the synthesis of the high molecular weight polypeptide L, but there is no direct evidence for this association.

Since at least two of the genome segments have been shown to code for multiple polypeptides, it becomes important to determine if the individual polypeptides are generated by transcription of multiple messages from a single genome segment or alternatively, if the polypeptides are derived by proteolytic cleavage of precursor polypeptides. We have undertaken studies of the RNA produced during infection of mammalian cells with Zagazig Hospital 501 strain of Rift Valley fever virus in order to determine whether one or both of these mechanisms are utilized during replication of this virus.

Copyright © 1984 by Academic Press, Inc.
All rights of reproduction in any form reserved.
ISBN 0-12-183501-4

VIRION STRUCTURAL COMPONENTS AND INTRACELLULAR PROTEIN SYNTHESIS

The RVF virus genome consists of three RNA segments. Compared to ribosomal RNA markers, the molecular weights of the genome segments determined are 0.7, 1.7 and 2.8 x 10^6 for S, M and L respectively (fig. 1). These genome segments have the eight conserved nucleotides at the 3' terminus of each segment, HO-UGUGUUUC (data not shown), reported previously for other members of the Phleboviruses (6).

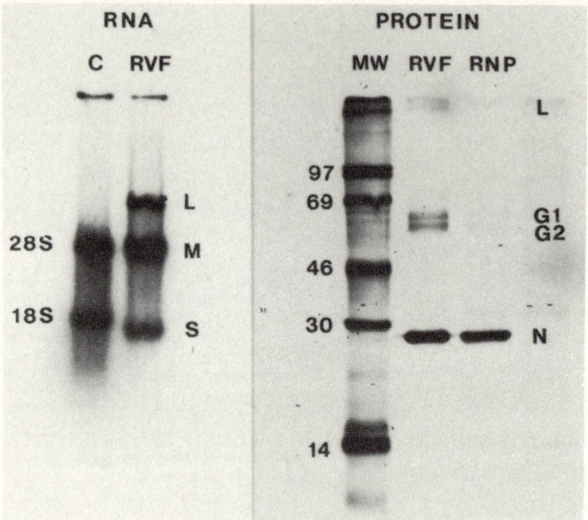

FIGURE 1. Structural components of Rift Valley fever virus. Infected Vero cells were labeled with 50 µCi/ml ^{32}P-orthophosphate or 25 µCi/ml ^{35}S-methionine (Amersham, 800 Ci/mMole) in Earles' minimal essential medium containing 2% fetal bovine serum. Clarified culture fluid was precipitated with saturated ammonium sulfate. The precipitated material was dissolved in 100 mM NaCl, 10 mM tris-HCl, pH 7.4, 1 mM EDTA and centrifuged on a 15-60% sucrose gradient. ^{32}P-orthophosphate labeled virus was extracted with phenol: chloroform. The RNA was ethanol precipitated and electrophoresed on 1.4% agarose gels containing 8 mM methyl mercuric hydroxide (7). ^{35}S-methionine labeled virus was disrupted with SDS and electrophoresed on 13% acrylamide gels according to Laemmli (8). Virion ribonucleoproteins were purified according to Leppert et al. (9). Vero cell marker RNA (C): Rift Valley fever virus (RVF); molecular weight markers (MW); the molecular weights are indicated x 10^{-3}; virion ribonucleoproteins (RNP).

Analysis of the virion structural polypeptides (fig. 1) revealed a nucleocapsid (N) protein of 28,000 daltons and the envelope proteins, G2 and G1, estimated at 58,000 and 65,000 daltons, respectively. The high molecular weight polypeptide at the top of the gel exhibits a molecular weight of approximately 180,000 daltons and presumably represents the virion associated transcriptase. G1 and G2 are glycoproteins as shown by their ability to incorporate ^{3}H-mannose and ^{3}H-glucosamine (not shown). Nucleocapsids obtained by equilibrium centrifugation in CsCl density gradients are composed primarily of the N protein (fig. 1, RNP).

Examination of the proteins synthesized during the course of infection, Figure 2, revealed a rapid decrease in host cellular protein synthesis with virus-specific proteins representing an increasing proportion of total protein

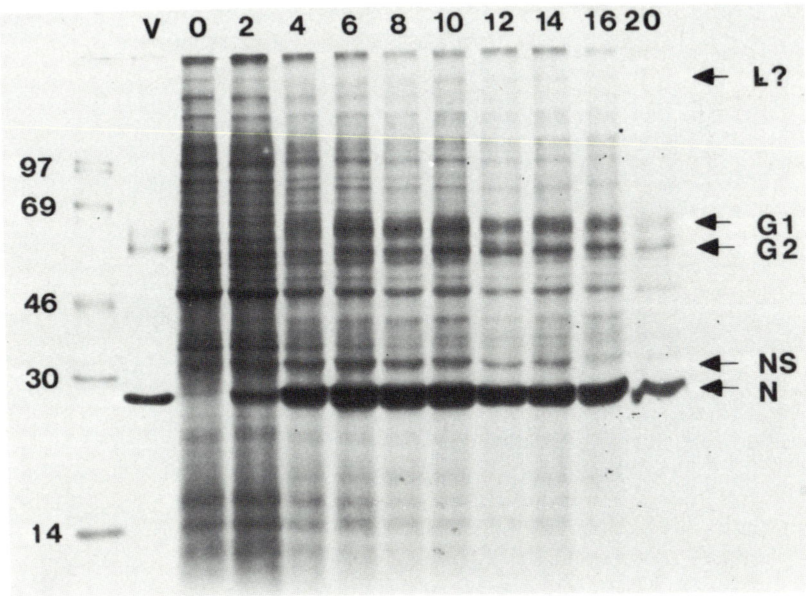

FIGURE 2. Time course of polypeptide synthesis during infection by Rift Valley fever virus. Vero cells were infected by a multiplicity of 10. At the hours indicated after infection, the cells were pulsed for two hours with 50 μCi/ml ^{35}S-methionine. Monolayers were lysed in 500 mM KCl, 20 mM tris-HCl, pH 7.2, 1% NP-40, 0.5% sodium deoxycholate. Electrophoresis was on 13% acrylamide gels according to Laemmli (8). The numbers at the left indicate the molecular weights of the marker proteins (x 10^{-3}). Polypeptides of purified virus (V).

synthesis. In addition to the structural polypeptides described above, RVF virus also directs the synthesis of a nonstructural polypeptide of 31,000 daltons. This protein has been found associated with the nuclear fraction of infected cells (3). We have evidence suggesting that this nonstructural polypeptide is phosphorylated at serine and threonine residues (not shown). Although other members of the Bunyaviridae have been found to direct the synthesis of at least one additional nonstructural polypeptide (2), we have not detected other virus specific polypeptides in lysates of RVF virus infected cells.

RNA FUNCTION

Agarose gel electrophoresis of radiolabeled RNA extracted from infected cells revealed an RNA of approximately 300,000 daltons, in addition to the three RNAs which migrate to the positions of the genome segments L, M and S (fig. 3).

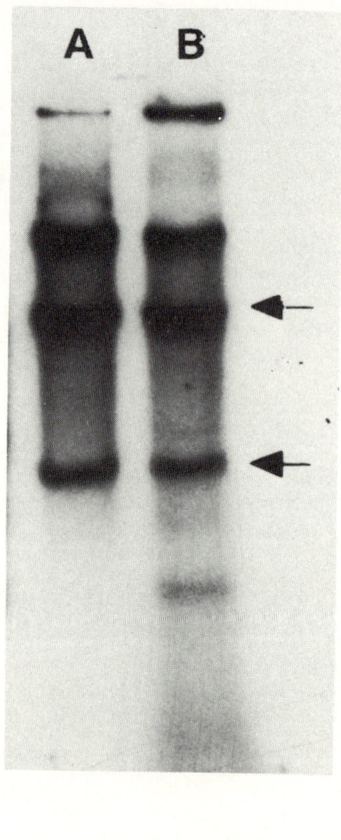

FIGURE 3. Rift Valley fever virus specific intracellular RNA. Infected Vero cells were treated with 10 μg/ml Actinomycin D at 4.5 hours postinfection and pulsed with 300 μCi/ml ^{32}P-orthophosphate from hours 5 through 7. Cells were lysed in 80 mM KCl, 20 mM tris-HCl, pH 7.4, 5 mM $MgCl_2$, 1% NP-40. The cytoplasmic RNA was extracted with phenol: chloroform. Radioactive virus was prepared as described in the legend to Figure 1. Electrophoresis was as described (7). Virion RNA (Lane A); Intracellular RNA (Lane B).

In order to further characterize these RNAs, lysates of infected cells were fractionated by CsCl centrifugation (8), yielding a pellet rich in ribosomal RNA and a peak of RNA at a density of 1.29, characteristic of virion ribonucleoproteins (RNP). Agarose gel electrophoresis of the RNA in the RNP fraction yielded only genome sized RNAs (fig. 4). The pellet also contained three virus-specific species of RNA. One migrated slightly ahead of the M segment found in the RNP, another migrated precisely with the S segment seen in the RNP fraction and the third was the 300,000 dalton RNA. Because the pelleted material consisted largely of polysomal associated RNA (9), it was expected that the virus-specific RNA in the pellet would be of positive polarity and would therefore hybridize to virion RNA. Unlabeled, RVF virus-infected cells were fractionated on CsCl and both RNP and pelleted RNA fractions were analyzed by Northern blotting. Probe was prepared by labeling a partial RNAse T1 digest of

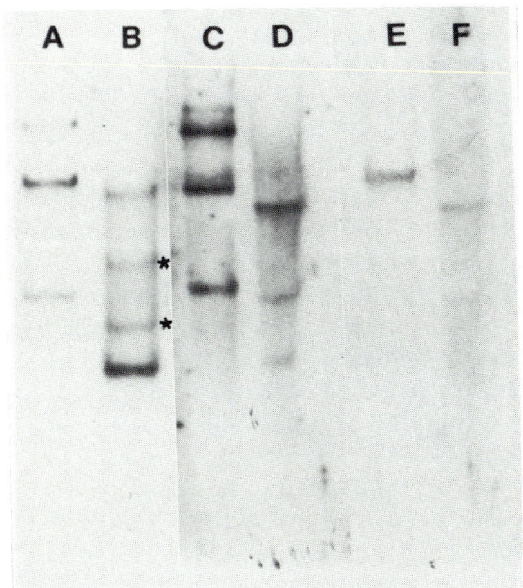

FIGURE 4. Northern blot analysis of virus specific intracellular RNA. Infected Vero cells were fractionated on CsCl (8) and electrophoresed on agarose gels containing 8 mM methyl mercuric hydroxide. The RNA was then electroblotted to ABM paper (BioRad Laboratories). ^{32}P-labeled RNP and pelleted RNA (lanes A and B), RNP and pelleted RNA probed with T1 fragments of virion RNA (lanes C and D); E and F, RNP and pelleted RNA probed with nick translated pRVF-34 (lanes E and F). Asterisk denotes two RNAs of nonviral origin which continued to be synthesized in the presence of Actinomycin D.

virion RNA with polynucleotide kinase and ^{32}P-ATP. Virion RNA probes hybridized to all three bands in the RNA extracted from intracellular ribonucleoproteins and to the three bands in the pelleted RNAs (fig. 4). An identical pattern of hybridization was obtained when RNA from the intracellular ribonucleoproteins was utilized as probe (data not shown). These results indicate that intracellular ribonucleoproteins contain RNA of both plus and minus polarity and that the CsCl gradient pellets contain plus stranded RNA species. When the same blot was probed using clone pRVF-34, which contains a DNA insert of approximately 2000 base pairs representing a portion of virus M segment near the 3' (Purchio et al., this volume), the resulting autoradiograph (fig. 4), illustrates that the middle band of the RNP RNA and the uppermost of the three bands in the pellet are homologous to sequences found in the M segment of the virus. The increased electrophoretic mobility of this band relative to the M segment of the genome suggests that this RNA is a subgenomic transcript and makes it a likely candidate for the glycoprotein messenger RNA.

In order to determine the polypeptides encoded by the mRNAs in the pelleted RNA, total pelleted RNA was fractionated by preparative agarose gel electrophoresis. Because early experiments indicated that the mRNA for the N and NS polypeptides sedimented at approximately 12-18S on sucrose gradients, the portion of the gel encompassing RNA of this size class was fractionated, the RNA eluted and translated in a reticulocyte extract. In Figure 5, the open arrow indicates the position of the S segment of vRNA in the agarose gel with fractions containing successively lower molecular weight RNA species to the right. The data in Figure 5 indicate that the maximum synthesis of N and NS polypeptides was directed by RNA in the molecular weight range of 300,000 daltons (lanes 6,7,8). It should be noted that the maximum of translational activity for N and NS did not exactly coincide suggesting that there are separate messages for each polypeptide. In fact, two bands of RNA in this region have been occasionally resolved on denaturing agarose gels. The data indicate that the NS polypeptide was translated from an mRNA slightly larger than that for N protein. A second region of messenger activity coincided with the location of the RNA corresponding to the full size S genome segment (lane 1). Because of the previous observation that the RNPs isolated from the cytoplasm contained positive stranded RNA, the RNA extracted from RNPs was also examined in an <u>in vitro</u> translation system. This RNA was also shown to yield N and NS polypeptides (lanes 11 and 12). However, while the relative amounts of the N and NS polypeptides synthesized in response to the 300,000 dalton mRNAs reflected the relative amounts of the two polypeptides synthesized *in vivo* (fig. 3) the yields of N and NS from the RNP RNA did not reflect this same ratio.

SUMMARY

The synthesis of multiple polypeptides from a single genome segment has precedent in the case of both influenza viruses and members of the family Bunyaviridae through proteolytic processing of precursor polypeptides and transcription of multiple messages from one genome segment (2,10,11). It appears that RVF virus accomplishes the synthesis of multiple polypeptides from two of its genome segments by different strategies.

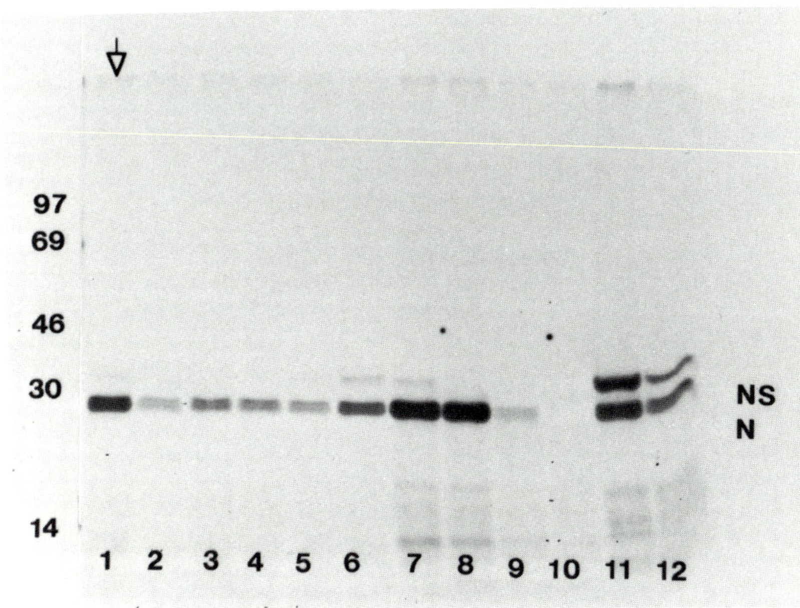

FIGURE 5. *In vitro* translation of fractionated intracellular mRNA. The CsCl pelleted RNA was electrophoresed on 1.4% agarose gels containing 8 mM methyl mercuric hydroxide (7). The gel was fractionated as described in the text, lanes 1-10. The arrow indicates the fraction containing the RNA which migrates as full length S segment. Translation products of RNA from intracellular RNP, lanes 11, 0.5 μg RNA; lane 12, 1 μg RNA. Molecular weights of marker polypeptides (x 10^{-3}) are shown at the left.

By analogy with other Bunyaviruses, it was assumed that the S segment would encode the N and probably the NS polypeptide. We have demonstrated this to be true by direct translation of polysomal RNA which is identical in size to the corresponding genome segment. The same polypeptides were translated from the RNA of intracellular RNPs which contain only genome size RNAs. In addition, the same polypeptides were translated from two separate, subgenomic mRNAs. In the absence of any additional polypeptides encoded by the S segment of RVF, it would appear that these data are in contrast to LaCrosse and snowshoe hare viruses which have substantially smaller S genome segments and utilize overlapping reading frames to encode two polypeptides (10,11). RVF virus may utilize separate reading frames to encode the N and NS polypeptides. The coding capacity necessary for these two polypeptides would account for essentially the entire S segment. Whether synthesis of the two subgenomic mRNAs is accomplished by internal initiation of transcription or some other mechanism is not known. Our observation that an RNA of S segment size can serve as messenger for two polypeptides which are also translated from two smaller RNAs poses an interesting question.

REFERENCES

1. Bishop, D. H. L., and Shope, R. E. (1979). In "Comprehensive Virology" (H. Fraenkel-Conrat and R. R. Wagner, eds.), p. 1. Plenum Press, New York.
2. Fuller, F., and Bishop, D. H. L. (1983). J. Virol. 41, 643.
3. Struthers, J. K., and Swanepoel, R. (1982) J. Gen. Virol. 60, 381.
4. Smith, J., and Pifat, D. (1982). Virology 121, 61.
5. Gentsch, J., and Bishop, D. H. L. (1979). J. Virol. 30, 767.
6. Clerx-Van Haaster, C.M., Clerx, J. P., Ushijimi, H., Akashi, H., Fuller, F., and Bishop, D. H. L. (1982). J. Gen. Virol. 61, 289.
7. Bailey, J. M., and Davidson, N. (1976) Anal. Biochem. 70, 75.
8. Laemmli, U. K. (1970). Nature 213, 1133.
9. Leppert, M. L., Rittenhouse, L., Summers, D. F., Kolakofsky, D. (1979). Cell 18, 735.
10. Patterson, J. L., Cabradilla, C., Holloway, B. P., Obijeski, J. F., and Kolakofsky, D. (1983). Cell 33, 791.
11. Cabradilla, C. D., Holloway, B. P., and Obijeski, J. F. (1983). Virology 128, 463.

TRANSCRIPTION OF THE S RNA SEGMENT OF GERMISTON BUNYAVIRUS[1]

M.Bouloy, N.Pardigon, S.Gerbaud, P.Vialat and M.Girard

Unité de Virologie Moléculaire, Institut Pasteur, 25 rue du Dr.Roux, 75724 Paris Cedex 15, France

STRUCTURAL ANALYSIS

Germiston virus is a member of the Bunyamwera serogroup of bunyaviridae(1). Its genome consists of three single-stranded RNA molecules of negative polarity: L (large), M (medium) and S (small). Each segment contains a unique nucleotide sequence and thus encodes different polypeptides (2,3). On agarose gels in the presence of methyl mercuric hydroxide the L, M and S RNA segments migrated with apparent molecular weights of 2.8×10^{6}, 1.6×10^{6} and 0.34×10^{6} daltons respectively(3).

Each Bunyavirus RNA subunit is associated with two polypeptides: a major one, the nucleoprotein N and a minor one, the large polypeptide L. In the case of Germiston virus, complementarity of the ends of the S RNA molecule has been demonstrated by direct sequencing and isolation of panhandle structures containing the 3' and 5' termini of the RNA molecules (4).

VIRUS-INDUCED PROTEIN SYNTHESIS

For all bunyaviruses studied so far, virus particles have been shown to contain four structural proteins two external glycoproteins G1 and G2, a major internal nucleoprotein N and a high molecular weight but minor protein L(2). These four structural proteins have been described in the case of Germiston virus(5).

We analyzed the proteins synthesized in BHK21 cells infected by Germiston virus after labeling with ^{35}S methionine from 4 to 6 hr after infection.

---

[1] This work was supported by grants n°PRC 134001 and CRL 811001 from Institut National de la Santé et de la Recherche Médicale

Copyright © 1984 by Academic Press, Inc.
All rights of reproduction in any form reserved.
ISBN 0-12-183501-4

The structural polypeptides L,G1 and N were clearly detected in infected cells (Fig.1)

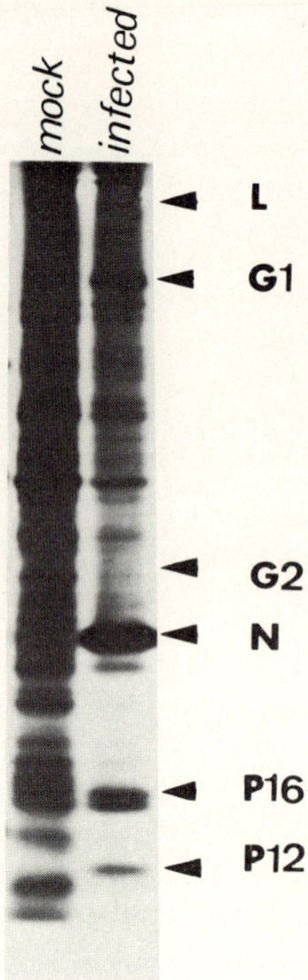

Fig.1:Polypeptides synthesized in Germiston virus infected cells or uninfected cells.Cells were labeled with ^{35}S methionine from 4 to 6 hr.Proteins were analyzed on a 14% polyacrylamide-SDS gel (Bouloy et al.,submitted for publication).

The glycoprotein G2 usually comigrated with a cellular protein so that the viral protein was visualized in the gel only late after infection when the cellular protein synthesis was shut off.Kinetics of viral protein synthesis indicated that N and G1 were present very early (1 to 1.5 hr after infection). In addition,several polypeptides which did not appear in uninfected cells and therefore were probably induced by the virus,have been observed. Two of them, the polypeptides of 12,000 and 16,000 daltons designated p16 and p12 are indicated in Fig.1. In other experiments,we observed the presence of polypeptides of 20,000, 60,000 and 72,000 daltons which were called respectively p20,p60 and p72.

SYNTHESIS OF INTRACELLULAR RNAs

Germiston virus infected BHK21 cells were exposed to (^3H) adenosine or (^3H) uridine in the presence of actinomycin D which does not block the replication of Bunyaviruses (6,7).Cytoplasmic extracts were prepared and the polysomes and viral ribonucleoproteins (RNP) were separated on sucrose gradients under the conditions previously described for Lumbo virus (8).The RNAs were analyzed by electrophoresis on a 2.8% polyacrylamide gel containing 6M urea (Fig.2a) or on 1% agarose gel containing 5mM methylmercuric hydroxide (Fig.2b).

Four RNA species were synthesized in infected cells,three of which (L,M and S1) had the same migration rate as the three genomic RNA segments (L,M and S respectively).The fourth RNA species,absent from the virion,migrated faster than the S segment and was designated S2.None of these RNAs could be bound to oligo(dT) cellulose (data not shown),suggesting that,contrary to snowshoe hare virus,the 3' end of the cRNAs of Germiston virus is not polyadenylated (9). RNA S2 was the most abundant species found in the infected cell polysomes (Fig.2a,lane 1) suggesting that it functioned as mRNA.By hybridization with viral RNAs,we showed that S2 RNA is a transcription product derived from the S segment of the genome.Transcripts from the L and M segments, the size of which was similar or identical to their respective templates were also present in the polysomes suggesting that full-length or almost full-length transcripts from the larger segments functioned as messengers.One should note that S1,the full-length transcript of the S segment was never found in the polysomal fraction (lane 1) but only in the RNPs (lane 4).Intracellular RNPs were resolved into three classes which contained respectively the L,M and S RNA (lanes 2,3 and 4 Fig.2 a and b). The L, M and S RNAs were composed of two types of molecules of opposite polarities,the virion-sense RNA and their equal size complementary molecules (antigenome).The presence of S2 in the S region of the RNPs (fig.2 a,lane 4),could be explained if the messenger RNA formed monosomes which cosedimented with the small RNPs.

The full-length transcript S1 and the messenger S2 were further analyzed by duplex formation with virion RNA.The size of the duplexes obtained

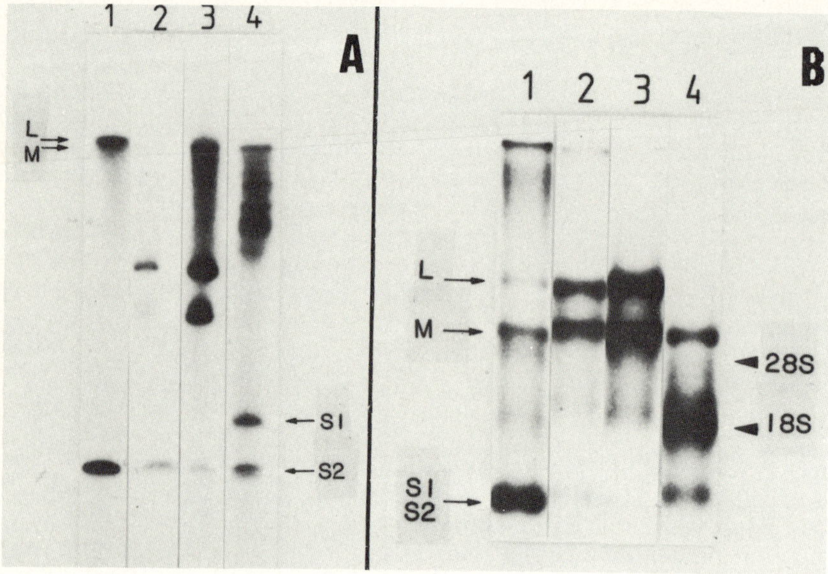

Fig.2: Analysis of cytoplasmic RNAs isolated from Germiston virus infected BHK21 cells labeled with ^{32}P (50 µCi/ml) from 4 to 6 hr in the presence of Actinomycin D (5 µg/ml). The cytoplasmic extract was fractionated on a sucrose gradient (8) and RNAs from the polysomes (lane 1), the L, M and S RNPs (lanes 2, 3, 4 respectively) were analysed on a 2.8% polyacrymide gel containing urea (a) or on a 1% agarose gel containing 5mM methylmercury (b).

after RNase T2 digestion was estimated by electrophoresis on a non-denaturing polyacrylamide gel. The duplex formed with S2 was about 100-150 bp shorter than the full length S1 RNA duplex, the size of which was about 1100 bp (data not shown). By hybrid formation with virion S RNA labeled at its 3 or its 5' end with 32P, we showed that transcription is iniatated exactly at the 3' end of the viron RNA and continues either to the 5' end of the template generating S1 or stops about 100 nucleotides before the 5' end, generating the molecule S2 (data not shown)

In order to determine whether the transcripts S1 and S2 are capped or not, we took advantage of of the capacity of the capping enzyme isolated from vaccinia virus to label specifically the 5' end of di- or triphosphate-terminated RNAs in the presence of ($\alpha^{32}P$) GTP and S-adenosyl methionine, to try to label the 5' end of RNA S2.As a control,we ran in a parallel reaction a RNA sample that had been chemically decapped by periodate oxidation followed by β elimination.S1 and S2 RNAs were purified from a cytoplasmic extract by sucrose gradient centrifugation. Because the fraction contained also cellular RNAs which would be labeled by the capping enzyme,we then annealed the sample with excess cold virion RNAs and analyzed the resulting RNase T2 resistant duplexes by polyacrylamide gel electrophoresis.Analysis of the RNase-resistant duplex molecules showed that S2 RNA,as well as S1 RNA,was labeled prior to the decapping treatment (fig.3),indicating that both transcripts were naturally uncapped. Labeling occured to about the same extent when the RNAs had been β eliminated before the capping reaction (data not shown).It should be noted that the cytoplasmic RNAs used for this experiment also contained the S genomic segment,which terminates with a 5' di-or triphosphate and could be capped (data not shown). Therefore,the label in the S1 duplex could be associated either with the 5' end of the antigenome or with that of the genome. However,because hybridization was performed with an excess of cold viral RNA,it is likely that the label was associated with the antigenome only. This problem does not apply to RNA S2 since the 5' end of the vRNA is not conserved in the duplexes formed with S2 RNA.

Added to the reticulocyte lysat, or L cell extracts,RNA S2 directed the synthesis of two proteins comigrating with nucleoprotein N and with the nonstructural protein p12 respectively.Both the N protein synthesized in vitro and the one labeled in vivo could be immunoprecipitated with an antivirion immuneserum but the same serum failed to immunoprecipitate polypeptide p12 whether synthesized in vitro or in infected cells (results not shown).Preliminary data on tryptic peptide analysis suggest that the two polypeptides synthesized in vitro are autentic N and p12 proteins.

Confirming the results obtained by Fuller and Bishop (10) with Lacrosse and snowshoe hare viruses these results indicate that the mRNA transcribed from the S segment of Germiston virus contains the genetic information for two proteins, the nucleoprotein and a virus coded nonstructural protein p12 which should be designated NSs according to their nomenclature.

DISCUSSION

The presence of two transcripts from the S RNA segment of Germiston virus was readily detected in infected BHK21 cells by electrophoresis of the RNAs in polyacrylamide-urea gels. The transcript S1 is a full length cRNA copy of the S segment,whereas transcript S2 is about 100 nucleotides shorter, probably missing part of the 3' end of transcript S1. The existence of similar RNAs has been reported in snowshoe hare and Uukuniemi virus infected cells (9,11,12).Here, we showed that viral transcription initiates at the exact 3' end of the viral RNA and either continues to the end of the template,generating S1 RNA or terminates before the 5' end of the template,generating S2 RNA.These results do not rule out the possible existence of minor transcription products which would have escaped detection.Multiple RNAs generated in a process involving cleavage or synthesis of leader RNAs have been described by Patterson et al. (13).

Translation directed by S2 RNA in the <u>in vitro</u> cell-free systems led to the synthesis of two proteins N and p12. This was unexpected since it is believed that translation of eukaryotic mRNAS begins with the 5' proximal AUG initiation codon (14,15). Germiston virus S2 RNA may therefore be an addition to the few exceptions violating the rule of the

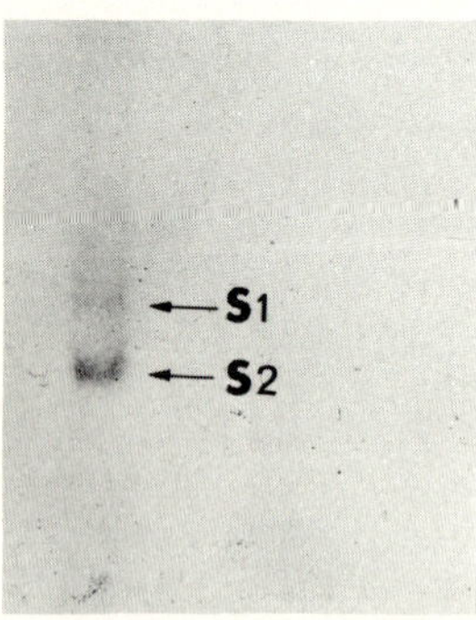

Figure 3. Analysis of the duplexes formed between viral RNA and the S RNA transcripts. The T2 RNase resistant duplexes were analyzed on a 3% polyacrylamide gel.
(Bouloy et al., submitted for publication).

scanning model.However,one cannot eliminate the possibility that the two polypeptides N and p12, might be translated from two different species of mRNA comigrating in the gel.These mRNA would be colinear except for their 5' sequences,each containing the first 5' proximal AUG triplet which precedes the open reading frame for the synthesis of the given polypeptide.It is also possible that in the cell free systems we used (reticulocyte lysate and L cell extracts)the S2 RNA was cleaved generating fragments where internal AUGs became exposed as 5' proximal initiation codons.Nucleotide sequencing of the S segment,which is presently in progress,will help to understand these problems.

REFERENCES

1.Berge,T.O. (1975) International Catalogue of Arboviruses U.S. DEW Publication n°CDC75.8301, Washington D.C.
2.Bishop,D.H.L. and Shope,R.E. (1979) In"Comprehensive Virology" (H.Fraenkel-Conrat and R.Wagner Eds) n°14, 1-15,Academic Press, New York
3.Pardigon,N.,Vialat,P.,Girard,M. and Bouloy,M. (1982) Ann.Virol.(Inst.Pasteur),133E, 461-472
4. Pardigon,N.,Vialat,P.,Girard,M. and Bouloy,M. (1982) Virology, 122, 191-197
5.Ozden,S. and Hannoun,C. (1980) Virology,103, 232- 234
6.Bouloy,M. et Hannoun,C.(1973) Ann.Microbiol.(Inst. Pasteur), 124 B, 547-553
7.Vezza,A.,Repik,P.,Cash,P. and Bishop,D.H.L. (1979) J.Virol., 31,426-436
8.Bouloy,M. and Hannoun,C. (1976) Virology, 71, 363-370
9.Cash,P.,Vezza,A.C.,Gentsch,J.R. and Bishop,D.H.L. J.Virol.,31,685-694
10.Fuller,F., and Bishop,D.H.L. (1982) J.Virol., 41, 643-648
11.Clerx-Van Haaster,C. and Bishop,D.H.L. (1980) Virology,105, 564-574
12.Ulmanen,I.,Sepälä,P. and Pettersson R,F.(1981) J.Virol., 37, 72-79
13.Patterson,J.L.,Cabradilla,J.C., Holloway,B.P., Obijeski,J.F. and Kolakofsky,D. (1983) Cell, 33, 791-799
14.Kozak,M. (1980) Cell,22,7-8
15.Kozak,M.(1981) Nucleic Acids Res., 9,5233-5252

THE S SEGMENT OF BUNYAVIRUSES CODES FOR TWO COMPLEMENTARY RNAs

Gordon Abraham and Asit K. Pattnaik

School of Science
Griffith University
Nathan, Queensland 4111
Australia

Members of the Bunyaviridae characteristically have three segments of RNA as their genetic material which provide the information for the coding of four structural proteins and at least two other non-structural proteins. This information transfer is undoubtedly mediated by messenger RNAs, but the literature concerning the number and properties of these RNAs is very limited.

Genetic and molecular evidence has shown that the intermediate sized RNA, segment M, contains the information for the coding of both the G1 and G2 glycoproteins found on the surface of the virion (1). The smallest segment, S, contains the information for the coding of the nucleocapsid protein, N, and it is believed that the minor protein, L., is coded by the largest (1) RNA segment. Recent studies show that two non-structural proteins, NS_s and NS_m, found in virus-infected cells are coded by the S and M genome segments respectively (2,3).

Direct confirmation of these relations by isolation of participating mRNAs has not been achieved as complementary RNAs from bunyavirus-infected cells are not polyadenylated (4, 5,6,7) although Cash *et al* (8) showed that both adenylated and non-adenylated fractions of RNA from infected cells could be translated to produce the nucleocapsid protein, N, of snow shoe hare virus. Thus mRNAs of bunyaviruses have been difficult to purify and to study. In addition, attempts to characterize fractionated RNAs from infected cells by translation in *in vitro* systems have not been completely successful (4,8).

Attempts to characterize mRNAs by specific hybridization to genome RNAs resulted in the detection of three different duplex RNA structures (8) in snow shoe hare virus infected cells, one of which corresponded to each of the genome segments L, M and S. This report shows that for both Akabane virus and Bunyamwera virus, four complementary RNA species can be identified by a similar approach and that two of them are transcribed from the S RNA segment.

FOUR COMPLEMENTARY RNAs IN BUNYAVIRUS-INFECTED CELLS

The synthesis of positive stranded RNAs in both Bunyamwera virus and Akabane virus-infected cells were studied by the

Copyright © 1984 by Academic Press, Inc.
All rights of reproduction in any form reserved.
ISBN 0-12-183501-4

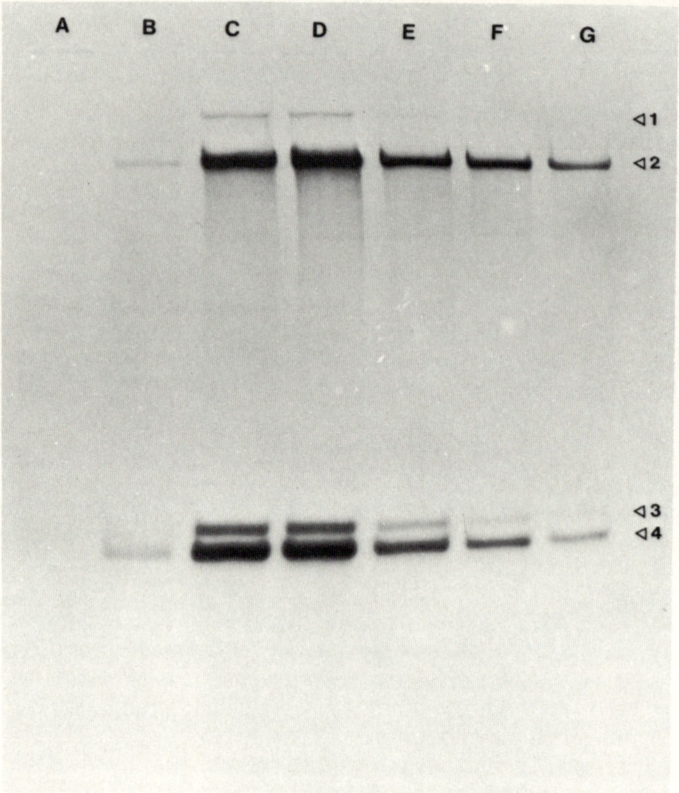

FIGURE 1. Polyacrylamide gel electrophoresis of virus-specific dsRNA hybrids. Parallel infected or mock infected cultures were incubated in the presence of actinomycin D at zero time post infection and labeled with ^{3}H-uridine for two hour periods thereafter. The RNA from each culture was extracted, hybridized with unlabeled vRNAs, and the duplexes formed were analyzed by electrophoresis as described elsewhere (6). Duplexes in lanes B to G were derived from infected cells labeled at 0, 2, 4, 6, 8 and 10 hours post infection, respectively. Lane A corresponds to a mock-infected culture labeled similarly. This figure is reproduced with the permission of the publishers of the Journal of Virology.

hybridization of total ^{3}H-uridine-labeled RNA extracted from cells with the negative stranded virion RNA. When unhybridized RNAs and single stranded regions were removed by digestion with S_1 nuclease, it was found that four double-stranded hybrid molecules could be resolved by gel electrophoresis.

This observation was extended by a kinetic study of the times at which each of the labeled cRNAs were synthesized following infection. Parallel cell cultures were infected with Akabane virus and the newly-synthesized RNAs were labeled with ^{3}H-uridine for two hour periods beginning from the time of infection. The RNAs extracted from each culture were separately hybridized with virion RNA and the hybrid molecules analyzed as described (Fig. 1). Arbitrarily numbering the duplex RNAs according to increasing electrophoretic mobility, it is seen that two of the virus specific RNAs, 2 and 4, are detected as early as two hours post infection (lane B). By 4 hours, all four duplex RNAs can be readily detected, and RNA synthesis reaches a peak at 4 to 6 hours post infection. After 8 hours, the synthesis of duplex RNA 1 falls to below detectable levels (lane F), while the synthesis of the other 3 duplexes continues in reduced amounts until at least 10 hours post infection (lane G). Similar experiments which analyzed protein synthesis in these conditions showed that the time of maximal synthesis (4-6 hours post infection) coincided exactly with the period of maximal RNA synthesis (data not shown). This observation is strongly suggestive that cRNA synthesis occurring at this time is indeed mRNA synthesis.

CODING ORIGINS OF THE FOUR cRNA SPECIES

Since four hybrid RNA species indicative of four cRNA species have been detected by this hybridization procedure, it was necessary to show that the virion RNA had, in fact, occurred. Therefore, viral RNA was prepared that was uniformly labeled with ^{32}P and the three characteristic RNA segments separated (Fig. 2) by denaturing gel electrophoresis. In this figure, Bunyamwera virus was used but similar results were obtained if Akabane virus was used alternatively. Individual L, M and S RNAs were eluted from the gel, separately hybridized with the total RNA extracted from Bunyamwera virus-infected cells and the resultant duplex molecules analyzed by electrophoresis (Fig. 2) as before. Lane T shows the four species of double-stranded-RNA (designated 1-4) obtained by hybridization of total ^{32}P-labeled RNA with RNA from infected cells. Using the L segment of vRNA produced a duplex RNA with an electrophoretic mobility corresponding principally to the largest of the four hybrid species (lane L). Similarly, the duplex RNA derived from segment M corresponded to hybrid 2. However, RNA segment S upon hybridization produced two hybrid species corresponding to hybrids 3 and 4 (lane S).

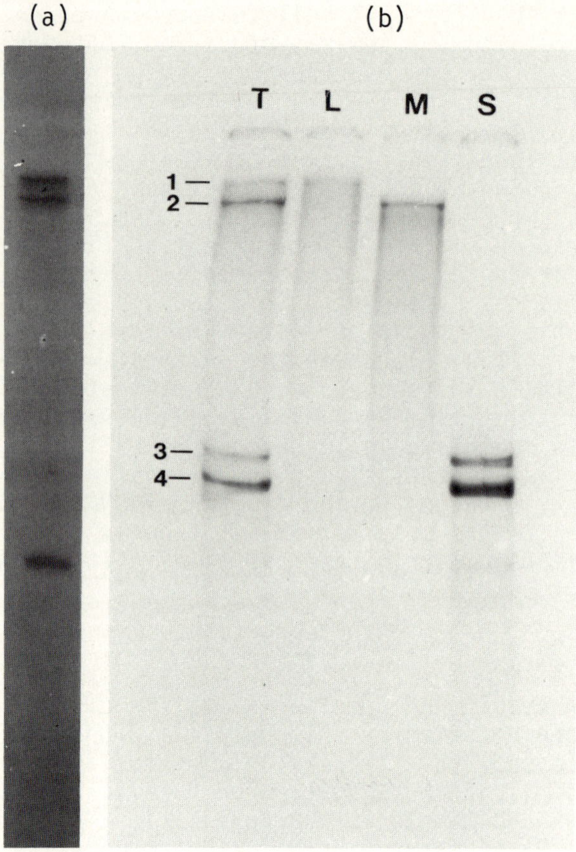

FIGURE 2. Polyacrylamide gel electrophoresis of single-stranded (a) or double-stranded (b) RNAs derived from Bunyamwera virus. Uniformly ^{32}P-labeled RNA extracted from Bunyamwera virus was analyzed by electrophoresis in denaturing conditions (a). The three radioactive bands located by autoradiography were cut out and the eluted RNA segments separately hybridized with the total RNA extracted from virus-infected cells. The hybrid molecules so formed were analyzed and detected as described elsewhere (5). Lane T shows duplex species detected following hybridization of total ^{32}P-labeled vRNA with complementary RNAs and identified arbitrarily as hybrids 1 to 4. Lanes L, M and S show hybrid products derived by hybridization of the individual genome segments with complementary RNAs. This figure is reproduced with the permission of the publishers of the Journal of General Virology.

Confirmation of the coding contained in each of the three vRNA segments was obtained by comparative oligonucleotide mapping. Individual ^{32}P-labeled L, M and S RNAs were digested with ribonuclease T_1 and the resulting oligonucleotides resolved by two-dimensional gel electrophoresis. Each RNA segment showed a characteristic fingerprint, but only that for RNA segment S is illustrated (Fig. 3).

Preparations of individual ^{32}P-labeled RNA hybrids 1 to 4 were obtained by elution from an electrophoretic separation similar to that shown in Fig. 2, lane T. Following heat denaturation, oligonucleotide maps for each hybrid species were prepared (Fig. 3). The unique oligonucleotides in the fingerprint prepared from hybrid 1 corresponded to those found in the L segment fingerprint, confirming the hybridization data that the cRNA found in hybrid 1 is transcribed from the L RNA segment (data not shown). Similarly, all of the unique oligonucleotides found in the fingerprint for hybrid 2 were represented in the fingerprint for RNA segment M (data not shown). The results for the analysis of hybrids 1 and 2 confirm the observations of Cash et al (8) that segments L and M of bunyaviruses each give rise to a single complementary RNA species.

Oligonucleotide maps prepared from hybrid RNA species 3 and 4 were compared with that obtained for RNA segment S (Fig. 3). All oligonucleotides except one in the fingerprint prepared from hybrid 3 were found also in the fingerprint of the S RNA segment, indicating that almost total protection for the S segment was provided by hybridization with this particular species of cRNA molecule. A similar fingerprint was prepared from RNA hybrid 4 except that at least five different oligonucleotides were absent when compared with the fingerprint of the S RNA segment. This result shows that a shorter complementary RNA species is present in RNA hybrid 4 which protects the RNA segment S from nuclease digestion to a lesser degree. However, the majority of the unique oligonucleotides were common to both hybrid RNA species 3 and 4, as well as to the parental molecule, segment S.

Inspection of the fingerprint for hybrid 3 (Fig. 3) shows that all but one of the unique oligonucleotides in the S RNA have been protected by hybridization, indicating that the cRNA species involved in this hybrid is almost a full-length copy of the genome segment. The fingerprint from hybrid 4 is somewhat less complex indicating the involvement of a smaller cRNA species. However, both fingerprints for hybrids 3 and 4 have a significant number of large oligonucleotides in common with each other and also with the genome segment S, indicating that considerable overlap occurs in the coding of these cRNAs. This information is consistent with previous reports (9,10) that the S segments of bunyaviruses have sequences that allow for the potential expression of two

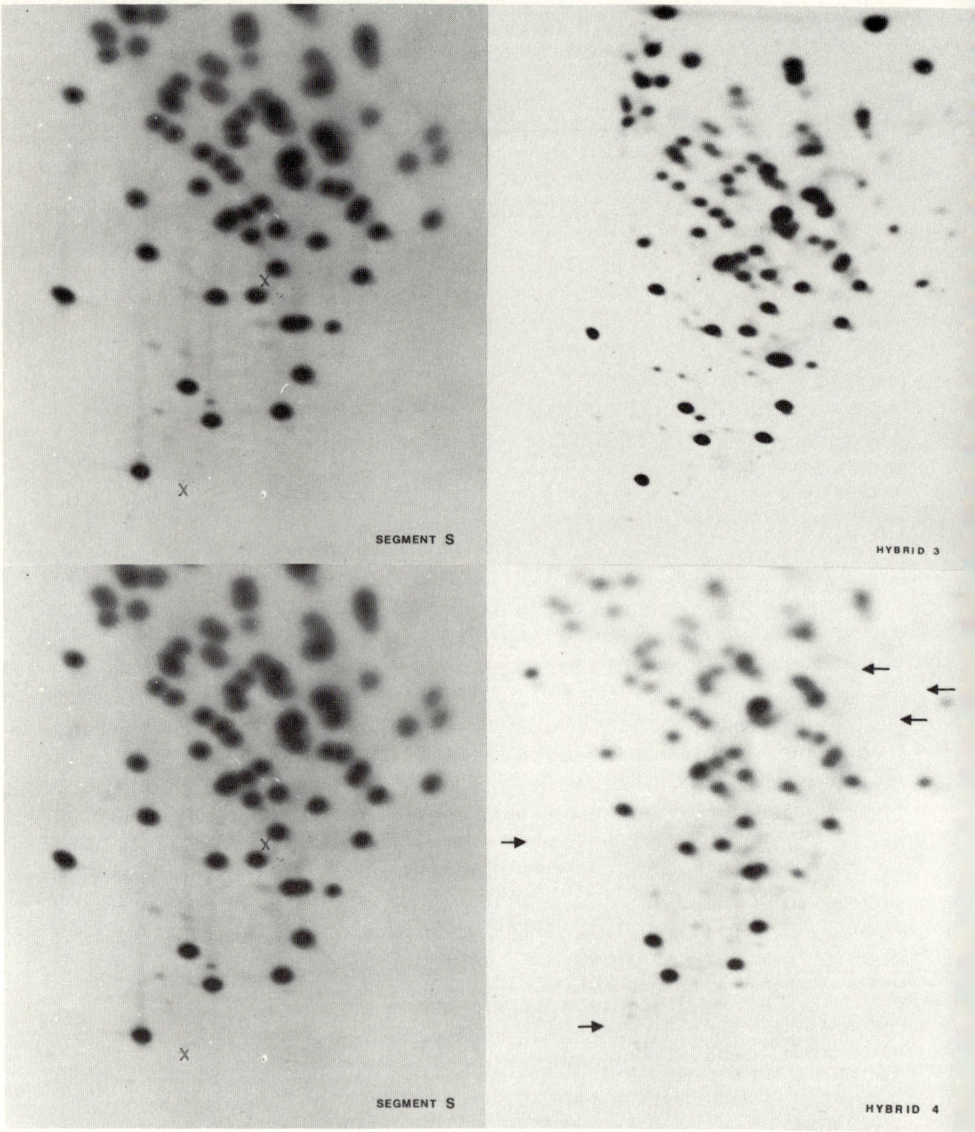

FIGURE 3. Oligonucleotide fingerprint analyses of uniformly labeled bunyaviral RNAs. Procedures for the preparation of the labeled RNAs and their digestion with ribonuclease T_1 are described elsewhere (5). The fingerprint for the S RNA segment is shown on the left hand side. Fingerprints for the hybrid RNAs 3 and 4 (see Figure 2) after denaturation are shown in the right hand panels. Arrows in the fingerprints on the right hand side indicate the corresponding positions of oligo-

gene products by using overlapping reading frames. In addition, it appears likely that both the N and NS proteins of snowshoe hare virus are unrelated, yet both are coded within the S RNA genome (3). If the cRNAs represented in hybrid RNAs 3 and 4 in this report are mRNAs, then they would have overlapping sequences which could allow the translation of two separate proteins by using different reading frames. Confirmation for the existence of two mRNAs coded by the S segment could conceivably be provided by in vitro translation studies but as yet, such data are not available.

An alternative explanation for the two transcripts from S segment RNA is that the larger represents a template used in S segment replication while the smaller is a functional mRNA. Such a situation exists, for example, in influenza virus replication (11). However, it is difficult to establish this point in bunyavirus systems as mRNAs cannot be selected on the basis of polyadenylation, nor can RNA transcription be satisfactorily restricted to that of the primary type only by using inhibitors of protein synthesis (5,6).

ACKNOWLEDGMENT

We thank Cecilia Devlin for excellent assistance.

REFERENCES

1. Gentsch, J. R., and Bishop, D. H. L. (1979). Journal of Virology 30, 767-770.
2. Gentsch, J. R., Robeson, G., and Bishop, D. H. L. (1979). Journal of Virology 31, 707-717.
3. Fuller, F., and Bishop, D. H. L. (1982). Journal of Virology 41, 643-648.
4. Ulmanen, I., Seppala, P., and Pettersson, R. F. (1981). Journal of Virology 37, 72-79.
5. Abraham, G., and Pattnaik, A. K. (1983). Journal of General Virology 64, 1277-1290.
6. Pattnaik, A. K., and Abraham, G. (1983). Journal of Virology 47, 452-462.
7. Patterson, J. L., Cabradilla, C., Holloway, B. P., Obijeski, J. F., and Kolakofsky, D. (1983). Cell 33, 791-799.
8. Cash, P., Vezza, A. C., Gentsch, J. R., and Bishop, D. H. L. (1979). Journal of Virology 31, 685-694.

---

nucleotides present in the single-stranded RNA segment but absent from the duplex analyses. This figure is reproduced with the permission of the publishers of the Journal of General Virology.

9. Clerx-Van Haaster, C. M., Akashi, H., Auperin, D. D., and Bishop, D. H. L. (1982). Journal of Virology 41, 119-128.
10. Bishop, D. H. L., Gould, K. G., Akashi, H., and Clerx-Van Haaster, C. M. (1982). Nucleic Acids Research 10, 3703-3713.
11. Hay, A. J., Abraham, G., Skehel, J. J., Smith, J. C., and Fellner, P. (1977). Nucleic Acids Research 4, 4197-4209.

RESTRICTION OF SUB-UNIT REASSORTMENT IN THE BUNYAVIRIDAE

C.R. Pringle, W. Clark, J.F. Lees
and R.M. Elliott

M.R.C. Virology Unit,
Institute of Virology, Church Street,
Glasgow G11 5JR, Scotland

In laboratory experiments exchange of genome sub-units among bunyaviruses has been limited to serologically related members of the genus Bunyavirus. For instance members of the Bunyamwera serogroup can exchange genome sub-units, as can members of the California encephalitis serogroup, but there appears to be no exchange of sub-units between groups (1,2, Elliott et al. in press). Within the Bunyamwera serogroup at least there are also restrictions to free exchange of sub-units, which cannot be attributed to geographical or ecological factors (2, Elliott et. al. in press).

Analysis of reassortment of genome sub-units has been limited by the methods available for determining genotype, primarily oligonucleotide fingerprinting which is definitve but time-consuming and suitable only for analysis of small numbers of samples. This paper describes a simple procedure for determining genotype by dot hybridisation, supplemented by polyacrylamide gel electrophoresis of virion proteins, which is both rapid and suitable for analysis of large numbers of isolates.

GENOTYPING BY DOT HYBRIDISATION

The molecular cloning of the genome of Bunyamwera virus has been described (Lees et al., this volume) and Table 1 lists the characteristics of recombinant plasmids containing Bunyamwera virus specific sequences which were screened for their potential as discriminatory probes. Table 1 indicates that some of the L and M gene probes, but none of the S gene probes, were discriminatory, i.e. they hybridised specifically with RNA extracted from

Copyright © 1984 by Academic Press, Inc.
All rights of reproduction in any form reserved.
ISBN 0-12-183501-4

TABLE 1
CHARACTERIZATION OF BUNYAMWERA VIRUS RECOMBINANT CLONES

Recombinant plasmid	Gene specific sequence	Method of cloning	Size of insert (bases)	% of gene sequence	Location of sequence on gene	Discriminatory in dot hybridization
pBUN 1	M	]EcoR1	1100	26	160-1260	ND[1]
2	M	]linker	2600	60	160-2760	-
5	M	]	360	8	ND	+
23	L	]	400	6	ND	-
61	L	]	1600	23	ND	+
62	M	]	900	21	ND	+
70	L	]GC	1200	17	ND	+
84	L	]tailing	1000	14	ND	+
92	S	]	175	19	ND	-
93	S	]	320	36	ND	-
95	S	]	190	21	ND	-
96	S	]	190	21	ND	-
133	M	blunt end into PvuII site	990	23	1-990	+

[1]ND = not determined

cells infected with Bunyamwera virus, but not with RNA from cells infected with heterologous bunyaviruses. Consequently it is possible to determine the parental origin of the L and M sub-units in progeny clones from crosses of Bunyamwera virus and any other bunyavirus. The dot hybridisation method is outlined in the legend to Figure 1, and requires only small amounts of unfractionated cytoplasmic RNA and nick-translated plasmid DNA. Figure 1 illustrates a typical experiment: there are 29 isolates which react with the pBUN70 probe only and 7 with the pBUN133 probe only, representing the parental genotypes. The six samples which reacted with both probes (B5,C6,C9,C10,D1,D11) and the five which reacted with neither (A3,A9,B12,D9,D10) represent recombinants. D12 is a mock infection.

Since it is known that the G1 polypeptide is encoded by the M sub-unit and the N protein by the S sub-unit (1), it is possible to establish the parental origin of the M and S sub-units from the characteristic electrophoretic mobility of the G1 and N proteins in polyacrylamide gels (2).

PROGENY ANALYSIS

Table 2 gives the genotypes of 151 non-ts recombinants from heterologous crosses of ts mutants. A striking feature of the data is that no more than one type of recombinant was obtained from each cross although two were expected (3), and the recombinants always received their L and S sub-unit from the same parent. Selection for the non-ts phenotype was not responsible for the coordinate segregation of L and S, since the same phenomenon was observed when the total progeny was examined. For example, in the BUN(II) x MAG(III) cross only one of the six possible genotypes was identified (Table 3). Altogether in these experiments the genotypes of 279 progeny clones were determined with no instance of independent segregation of the L and S sub-units.

Table 4, however, shows that there is no incompatibility between these heterologous sub-units. When the parents are recombinants with heterologous combinations of genome sub-units the coordinate segregation of the L and S sub-units was no longer observed and all six possible recombinant genotypes were recovered (Table 4). The absolute

TABLE 2

ANALYSIS OF THE GENOTYPE OF WILD-TYPE PROGENY FROM CROSSES OF *ts* MUTANTS OF BATAI VIRUS (BAT), BUNYAMWERA VIRUS (BUN) AND MAGUARI VIRUS (MAG).

Cross*			Progeny Analysis							
Bun Parent		Heterologous Parent	RNA**		Protein***		Inferred Genotype			No of Recombinant Clones
			L	M	Gl	N	L	M	S	
BUN(l)	x	MAG(ll)	-	+	BUN	MAG	MAG	BUN	MAG	28
BUN(l)	x	BAT(ll)	-	+	BUN	BAT	BAT	BUN	BAT	21
BUN(ll)	x	MAG(l)	+	-	MAG	BUN	BUN	MAG	BUN	33
BUN(ll)	x	BAT(l)	+	-	BAT	BUN	BUN	BAT	BUN	43
BUN(l)	x	MAG(lll)	x	x	x	x	x	x	x	none
BUN(ll)	x	MAG(lll)	+	-	MAG	BUN	BUN	MAG	BUN	26
Total										151

* Data pooled from several crosses of different parental *ts* mutants; recombination group in brackets.

** Dot hybridization with BUN cDNA probes

*** PAGE of ^{35}S-met labelled virion proteins

TABLE 3
TOTAL PROGENY ANALYSIS FROM BUNII x MAGIII CROSS

RNA		Protein		Inferred Genotype			No of	
L	M	Gl	N	L	M	S	Clones	
+	+	BUN	BUN	BUN	BUN	BUN	42	
-	-	MAG	MAG	MAG	MAG	MAG	10	
+	+	BUN	MAG	BUN	BUN	MAG	0	
-	-	MAG	BUN	MAG	MAG	BUN	0	
+	-	MAG	MAG	BUN	MAG	MAG	0	14.8%
-	+	BUN	BUN	MAG	BUN	BUN	0	
+	-	MAG	BUN	BUN	MAG	BUN	9	
-	+	BUN	MAG	MAG	BUN	MAG	0	
Total							61	

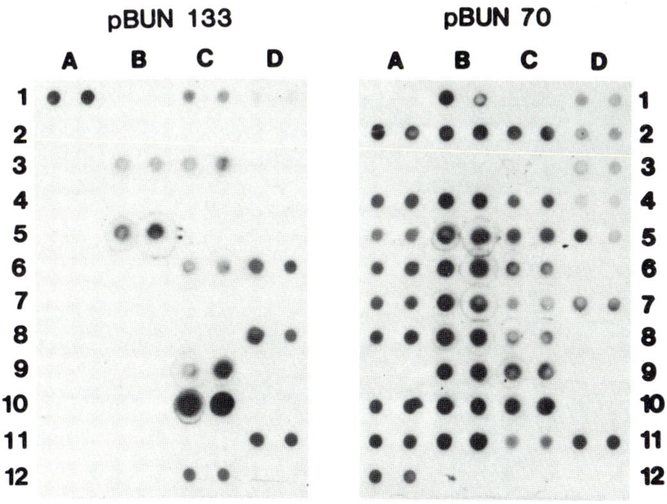

FIGURE 1. Dot hybridisation filters showing reaction of the pBUN70 L sub-unit probe and the pBUN133 M sub-unit probe with cytoplasmic RNA extracts from 47 virus-infected (A1 - D11) cultures and a mock-infected (D12) culture.

Total cytoplasmic RNA was extracted (4) from 2 x 10^6 BHK-21 cells infected at a m.o.i. of 10-20. Each extract was spotted in duplicate onto replicate nitrocellulose filters. Hybridisation to nick-translated denatured plasmid DNA was performed in 50% formamide, 5 x SSC at 42°C for 16-20 h. The filters were washed, dried and exposed to Kodak X-ray film with intensifyimg screens at -70°C for 48h. See text for explanation of the code.

TABLE 4
Rec:BUN/MAG/BUN x Rec:BAT/BUN/BAT - PROGENY ANALYSIS

RNA		Protein		Inferred Genotype			No of Clones	
L	M	Gl	N	L	M	S		
+	-	MAG	BUN	BUN	MAG	BUN	26	
-	+	BUN	BAT	BAT	BUN	BAT	5	
+	+	BUN	BAT	BUN	BUN	BAT	1	
-	-	MAG	BUN	BAT	MAG	BUN	2	
+	-	MAG	BAT	BUN	MAG	BAT	3	31.1%
-	+	BUN	BUN	BAT	BUN	BUN	1	
+	+	BUN	BUN	BUN	BUN	BUN	4	
-	-	MAG	BAT	BAT	MAG	BAT	3	
Total							45	

frequency of recombinants in these experiments was also higher, in line with the relaxation of restriction.

These results imply that the specificity of assembly of the multipartite genotype of bunyaviruses is determined at the gene product level. An experiment is in progress to determine whether restriction is restored in crosses of viruses with homologous combinations of genome sub-units derived by recombination.

ACKNOWLEDGEMENT

We thank Margaret Wilkie for assistance with many of these experiments.

REFERENCES

1. Bishop, D.H.L., Clerx, J.P.M., Clerx-Van Haaster, C.M., Robeson, G., Rozhon, E.J., Ushijima, H., and Veerisetty, V. (1981). In "The Replication of Negative Strand Viruses" (D.H.L. Bishop and R.W. Compans, eds.), p.135. Elsevier/North Holland.
2. Iroegbu, C.U., and Pringle, C.R. (1981). J. Virol. 37, 383.
3. Pringle, C.R., and Iroegbu, C.U. (1982). J. Virol. 42, 873.
4. White, B.A., and Bancroft, F.C. (1982). J. Biol. Chem. 257, 8569.

GENETIC ANALYSES OF PICHINDE AND LCM ARENAVIRUSES; EVIDENCE FOR A UNIQUE ORGANIZATION FOR PICHINDE S RNA[1]

David D. Auperin, Victor Romanowski, and David H.L. Bishop

Department of Microbiology
University of Alabama in Birmingham
Birmingham, AL 35294

VIRUS GENOME RNA AND CODING STRATEGY

The genome of arenaviruses consists of two single-stranded, negative-sense, RNA molecules designated L and S (1,2). For Pichinde virus the L RNA has an estimated size of 2.6×10^6 daltons, the S RNA an estimated size of 1.3×10^6 daltons (3). It has been shown that the Pichinde L and S RNA species each have unique nucleotide sequences (2,4,5). The S RNA has been shown to encode the viral nucleocapsid protein (N) and the surface glycoproteins (G1 and G2); in the cell the glycoproteins are synthesized in a precursor form named GPC (1,6-9). The L RNA encodes a 200×10^3 dalton polypeptide which may be a component of the virion transcriptase (6). It is not known if other proteins (structural or non-structural) are coded by either the L and, or S RNA species. RNA nucleotide sequence analyses have demonstrated that the 3' terminal 19 nucleotides of the L and S viral RNA species of Pichinde, lymphocytic choriomeningitis (LCM, Armstrong strain) and Tacaribe arenaviruses are conserved (10,11).

Although the genome coding assignments for the three primary gene products of Pichinde viruses are known, the gene order of the S encoded proteins and their transcriptional strategies are unknown. In order to begin to address these questions, we have undertaken a study to determine the sequence of the Pichinde S genome RNA segment.

MOLECULAR CLONING PICHINDE S RNA SPECIES

The strategy employed for generating cDNA transcripts of the (approx.) 4,000 nucleotide long Pichinde S viral RNA species was as follows. A 5' phosphorylated oligonucleotide complementary to the 3' terminus of Pichinde S RNA segment (10, 11) was synthesized and used

---

[1]This work was supported by NIH Grant AI-14183. D.A. is the recipient of an NIH postdoctoral fellowship (AI-06474).

Copyright © 1984 by Academic Press, Inc.
All rights of reproduction in any form reserved.
ISBN 0-12-183501-4

to prime cDNA transcription using RNA templates obtained from purified virions. The single-stranded cDNA transcripts were converted to double-stranded molecules and cloned into competent E. coli MC1061 cells. The resulting transformants were then analysed for the presence of Pichinde S cDNA sequences and their cDNA inserts sized. The largest clone obtained was estimated to be about 1600 base pairs in length (clone A3-89). After sequencing (12), a second set of DNA clones was obtained using a 19 base oligonucleotide complementary to an internal region of the deduced S RNA sequence (nucleotide positions 1139-1157). One of these clones (2-79), estimated to be about 1000 nucleotides in length, was also sequenced (12).

The consensus nucleotide sequence data confirmed the sequence information obtained previously for the 3' proximal region of Pichinde S RNA using direct RNA sequencing methods (11).

NUCLEOTIDE SEQUENCES AND PREDICTED GENE PRODUCTS

Figure 1 presents the predicted 3' proximal Pichinde S RNA gene product which derives from an open reading frame in the viral complementary RNA originating with a methionine codon at a position complementary to the viral (3') nucleotide residues 84-86 and terminating with an opal codon corresponding to residues 1767-1769 (12). The protein encoded by this reading frame has a calculated size of 62,911 daltons and is 561 amino acids in length (Figure 1).

```
        10         20         30         40         50         60
MSDNIPSFRW VQSLRRGLSN WTHPVKADVL SDTRALLSAL DFHKVAQVQR MVRKDKRTDS

        70         80         90        100        110        120
DLTKLRDMNK EVDALMNMRS VQRDNVLKVG GLAKEELMEL ASDLDKLRKK VTRTEGLSQP

       130        140        150        160        170        180
GVYEGNLTNT QLEQRAEILR SMGFANARPA GNRDGVVKVW DIKDNTLLIN QFGSMPALTI

       190        200        210        220        230        240
ACMTEQGGEQ LNDVVQALSA LGLLYTVKFP NMTDLEKLTQ QHSALKIISH EPSALNISGY

       250        260        270        280        290        300
NLSLSAAVKA AACMIDGGNM LETIQVKPSM FSTLIKSLLQ IKNREGMFVS TTPGQRNPYE

       310        320        330        340        350        360
NLLYKICLSG DGWPYIGSRS QVQGRAWDNT TVDLDSKPSA IQPPVRNGGS PDLKQIPKEK

       370        380        390        400        410        420
EDTVVSSIQM LDPRATTWID IEGTPNDPVE MAIYQPDTGN YIHCYRFPHD EKSFKEQSKY

       430        440        450        460        470        480
SHGLLLKDLA DAQPGLISSI IRHLPQNMVF TAQGSDDIIR LFEMHGRRDL KVLDVKLSAE

       490        500        510        520        530        540
QARTFEDEIW ERYNQLCTKH KGLVIKKKKK GAVQTTANPH CALLDTIMFD ATVTGWVRDQ

       550        560
KPMRCLPIDT LYRNNTDLIN L
```

FIGURE 1. Predicted 3' proximal Pichinde S RNA gene product.

Assuming that lysine and arginine residues each have a +1 charge at neutral pH, histidine +0.5 and aspartic acid and glutamic acid each -1, the net charge of the protein is +9. A second reading frame originating with a methionine codon at position 1827-1829 of the viral complementary S RNA sequence remains open for the duration of the sequence present in the clone (i.e., incomplete, 72 amino acids, Figure 2).

```
         10         20         30         40         50         60
MSEAELIFPT LPICKVSLVF IIPAATQGVS CNPVWFRATI TNDVPMSRYS NYVEKYCNGV

         70
KNQRPEAYIC QCQRGFF
```

FIGURE 2. Predicted amino terminal portion of the second Pichinde S RNA gene product (incomplete, GPC?).

No other open reading frame of considerable length (i.e., greater than 80 amino acids was identified in either of the other 2 possible reading frames of the viral complementary sequence, or greater than 93 amino acids in length in the viral RNA sequence (Figure 3).

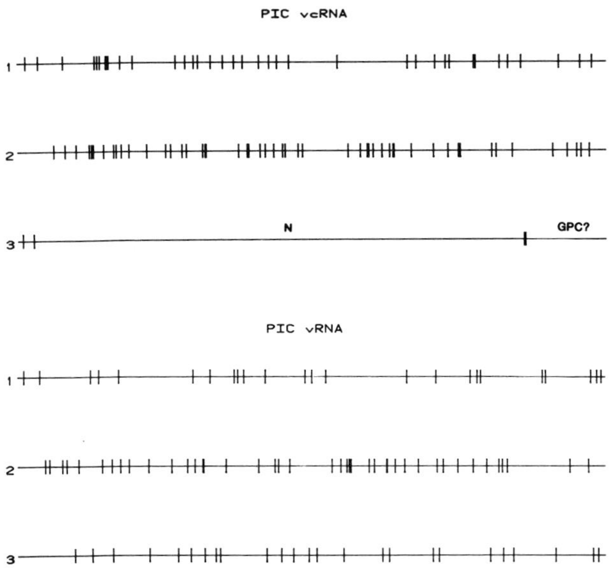

FIGURE 3. Termination codons in the (incomplete) viral (v) and viral complementary (vc) sequences of Pichinde S RNA.

The size and character of the first S RNA gene product (residues 84-1767 in the viral complementary sequence) is compatible with it coding for the viral nucleocapsid protein, N. The second reading frame (residues 1827-?) may code for the amino terminal portion of the precursor to the viral glycoproteins, GPC. The hydrophobic character of its amino proximal sequences could represent a signal, or transmembranal, region in common with other membrane, or secreted proteins (13-15). Thus the sequence data support the hypothesis that the gene order in the S RNA is 3'-N-GPC-5'. However, we cannot as yet rule out the possibility that the second reading frame may represent another gene product. Note, the estimated size of the S RNA species (i.e., 1.3×10^6 daltons) is sufficient to code for both N and GPC in consecutive, non-overlapping, reading frames. Of particular interest from the sequence data is the fact that there are termination codons at the end of the N coding region (12). Unless they are removed by splicing, this observation suggests that N is translated as a unique product. There is no U rich sequence following the N gene that could serve as a polyadenylation signal if a monocistronic N message is synthesized as, for example, in the case of rhabdoviruses (16).

THE INTERGENIC REGION OF PICHINDE S RNA

Figure 4 is a schematic representation of the deduced coding strategy in the viral complementary sequence of Pichinde S RNA (12).

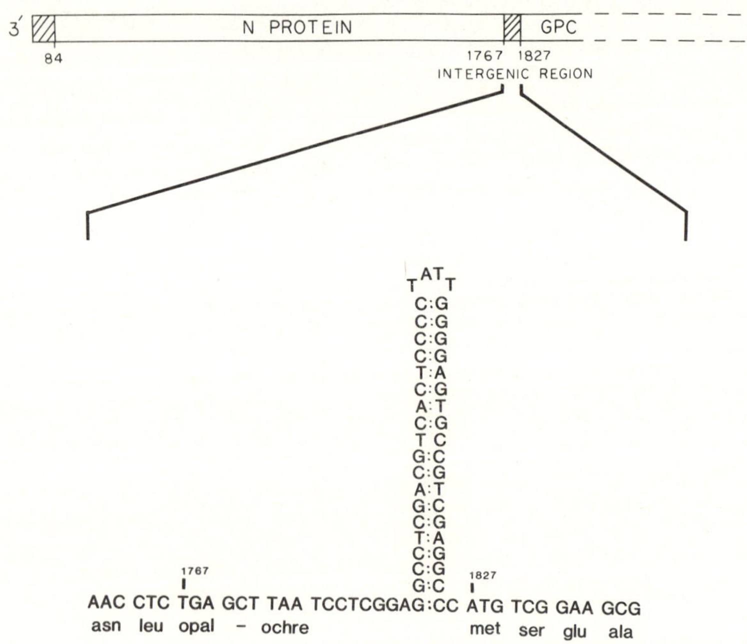

FIGURE 4. Schematic of the S vcRNA organization.

The nucleotide sequence within the intergenic region predicts a hairpin structure stabilized by 14 G-C and 4 A-T base pairs. The function of this structure is unknown. Possibly it is involved in the termination of mRNA transcription and the formation of a monocistronic (unpolyadenylated?) N mRNA species. Alternatively if there is a polycistronic messenger RNA species, it may be important in ribosome scanning, or initiation of translation of the second gene product. Clearly the mRNA strategies of the S RNA species need to be elucidated in order to determine the role of the unique intergenic region.

GENETIC ATTRIBUTES OF PICHINDE AND LCM VIRUSES

It has been demonstrated that arenaviruses can form intertypic recombinant viruses by RNA segment reassortment (7,9,17,18). As demonstrated in Table 1 and confirmed by oligonucleotide fingerprint analyses, recombinant viruses genetically diploid with respect to the S RNA species have been obtained using distinguishable genotypes of LCM virus (WE, ARM) when ts mutants representing alternate S coded gene products (N, GPC?) are used in dual virus infections and recombinant viruses identified by a wild-type plaquing phenotype (18).

TABLE 1
RECOMBINATION OF LCM ARM AND WE TS MUTANTS

	% Wild-type recombinants[a]							
	ARM ts mutants				WE ts mutants			
Mutant	Group I	Group II			Group I			Group II
	A1	A2	A8	A13	W3	W4	W6	W20
W2	12	58	50	50	3	4	4	2[b]

[a]The % wild-type recombinants obtained from 35°C ARM (A) and WE (W) ts mutant virus crosses were calculated as described previously (5). Group I mutants have S RNA mutations, Group II L RNA mutations (18).

[b]The % wild-type recombinants obtained from 40°C crosses were 39-68%.

The high frequency segregation of ts progeny from these genetically diploid viruses agrees with the conclusion that the wild-type phenotype is maintained by gene product complementation.

We have found that the yields of ts mutants of Pichinde and LCM viruses following conventional methods of mutagenesis (e.g., growth in the presence of high concentrations of the mutagen 5-fluorouracil 5FU) are invariably low when compared to the results obtained with other RNA viruses (e.g., bunyaviruses). Although this could be explained in part by the prevelance of genetically diploid viruses (Romanowski and Bishop, 1983), another contributing factor could be more faithful copying by their replicases. Typical recoveries of ts progeny having mutations located in LCM L or S RNA species following growth in the presence of 100-150 ug 5FU per ml of growth medium are given in Table 2. Compared to these data are the yields of ts mutants obtained in the absence of a chemical mutagen from coinfections of wild-type LCM WE virus (WE/WE) with an LCM ARM ts mutant (A2) having an L RNA mutation (ARM*/ARM). Relatively large numbers of ts mutants ($40^{o}/35^{o}C$ efficiencies of plating of less than 0.1%) were recovered having mutations located in RNA segments other than the ARM L RNA. We interpret these data to indicate that the A2 mutant functions as a "mutator" virus (Table 2). An explanation which is under investigation is that the mutant A2 viral replicase is less faithful in its replicative role than the wild-type enzyme.

TABLE 2
RECOVERY OF TS MUTANTS FOR LCM VIRUSES

Location of ts mutation[a]	Mutagen	
	5-Fluorouracil[b]	A2 ts mutant[c]
ARM L	7.4%	-
ARM S	0.35%	2%
WE L	0.1%	1%
WE S	0.4%	ND

[a]Genotype analyses of derived ts mutants and determination of the location of the ts lesion were undertaken by reassortment and oligonucleotide fingerprinting analyses (Romanowski and Bishop, 1983)

[b]5-Fluorouracil was used as a mutagen at concentrations of 100-150 ug/ml.

[c]Clonal analyses of progeny ts mutants obtained from coinfections of ts A2 (ARM*/ARM) and 10-fold higher multiplicities of wild-type WE virus (WE/WE) were undertaken by reassortment and oligonucleotide fingerprinting (Romanowski and Bishop, 1983). ND = none detected.

REFERENCES

1. Rawls, W.E. and Leung, W.-C. (1979). Comp. Virol. 14, 157.
2. Vezza, A.C., Clewley, J.P., Gard, G.P., Abraham, N.Z., Compans, R.W., and Bishop, D.H.L. (1978). J. Virol. 26, 485.
3. Ramsingh, A., Dimock, K., Rawls, W.E., and Leung, W.-C. 1980. Intervirol. 14, 31.
4. Leung, W.-C., Ramsingh, A., Dimock, K., Rawls, W.E., Petrovich, J., and Leung, M. (1981). J. Virol. 37, 48.
5. Vezza, A.C. and Bishop, D.H.L., (1977). J. Virol. 24, 712.
6. Harnish, D.G., Dimock, K., Bishop, D.H.L., and Rawls, W.E. (1983). J. Virol. 46, 638.
7. Harnish, D.G., Leung, W.-C., and Rawls, W.E. (1981). J. Virol. 38, 840.
8. Leung, W.-C., Harnish, D., Ramsingh, A., Dimock, K., and Rawls, W.E. (1981). In "Replication of Negative Strand Viruses" (D.H.L. Bishop and R.W. Compans, eds.), p 51. Elsevier North-Holland, N.Y.
9. Vezza, A.C., Cash, P., Jahrling, P., Eddy, G., and Bishop, D.H.L. (1980). Virology 106, 250.
10. Auperin, D.D., Compans, R.W., and Bishop, D.H.L. (1982). Virol. 121, 200.
11. Auperin, D.D., Dimock, K., Cash, P., Rawls, W.E., and Bishop, D.H.L., (1982). Virol. 116, 363.
12. Auperin, D.D., and Bishop, D.H.L. (1983). J. Virol. Submitted for publication.
13. Blobel, G., and Dobberstein, B. (1975). J. Cell Biol. 67, 835.
14. Rose, J.K., Welch, W.K., Sefton, B.M., Esch, F.S., and Ling, N.C. (1980). Proc. Natl. Acad. Sci. USA 77, 3884.
15. Sabatini, D.D., Kreibich, G., Morimato, T., and Adenich, M. (1982). J. Cell Biol. 92, 1.
16. Rose, J.K. (1980). Cell 19, 415.
17. Kirk, W.E., Cash, P., Peters, C.J., and Bishop, D.H.L. (1980). J. Gen. Virol., 51, 213.
18. Romanowski, V., and Bishop, D.H.L. (1983). Virology, 126, 87.
19. Gentsch, J., and Bishop, D.H.L. (1976). J. Virol. 20, 351.

STUDIES ON THE MOLECULAR BIOLOGY OF LYMPHOCYTIC CHORIOMENINGITIS VIRUS[1]

Peter J. Southern and Michael B. A. Oldstone

Department of Immunology
Scripps Clinic and Research Foundation
La Jolla, California 92037

RATIONALE FOR MOLECULAR STUDIES

Lymphocytic Choriomeningitis Virus (LCMV) has been studied for many years with respect to the biology and pathology of virus infection (1). However, until very recently, there has been only limited information available relating to the molecular biology of LCMV infections. The genome of LCMV comprises two single-stranded RNA segments of negative polarity (RNAs designated L and S, approximate lengths 9 and 4 kb, respectively) which exist as ribonucleoprotein (RNP) complexes within both purified virions and infected cells. The LCMV virion also contains varying numbers of host ribosomes but it has been suggested that these ribosomes are not required for normal virus replication (2). We and other groups (see the papers from the groups of D.H.L. Bishop and W-C. Leung, this volume) have initiated DNA cloning experiments: 1) to evaluate the full genetic potential of LCMV, 2) to generate hybridization probes for the study of LCMV gene expression and gene regulation, 3) to examine sequence relatedness between different LCMV strains and other arenaviruses, and 4) to investigate the molecular basis of LCMV induced disease.

STRATEGY FOR CLONING LCMV VIRION RNAS

The strategy that we have chosen for cloning the LCMV virion RNAs was designed to take account of two particular

---

[1]This work was supported by NIH grants NS-12428, AI 09484 and by USAMRIID Contract C-3013. The findings in this report are not to be construed as an official Department of the Army position unless so designated by other authorized documents.

Copyright © 1984 by Academic Press, Inc.
All rights of reproduction in any form reserved.
ISBN 0-12-183501-4

problems: 1) the negative polarity of the virion RNAs and 2) the presence of contaminating ribosomal RNAs in samples of LCMV RNA prepared from purified virions. Initially, virion RNA preparations were treated with polynucleotide phosphorylase (3) to add a stretch of A residues at the 3' termini. The "poly A^+" RNA was isolated by chromatography on oligo dT cellulose and then a synthetic dT primer (dT 8-12) was used to initiate reverse transcription. More recently, we have used a calf thymus primer (4) to initiate reverse transcription at random positions along the RNA templates. Double stranded cDNAs were constructed by standard procedures and were then cloned into the *PstI* restriction site of pBR322 by homopolymer tailing (5). Independent bacterial colonies were isolated after transformation of competent *E. coli* HB101 cells under selection for tetracycline resistance.

IDENTIFICATION OF LCMV cDNA CLONES

Bacterial colonies containing cDNA inserts were screened by colony hybridization (6) using purified and partially hydrolysed LCMV L or S virion RNAs as probes - the RNAs were labelled at 5' ends with polynucleotide kinase and $\gamma\ ^{32}P$ ATP (Fig. 1). Following a brief wash in dilute alkali, the filters could be successfully re-hybridized with a second probe. Our analyses indicate that, under suitable hybridization conditions, the intensity of signal on the autoradiograph correlates closely with the size of the inserted cDNA segment. Plasmid DNA was purified from selected positive clones, labelled *in vitro* by nick translation and then used in hybridization studies. In order to confirm the authenticity of LCMV L and S cDNA clones, we have hybridized to unlabelled LCMV infected and control uninfected total cellular RNA preparations (7) (Fig. 2). In every case the hybridization was specific to the LCMV infected RNA sample.

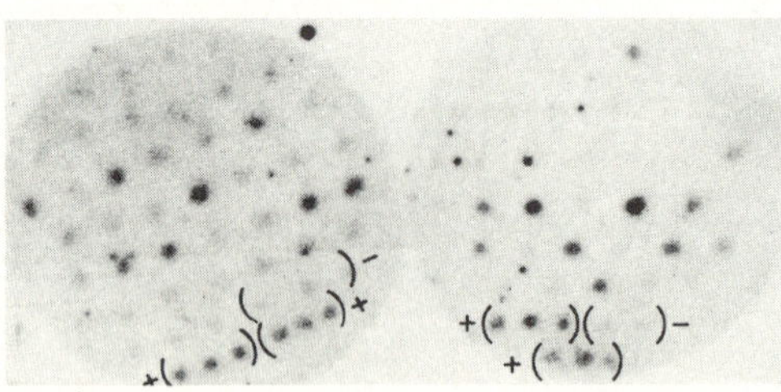

FIGURE 1. Colony hybridization to detect LCMV specific cDNA clones. These filters were hybridized with 5' end labelled LCMV S RNA. Positive (+) and negative (-) control colonies are indicated.

We have also used DNA-DNA hybridization to establish sequence overlaps between independent cDNA clones and have identified several clones that were derived from the 3' portion of the S RNA (Fig. 3).

IDENTIFICATION OF 3' SPECIFIC cDNA CLONES AND NUCLEOTIDE SEQUENCING

The 3' terminal sequence (approximately 50-150 bases) of several different arenavirus virion RNAs has been determined by D.H.L. Bishop's group (8,9). Using the published sequence for LCMV S RNA, we converted this sequence into DNA, scanned the DNA sequence for restriction enzyme recognition sites and then searched for these predicted sites in the cDNA clones. One clone, S9 (Fig. 3) contained a Bam HI site very close to the cDNA:pBR322 junction and this became a strong candidate for a 3' specific S clone (Fig. 4). Other S clones have now been positioned relative to S9 (Fig. 3).

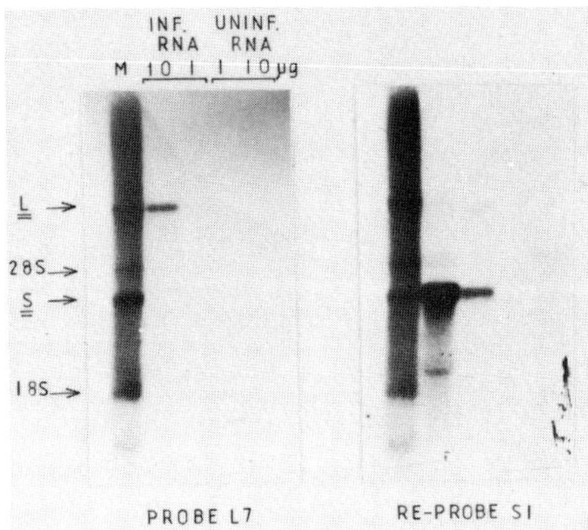

FIGURE 2. Detection of LCMV RNAs by hybridization with radioactively labelled cDNA probes. Total cellular RNAs were denatured with glyoxal, fractionated on the basis of size by agarose gel electrophoresis, transferred to nitrocellulose and then hybridized with ^{32}P cDNA probes that had been labelled by nick translation. The left panel shows the hybridization pattern with an L specific probe and then the filter was washed extensively and re-hybridized with an S specific probe (right panel). Note that neither the L nor S probe hybridizes to uninfected cell RNA and that there is no apparent homology between the L and S RNAs shown by this pair of probes. The track labelled M represents a sample of *in vivo* labelled LCMV virion RNA and shows ribosomal 28S and 18S and LCMV L and S RNA bands as markers.

We have initiated nucleotide sequencing studies (10) and have deduced the sequence of virtually all of the first 1300 bases 3'→5' of the S RNA including a continuous stretch of 620 bases starting from the 3' end. Inspection of the 3' end sequence reveals a translation initiation codon at nucleotide 247 (nucleotides are numbered in an increasing sense from the 3' terminus) and this translation frame remains open until the end of the known sequence. There are 3 other initiation codons that lie closer to the end of the S RNA but each of these is followed relatively quickly by a translation termination signal; the longest potential translation product would contain 46 amino acids.

COMPARISON OF LCMV S GENOMIC SEQUENCES

The DNA sequence deduced from the S DNA clone (S9) matches very closely with the sequence obtained from the S RNA (Fig. 4) for the first 45 nucleotides. However, beyond this point there appears to be total sequence divergence including differences in the positions of translation initia-

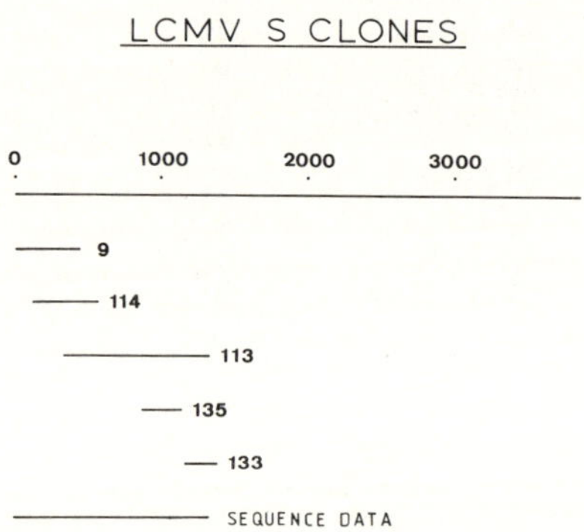

FIGURE 3. Diagrammatic representation of overlapping clones extending from the 3' end of the LCMV S virion RNA. These overlaps were established by DNA-DNA hybridization (Southern Blotting) experiments. Clone S9 is lacking only 5 bases from the 3' end of the virion S RNA as inferred from direct RNA sequencing experiments (9). Clone S9 was isolated from 3' poly A addition and dT primed cloning reactions whereas the other clones were derived from random primed reverse transcription reactions.

Bam HI

5'(dGTail)TGTGGATCCTAGGCTTTTTGGATTGCGC_TTTCCTCTAGATC

CGCACAGTGGATCCTAGGCATTT GATTGCGCATTTG TCTTGAGA

FIGURE 4. Comparison of S RNA 3' terminal sequences for LCMV strains. The upper line shows the sequence from cDNA clone S9 for ARM 1371 and the lower line shows the DNA sequence deduced from the published E-350 RNA sequence (10). Gaps have been left to allow the maximum alignment between the two sequences; bases that differ are underlined. The Bam HI restriction endonuclease recognition sequence is also shown.

tion codons. We used LCMV ARM 1371 for cloning and sequencing whereas Bishop's group used LCMV E-350. Although passage of the virus in different laboratories may have resulted in sequence changes in a non-coding region of the viral genome, oligonucleotide fingerprinting (11) had indicated that these LCMV strains showed extensive sequence similarity for the S RNAs. We are unable to provide an explanation for these unexpected differences between two closely related strains. This discrepancy may be resolved in terms of RNA analyses involving populations of molecules whereas the cDNA clone represents the sequence of an individual molecule within the population. Experiments are now in progress to reconcile this apparent inconsistency.

CONCLUDING REMARKS

We have generated partial cDNA clones to the L and S genomic RNA segments of the ARM 1371 strain of LCMV. These cDNA clones are being used as hybridization probes and also for nucleotide sequencing. We are interested in examining sequence divergence of LCMV strains and using this information to understand the molecular basis of LCMV induced disease.

ACKNOWLEDGMENTS

This is Publication No. 3195-IMM from the Department of Immunology, Scripps Clinic and Research Foundation, La Jolla, California 92037. The authors thank Frank Dutko for helpful discussions in the early stages of this work. We thank Lisa A. Flores for skillful preparation of the manuscript.

REFERENCES

1. Buchmeier, M. J., Welsh, R. M., Dutko, F. J., and Oldstone, M. B. A. (1980). Adv. Immunol. 30, 275.
2. Leung, W-C., and Rawls, W. E. (1977). Virol. 81, 174.
3. Engel, J., and Davidson, N. (1978). Biochem. 17, 3883.
4. Taylor, J., Illmensee, R., and Summers, J. (1976). Biochem. Biophys. Acta 442, 324.
5. Roychoudhury, R., Jay, E., and Wu, R. (1976). Nucleic Acid. Res. 3, 863.
6. Grunstein, M., and Hogness, D. (1975). Proc. Natl. Acad. Sci. USA 72, 3961.
7. Thomas, P. S. (1980). Proc. Natl. Acad. Sci. USA 77, 5201.
8. Auperin, D., Dimock, K., Cash, P., Rawls, W. E., Leung, W-C., and Bishop, D. H. L. (1982). Virol. 116, 363.
9. Auperin, D., Compans, R. W., and Bishop, D. H. L. (1982). Virol. 121, 200.
10. Maxam, A., and Gilbert, W. (1980). In "Methods in Enzymology" (L. Grossman and K. Moldave, eds.), Vol. 65, p. 499. Academic Press, New York.
11. Dutko, F. J., and Oldstone, M. B. A. (1983). J. Gen. Virol., in press.

SEQUENCE REARRANGEMENTS IN INFLUENZA VIRUS RNA AND RIBONUCLEOPROTEIN STRUCTURE

James S. Robertson[1,2], Philip A. Jennings[3], John T. Finch[3] and Greg Winter[3].

[1]Division of Virology
University of Cambridge
Addenbrooke's Hospital
Cambridge, U.K.
[3]Laboratory of Molecular Biology
Medical Research Council Centre
Cambridge, U.K.

SUBGENOMIC RNAs

Influenza virions often contain small sub-genomic RNA species (sgRNAs) in addition to the usual eight RNA segments. These sgRNAs are abundant in defective-interfering virus preparations (1,2) and have occasionally been found in standard virus (3,4). To investigate the generation of these sgRNAs from their parental segment, we carried out complete nucleotide sequencing of thirty-five different sgRNAs (1).

The results of this study led us to propose that the tertiary structure of the viral ribonucleoprotein (RNP) is responsible for the sequence rearrangements in sgRNAs and we present a model describing the path of the RNA within the RNP which illustrates how the sgRNAs may be generated during RNA replication.

CLASSIFICATION OF sgRNAs

Sequence analyses of sgRNAs by us and others (1,3-6) indicate several different structural classes of sgRNAs based on their relationship with the progenitor segment (Fig. 1). The most abundant is the simple deletion sgRNA (36 out of 41)

---

[2]Present address: National Institute for Biological Standards and Control, Hampstead, London, U.K.

J.S.R. was supported by an MRC programme grant (G979/140) and P.A.J. by a CSIRO postdoctoral award.

Copyright © 1984 by Academic Press, Inc.
All rights of reproduction in any form reserved.
ISBN 0-12-183501-4

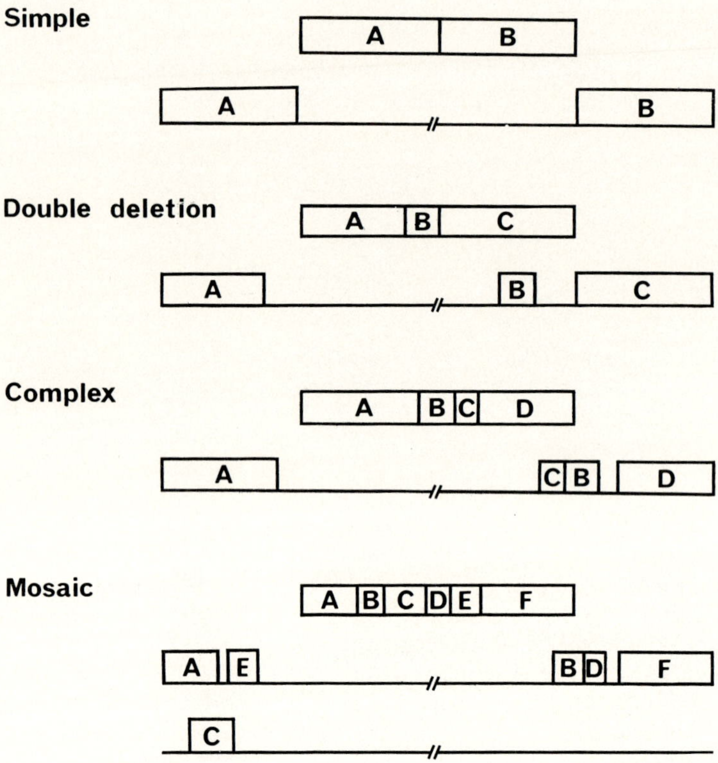

FIGURE 1. Classification of sgRNAs. Defined regions of a sgRNA (upper bars) are derived from specific locations within the progenitor segment as indicated (lower bars).

in which a large internal deletion of the progenitor segment gives rise to a sgRNA containing the original 5' and 3' terminal regions. Three double deletion sgRNAs have been sequenced (1,6). This type contains a small deletion close to the 3' end (mRNA sense) in addition to the large deletion as in simple sgRNAs. In our study we also described a complex sgRNA in which a region from the progenitor segment has been transposed to a region in the sgRNA downstream from its original location. The mosaic sgRNA (4,5) contains sequences derived from two progenitor segments and was apparently generated by a series of deletions and insertions.

Fig. 1 is reproduced with permission from ref. (1).

A JUMPING POLYMERASE?

It has been suggested that sgRNAs are generated by a jumping viral polymerase during RNA replication (3,7,8). Analysis of the nucleotide sequences at the sites of transition of 35 unique influenza virus sgRNAs (1) reveals no obvious explanation for the generation of these sgRNAs. However since the replication of viral RNA presumably occurs on a scaffold of RNP, it is likely that the structure of the RNP may guide the jumps of the polymerase. This would confine transitions to regions of RNA template brought into three-dimensional proximity and maintained in a stable tertiary configuration by the nucleoprotein. We investigated RNP structure using electron-microscopy and combining the data with the observed sequence rearrangements we propose a three-dimensional model of the RNP in which a jumping polymerase could successfully generate all sgRNA structures (Fig. 2).

ELECTRON MICROSCOPY OF RNPs

Electron micrographs of influenza virus RNPs show a characteristic structure of a strand of subunits folded back on itself and twisted into a double helical configuration (9,10). In Fig. 2 these subunits are quite evident especially in the end loops and from our own measurements and previously published data we have been able to correlate the visible subunits with NP monomers. An outer helical path of 40 nm can be calculated from the pitch of the double helix (12.5-15 nm) and the outer diameter of the rod-stem of the RNP (13 nm). The spacing of the subunits at the outer radius of small loops is 7 nm and indicates that each turn of a single helix contains ~5½ subunits (~11 subunits per turn of the double helix or 0.7-0.9 subunits per nm of rod). From the measurements of Compans et al (10) we estimate that 1 nm of rod length is associated with 21 nucleotides of RNA. Thus each visible subunit is associated with 23-30 nucleotides. It can be deduced also, from the NP content of the RNP, that each NP is associated with 22-26 nucleotides. Thus we suggest that the visible subunits are NP monomers. This was confirmed by counting the number of subunits in segment 8 RNPs. The predicted number of NPs per segment 8 RNP (34-40) was comparable with the number of visible subunits in segment 8 RNPs (30-40) partially unwound by raising the salt concentration (Fig. 2b).

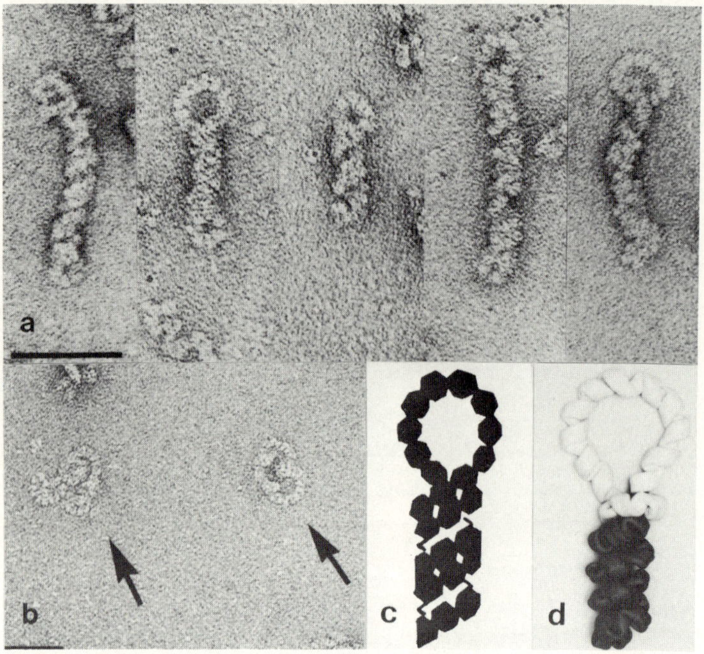

FIGURE 2. Electron micrographs of influenza virus RNPs. (a) Virion RNPs in 0.1M NaCl were negatively stained with uranyl acetate. The RNPs display a characteristic double helical ribbon structure with a variable sized loop at one or both ends. Individual subunits comprising the RNP structure are evident, each subunit corresponding to an NP monomer (see text). (b) Segment 8 RNPs opened out by raising the salt concentration to 0.5M. The RNPs were verified as representing >95% segment 8 by analysis of their RNA content (data not shown). (c) A diagrammatic representation of a segment 8 RNP particle consisting of a string of ~35 NP subunits folded to form a double ribbon and twisted into a right-handed helix with a loop at one end. Approximately 25 nucleotides of RNA are associated with each NP subunit and about 6 NP subunits per turn of a single helix. (d) A low resolution model of the path of the RNA in an RNP. Each coil represents 25 nucleotides of RNA associated with an NP subunit. The regions in white indicate sequences transcribed during the generation of the complex sgRNA illustrated in detail in Fig. 5. This and subsequent models are equivalent to a segment 8 RNP and are readily extrapolated to the other segments by increasing the length of the stem. Bar=50 nm.

Fig. 2 is reproduced with permission from ref. (1).

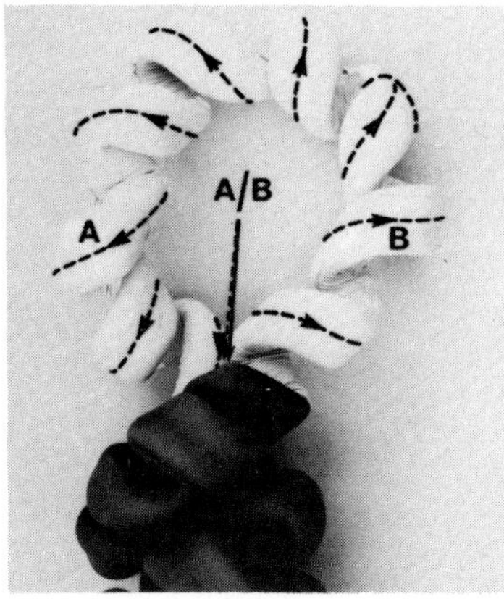

FIGURE 3. Generation of a simple sgRNA. The 5' and 3' ends of the vRNA in this and all subsequent models are located in the terminal loop and indicated by (----). Transition of the polymerase from the 3' region of the vRNA to the 5' region at the junction A/B gives rise to a simple deletion sgRNA. The regions transcribed into the sgRNA are shown in white and the direction of synthesis of the nascent RNA by ---->.

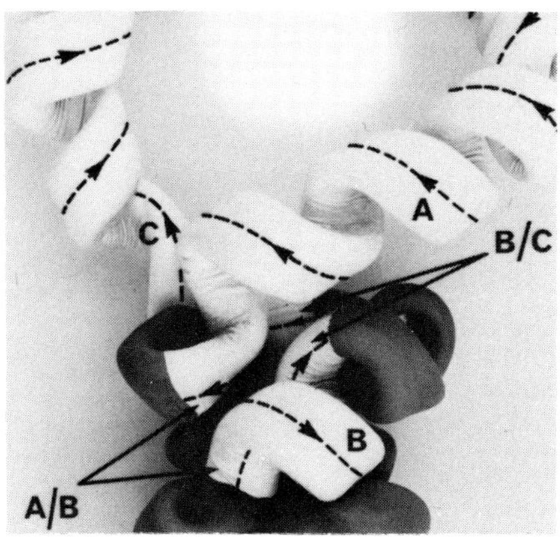

FIGURE 4. Generation of a double deletion sgRNA. A posterior view of the RNP structure shows details of the proposed transitions. The first transition of the polymerase from the 3' region of the vRNA to the 5' region (A/B) deletes the bulk of the progenitor gene. A second smaller deletion of 40-60 nucleotides occurs when the polymerase jumps from region B to C (B/C).

Figs. 3 & 4 are reproduced with permission from ref. (1).

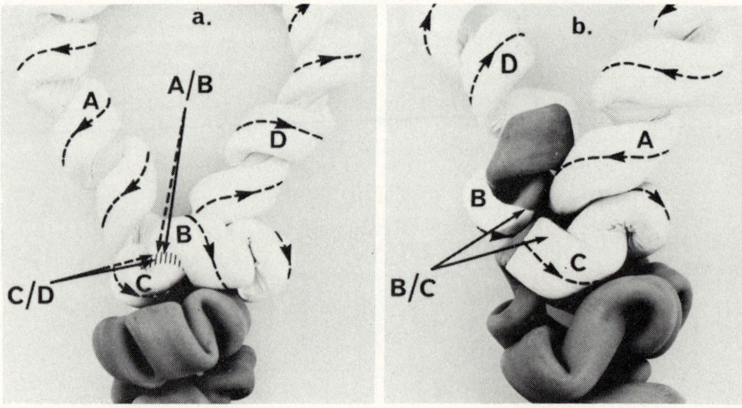

FIGURE 5. Generation of the complex sgRNA derived from segment 1. The detailed transitions are marked in a front view (a) and a back view (b) of the RNP structure. After the transitions from A to B (A/B) and from B to C (B/C), the polymerase continues until it encounters the nascent RNA strand still associated with region B (the C-B junction is indicated by |||||||||). As further synthesis on this portion of the template is blocked by the nascent strand, the polymerase makes a transition to an adjacent region D. (The critical region of D is behind C and is indicated by a dotted arrow). The end result is a transposition of region C which is normally upstream of B in the progenitor to a location downstream in the sgRNA.

MODEL FOR RNP STRUCTURE

We propose that the path of the RNA within the RNP follows the strand of NP subunits. As the length of the RNA associated with a subunit (11-13 nm) is considerably greater than the width of a subunit (5 nm), the RNA must presumably be gathered or coiled. In our model we thus present the path of the RNA as a coil; each loop of the coil corresponds to 25 nucleotides with a pitch equal to the subunit spacing. The coil is folded back on itself and twisted into a double helical structure with a pitch of 13 nm and 5½ loops per (single) turn of the helix (Fig. 2d).

For generation of a simple deletion sgRNA we require that a transition of the polymerase occurs between the 5' and 3' terminal regions of the progenitor segment. We propose that the 5' and 3' ends of the vRNA are located in the terminal loop. The transition from open loop to supercoiled

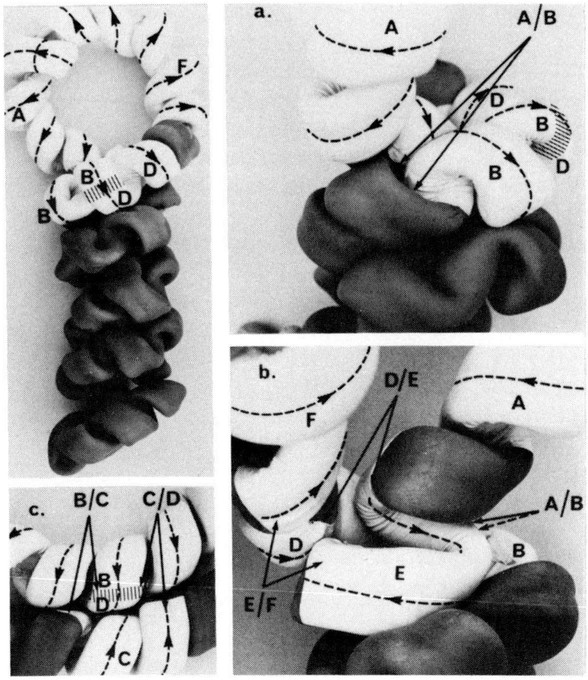

FIGURE 6. Generation of the mosaic sgRNA (4,5) derived mainly from segment 3. The B/D junction (||||||||) is the locus for the insertion from segment 1. A side view (a) of the RNP illustrates the first transition A/B and a back view (b) illustrates the D/E and E/F transitions. This would generate a sgRNA (A,B,D,E,F) derived entirely from segment 3. The insertion of a 37 nucleotide region from segment 1 (region C) is accomplished by the transitions B/C and C/D (c).

structure may impede polymerase progress and result in polymerase 'jumping' (Fig. 3). The range of loop sizes observed using close to in vivo conditions (8-15 subunits, corresponding to 200-400 nucleotides) is compatible with the modal size of the sgRNAs (350-450 nucleotides).

The double deletion sgRNAs would arise as do the simple sgRNAs but the second smaller deletion would appear as a consequence of transitions across a short region that is effectively 'looped out' (Fig. 4). The RNA loop size of ~25 nucleotides accomodates the required transitions in the double deletion sgRNAs and an essentially similar mechanism can also explain the necessary transpositions in the complex (Fig. 5) and the mosaic (Fig. 6) sgRNAs.

CONCLUSION

The models described in Figs. 2-6 for the generation of sgRNAs encompass all observations from currently available nucleotide sequence data (1). These observations are:

- virtually all viral segments can give rise to sgRNAs,
- there are no universal sequence features to account for their generation,
- the sgRNAs are derived from regions near the 5' and 3' termini of the parental segments,
- these regions are in the same sense in the sgRNAs,
- sequences from the progenitor regions may be alternated or transposed in the sgRNAs,
- all but one of the sgRNAs derive from one segment only and in this exceptional case two different segments contribute to the final structure of the sgRNA.

Our proposal of a mechanistic explanation for the generation of sgRNAs reconciles two distinct sets of data, nucleotide sequence and electron microscopic, and provides a basis for further experimentation.

ACKNOWLEDGMENTS

We would like to thank Drs. B.W.J. Mahy, J.W. McCauley and A. Klug for their helpful criticisms.

REFERENCES

1. Jennings, P.A., Finch, J.T., Winter, G., and Robertson, J.S. (1983). Cell 34, 619.
2. Nayak, D.P. (1980). Ann. Rev. Microbiol. 34, 619.
3. Winter, G., Fields, S., and Ratti, G. (1981). Nucleic Acids Res. 9, 6907.
4. Moss, B.A., and Brownlee, G.G. (1981). Nucleic Acids Res. 9, 1941.
5. Fields, S., and Winter, G. (1982). Cell 28, 303.
6. Sivasubramanian, N., and Nayak, D.P. (1983). Virology 124, 232.
7. Leppert, M., Kort, L., and Kolakofsky, D. (1977). Cell 12, 539.
8. Perrault, J., and Leavitt, R.W. (1977). J. Gen. Virol. 38, 35.
9. Pons, M.W., Schulze, I.T., Hirst, G.K., and Hauser, R. (1969). Virology 39, 250.
10. Compans, R.W., Content, J., and Duesberg, P.H. (1972). J. Virol. 10, 795.

MOLECULAR TOPOGRAPHY OF THE INFLUENZA VIRUS P PROTEIN COMPLEX DURING CAPPED RNA-PRIMED MESSENGER RNA SYNTHESIS[1]

Janet Braam, Ismo Ulmanen,[2] and Robert M. Krug

Molecular Biology and Virology Program
Graduate School of Medical Sciences
Memorial Sloan-Kettering Cancer Center
New York, NY 10021

INTRODUCTION

Influenza viral nucleocapsids (or cores) catalyze the unique process of influenza viral messenger RNA (mRNA) synthesis (Plotch, et al., 1981; Ulmanen, et al., 1981), which requires primers generated from heterologous cap 1 ($m^7GpppNm$)-containing RNAs (Bouloy, et al., 1978; Plotch, et al., 1979; Plotch, et al., 1981). These RNAs are cleaved by a viral endonuclease at a purine residue 10 to 13 nucleotides from the cap, and the resulting capped fragments serve as primers (Plotch, et al., 1981). Transcription is initiated by the addition of a G residue onto the 3' end of the capped primer fragments, directed by the penultimate C residue of the virion RNA (vRNA) templates (Plotch, et al., 1981). The viral mRNA chains are then elongated.

Four virus-specific proteins are associated with each of the eight vRNA segments in the viral nucleocapsids (Inglis, et al., 1976; Ulmanen, et al., 1981). The nucleocapsid (NP) protein, which is the predominant species (representing over 90% of the total protein), presumably serves largely a structural role. The other three viral proteins, the P proteins, distribute during two-dimensional gel electrophoresis into two basic species, PB1 and PB2, and one acidic species, PA (Horisberger, 1980; Ulmanen, et al., 1981). Previous experiments have established that one of the P proteins, the PB2 protein, is the cap 1-recognizing protein. Ultraviolet (UV) light-induced crosslinking studies

---

[1]This work was supported by U.S. Public Health Service grants; and by a European Molecular Biology Organization fellowship to I.U.

[2]Present address: Recombinant DNA Laboratory, University of Helsinki, Helsinki, Finland

Copyright © 1984 by Academic Press, Inc.
All rights of reproduction in any form reserved.
ISBN 0-12-183501-4

employing a capped RNA labeled with ^{32}P only in its cap structure showed that the PB2 protein is closely associated with the cap during the endonuclease reaction (Ulmanen, et al., 1981). Experiments using temperature-sensitive virus mutants established that the physical association of the PB2 protein with the cap 1 structure seen in crosslinking studies reflects a functional association. Temperature-sensitive mutants with a defect in the PB2 protein were shown to exhibit temperature-sensitivity in cap recognition during viral mRNA synthesis in vitro (Ulmanen, et al., 1983).

The ability of UV-crosslinking experiments to discern the function of the PB2 protein prompted us to use this methodology to delineate the functions and movements of the three viral P proteins during viral mRNA synthesis (Braam et al., 1983). An earlier UV-crosslinking experiment indicated that the PB1 protein catalyzes the initiation of transcription, as this protein is found in close association with the initial guanosine (G) residue coincident with its addition onto the 3' end of the primer fragment (Ulmanen, et al., 1981). We have now been able to follow the movements of the three P proteins during both the initiation of transcription and the elongation of the viral mRNA chains (Braam, et al., 1983). As a result of these studies, we can now present a coherent model for the functions and movements of the viral P proteins as they transcribe the influenza vRNA segments into viral mRNAs (Braam, et al., 1983).

RESULTS

Identification of the Elongation Protein in Viral Cores. Following the initiation of transcription by the incorporation of a G residue onto the 3' ends of capped primer fragments, influenza viral mRNA chains are elongated. A protein responsible for catalyzing this chain elongation would be expected to be situated at the very 3' end of the growing transcript, coincident with the addition of the 3' terminal residue. To identify the elongation protein, we performed transcription reactions that result in the synthesis of incomplete mRNA chains containing ^{32}P label essentially only in the 3' terminal nucleotide, and used UV light irradiation to covalently crosslink this labeled nucleotide to the protein(s) closely associated with this nucleotide (Braam, et al., 1983).

As shown in Figure 1, the initial part of the viral-coded region of each mRNA chain lacks uridine (U) residues (Robertson, 1979). Our strategy then was to incubate viral cores with a capped RNA primer (alfalfa mosaic virus RNA 4 [AlMV RNA 4]) and unlabeled GTP, CTP and

ATP and then to isolate the viral cores containing the nascent partial mRNAs free from unincorporated triphosphates. Transcription was then resumed by the addition of (α-^{32}P) UTP as the only ribonucleoside triphosphate. Analysis of the RNA in an aliquot from this reaction is shown in Figure 2A, lane 3. As predicted, the size of the RNA is 1-3 nucleotides longer than the partial transcripts terminated prior to the first U residue (lane 2) (Bramm et al., 1983).

To identify the protein(s) closely associated with these labeled terminal U residues, the reaction mixture was exposed to UV light, and after nuclease treatment, the proteins were analyzed by two-dimensional gel electrophoresis. PB1, the larger of the two basic P proteins, is essentially the only P protein labeled under these conditions (Figure 2B) (Braam, et al., 1983). The predominant labeling of PB1 was consistently observed in several experiments. In contrast, the amount of the NP that was labeled varied substantially and often was less than that shown in Figure 2B. Previously, we showed that NP crosslinks nonspecifically to labeled primer fragments and product RNAs (Ulmanen, et al., 1981). Our results thus indicate that PB1 is the protein closely associated with the U residues at the growing end of the mRNA chains, and therefore most likely catalyzes these 3' terminal nucleotide additions and hence functions to elongate viral mRNA chains (Braam, et al., 1983).

MRNA SEGMENT	PRIMER REGION	VIRAL-CODED REGION
1, 3	m^7GpppGm. . . A	G C G A A A G C A G G (U)
7	m^7GpppGm. . . A	G C A A A A G C A G G (U)
5, 8	m^7GpppGm. . . A	G C A A A A G C A G G G (U)
4	m^7GpppGm. . . A	G C A A A A G C A G G G G (U U)
6	m^7GpppGm. . . A	G C A A A A G C A G G A G (U U)
2	m^7GpppGm. . . A	G C G A A A G C A G G C A A A C C A (U U U)

FIGURE 1. The 5' Terminal Sequences of the Eight Viral mRNA Chains. The viral mRNA chains or segments are numbered 1 to 8 in order of decreasing molecular weight. The sequence of the 5' viral-coded region of the mRNAs is from Robertson (1979) (Braam, et al., 1983).

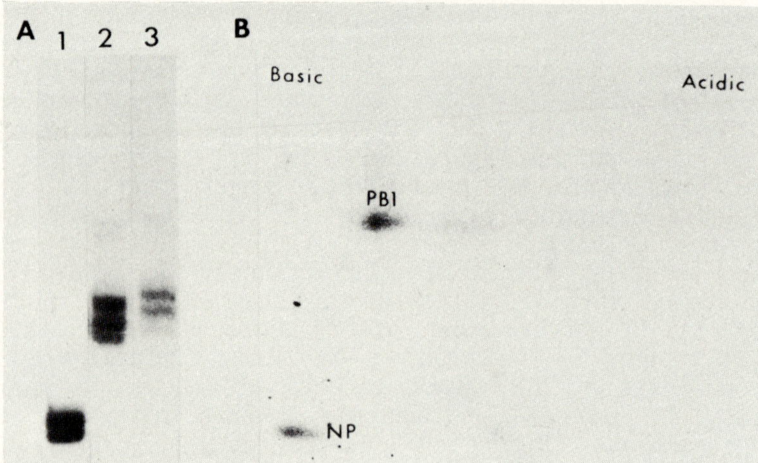

FIGURE 2. Identification of the P Protein That Catalyzes the Addition of the 3' Terminal Nucleotide Onto Growing Viral mRNA Chains. Viral cores were incubated for 20 minutes at 31C in a 100μl reaction mixture containing AlMV RNA 4 and three unlabeled ribonucleoside triphosphates (GTP, CTP and ATP), each at 100μM. These cores with the nascent chains were freed of ribonucleoside triphosphates by spin chromatography, and were then incubated for 20 minutes at 31C in the presence of (α-^{32}P) UTP (10μM, 300Ci/mmole) as the only ribonucleoside triphosphate. (A). The labeled RNA products of this reaction were electrophoresed on a 20% acrylamide/7M urea gel (lane 3). For size comparison, the RNAs synthesized in two other 100μl reactions were analyzed on the same gel: viral cores incubated for 20 minutes at 31C with AlMV RNA 4 and either (α-^{32}P) GTP alone (10μM 15Ci/mmole) (lane 1) or three ribonucleoside triphosphates (α-^{32}P) GTP, CTP and ATP (lane 2). (B). After UV irradiation and nuclease digestion, the labeled proteins were analyzed on a two-dimensional gel. From Braam, et al., (1983).

The PB1 protein has previously been shown to be associated with the initial G residue added to the capped primer fragments (Ulmanen, et al., 1981). Because of the demonstration that during elongation PB1 is found at the 3' end of the growing chains (Braam, et al., 1983), it would be predicted that the PB1 protein initially associated with the initial G residue moves away from this residue as soon as elongation begins. To test this prediction, cores were first incubated with a primer RNA and (α-^{32}P) GTP as the only ribonucleoside triphosphate, and these initiated chains were then elongated in a cold chase either with three triphosphates (GTP, CTP and ATP) to generate interrupted chains, or all four triphosphates to generate complete mRNA chains (Braam, et al., 1983). Figure 3 shows that the close association of PB1 with the initial labeled G residue - lanes 1 and 3 - disappears after the chains are elongated in the presence of three triphosphates - lane 2 - or in the presence of four triphosphates - lane 4. Therefore, after functioning in the initiation of transcription, the PB1 protein moves away from the initiating G residue to the next and subsequent downstream residues in order to elongate the viral mRNA chains (Braam, et al., 1983).

The Three P Proteins Move Down the Growing Viral mRNA Chains as a Complex. To determine whether either of the other two P proteins are also associated with the growing mRNA chains, we synthesized uniformly labeled partial transcripts terminated before the first viral-encoded U residue and exposed this reaction mixture to UV light (Braam, et al., 1983). As shown in Figure 4, all three P proteins were associated with the uniformly labeled partial transcripts. The relative amounts of the three P proteins that crosslinked to these partial transcripts varied in different experiments; consistently, however, all three of the P proteins were crosslinked, suggesting that the three P proteins are associated with the growing viral mRNA chains as a complex.

If they are subunits of a polymerase complex, the three P proteins should move together from this initial region of the viral mRNA chains to downstream regions as the mRNA chains are further elongated. These initial partial chains can be extended to various lengths by a second incubation with unlabeled UTP and other unlabeled triphosphates. Figure 5A shows that a chase with unlabeled UTP, ATP and GTP effectively elongates the initial partial transcripts (compare lanes 1 and 2) (Braam, et al., 1983). Only a small fraction (about 10-20%) of the initial partial transcripts are apparently not extended during the chase. The close association of the three P proteins with the labeled initial partial chains (Figure 5B, lanes 1 and 3) was gradually reduced as the partial transcripts were further elongated in a chase with unlabeled UTP and ATP (lane 4) and with unlabeled UTP, ATP and GTP (lanes 2 and 5) (Braam, et al., 1983). No detectable reduction in crosslinking occurred when ribonucleoside triphosphates were not added to the second incubation (Figure 5B, lanes 1 and 3), indicating that chain elongation was required. Thus, three P proteins

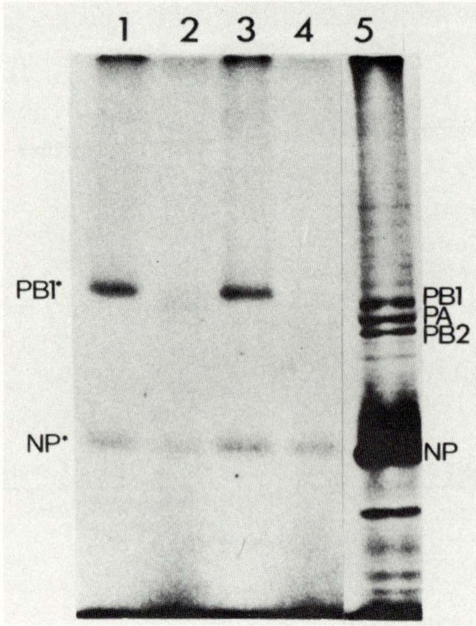

FIGURE 3. Movements of the PB1 Protein During Elongation. Two reaction mixtures (100μl), each containing viral cores, AlMV RNA 4 and (α-^{32}P) GTP (2.5μM, 400Ci/mmole) as the only ribonucleoside triphosphate, were incubated for 20 minutes at 31C. Each reaction mixture was divided in half. One half of each was then incubated for another 20 minutes at 31C without any further addition (lanes 1 and 3). The other half of the reaction mixture was incubated either in the presence of three unlabeled ribonucleoside triphosphates, GTP, CTP and ATP, each at 2mM (lane 2), or in the presence of four unlabeled ribonucleosde triphosphates, each at 2mM (lane 4). Following UV irradiation and nuclease treatment, the proteins were analyzed on one-dimensional gels. Lane 5 shows the gel pattern of the (^{35}S) methionine-labeled proteins in purified virions. Crosslinked proteins are denoted with a (*) superscript, as they migrate slightly slower than the noncrosslinked marker proteins because of the presence of a few covalently linked nucleotides (Ulmanen, et al., 1981). From Braam, et al., (1983).

move away from the initial labeled portion of the partial chains after subsequent elongation (Braam, et al., 1983).

To demonstrate directly that the three P proteins moved to the extended regions of the chains, we synthesized the initial partial chains in the presence of unlabeled ATP, CTP and GTP, and then carried out an elongation with three (α-^{32}P) labeled ribonucleoside triphosphates, UTP, ATP and GTP (Braam, et al, 1983). Under these conditions, only the extended regions of the chains are labeled (Figure 5A, lane 3). As shown in Figure 5B (lane 6), the three P proteins are closely associated with the labeled residues in the extended region of the chains. Two-dimensional gel electrophoresis confirmed that all three P proteins were present (data not shown). Consequently, the three P proteins move down the mRNA chains together during elongation (Braam, et al., 1983).

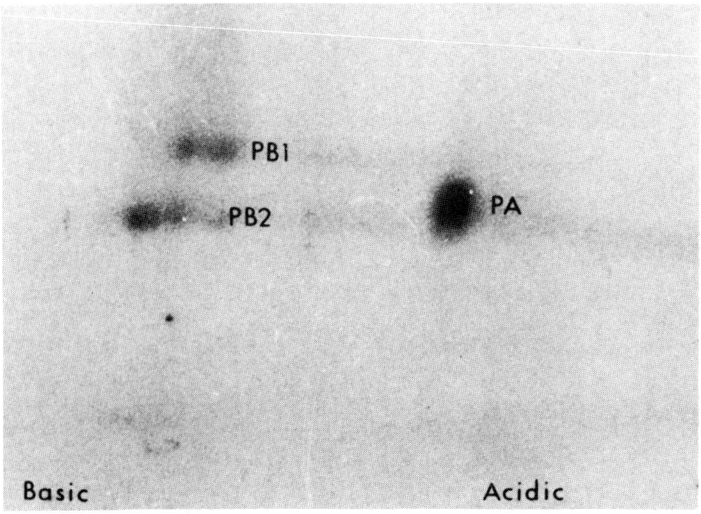

FIGURE 4. Identification of the P Proteins Associated with the Growing Viral mRNA Chains. Viral cores were incubated for 20 minutes at 31C with AlMV RNA 4 and three (α-^{32}P) labeled ribonucleoside triphosphates, GTP, CTP and ATP, each at 20 M and 100Ci/mmole, in a final volume of 100 μl. The labeled triphosphates were removed by spin chromatography, and after UV irradiation and nuclease digestion, the proteins were analyzed by two-dimensional gel electrophoresis. From Braam, et al., (1983).

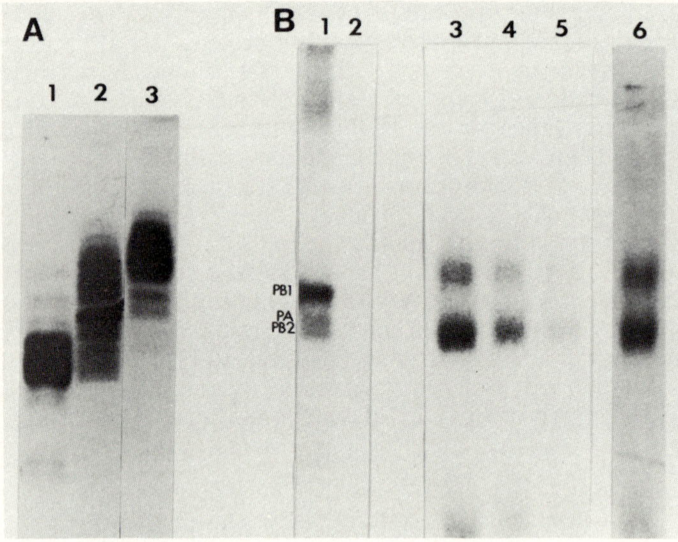

FIGURE 5. Movement of the Three P Proteins Together During Chain Elongation. Two reaction mixtures, one of 200 μl and one of 300μl, each containing viral cores, AlMV RNA 4 and three (α -^{32}P) labeled ribonucleoside triphosphates, GTP, CTP and ATP, each at 20 μM and 100Ci/mmole, were incubated for 20 minutes at 31C. The labeled triphosphates were removed by spin chromatography, and each of the two reaction mixtures was divided into 100μl aliquots. One 100 μl aliquot of each reaction mixture was incubated for an additional 20 minutes at 31C in the absence of ribonucleoside triphosphates. The labeled RNA was analyzed on a 20% gel (A, lane 1) and, after UV irradiation and nuclease digestion, the crosslinked proteins were analyzed on a one-dimensional gel (B, lanes 1 and 3). Another 100μl aliquot of one of the reaction mixtures was incubated for an additional 20 minutes at 31C in the presence of unlabeled UTP and ATP, each at 100μM, and the crosslinked proteins were analyzed on a one-dimensional gel (B, lane 4). Another 100μl aliquot of each reaction mixture was incubated for an additional 20 minutes at 31C in the presence of unlabeled UTP, ATP and GTP, each at 100μM, and the labeled RNA (A, lane 2) and crosslinked proteins (B, lanes 2 and 5) were analyzed as described above.

A separate reaction mixture (100 μl), containing viral cores, AlMV RNA 4 and three unlabeled ribonucleoside triphosphates, GTP, CTP and ATP, each at 100 μM, was incubated for 20 minutes at 31C. After removal of unlabeled

Dissociation of the PB2 Protein From the Cap During mRNA Synthesis. Since the PB2 protein is part of this complex moving down the chain, we determined whether the PB2 protein remains with, or dissociates from, the cap during chain elongation (Braam, et al., 1983). We used a capped RNA primer (AlMV RNA 4) labeled with ^{32}P only in its cap and carried out transcription reactions which yielded mRNA chains extended to various lengths. As before, UV-crosslinking was employed to monitor the association of the PB2 protein with the labeled cap. The controls for these experiments were the reaction conditions under which the PB2 protein has been demonstrated previously to be closely associated with the cap: the endonuclease reaction in the absence of ribonucleoside triphosphates (Figure 6A, lane 1); and the initiation step in the presence of GTP (lane 2) (Ulmanen, et al., 1981). The association of PB2 with the cap was maintained even after the mRNA chains were elongated up to the first viral-coded U residue (lane 3) (Braam, et al., 1983). In contrast, the association of PB2 with the cap largely disappeared when complete viral mRNA chains were synthesized in the presence of all four ribonucleoside triphosphates (lane 4). This result indicates that the PB2 protein dissociates from the cap of the viral mRNA chains at some stage of their elongation. As shown in Figure 6B, this dissociation occurs soon after elongation proceeds past the first viral encoded U residue. When the initial transcripts terminated prior to the first U residue were extended by a second incubation in the presence of UTP, ATP and GTP, the PB2 protein was no longer associated with the cap of the nascent mRNA chains (lane 2). A control reaction in which ribonucleoside triphosphates were omitted from the second incubation established that the dissociation of PB2 from the cap required further chain elongation (lane 1). Thus, PB2 dissociates from the cap soon after more than 11 to 15 nucleotides have been added to the capped primer fragments (Braam, et al., 1983).

---

triphosphates by spin chromatography, three (α-^{32}P) labeled triphosphates, UTP, ATP and GTP, each at 20 μM and 100Ci/mmole, were added, and the reaction mixture was incubated for an additional 20 minutes at 31C. After removal of the labeled triphosphates by spin chromatography, the labeled RNA (A, lane 3) and crosslinked proteins (B, lane 6) were analyzed as described above. From Braam, et al., (1983).

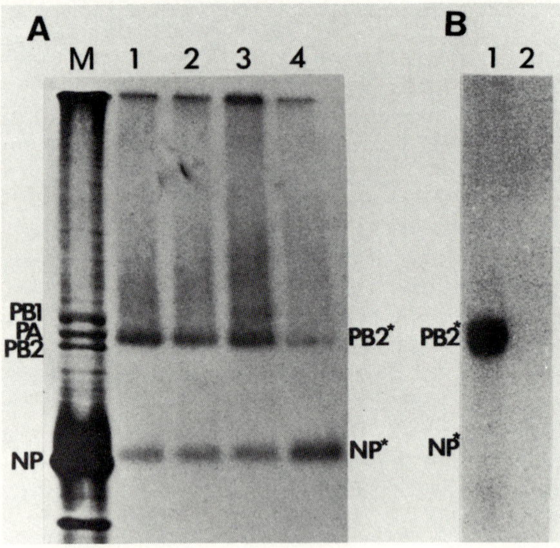

FIGURE 6. Dissociation of the PB2 Protein from the 5' Terminal Cap During Chain Elongation. (A). AlMV RNA 4 containing a ^{32}P-labeled cap 1 structure (10^5 cpm per each 300μl reaction mixture) was incubated with viral cores in the absence of triphosphates (lane 1), in the presence of GTP (10 M) alone (lane 2), in the presence of three triphosphates, GTP, CTP and ATP, each at 100μM (lane 3), or in the presence of four triphosphates, each at 100μM (lane 4). After UV irradiation and nuclease digestion, the proteins were analyzed by one-dimensional gel electrophoresis. Lane M (marker) shows the gel pattern of (^{35}S) methionine labeled proteins from purified virions. The crosslinked proteins denoted by a (*) superscript run slightly slower than the marker noncrosslinked proteins (Ulmanen, et al., 1981). (B). After an initial incubation of viral cores with ^{32}P cap-labeled AlMV RNA 4 and three ribonucleoside triphosphates, GTP, CTP and ATP, each at 100μM, the triphosphates were removed by spin chromatography. The cores were then divided in half. One half was incubated for 20 minutes at 31C in the absence of triphosphates (lane 1), and one half was incubated in the presence of UTP, ATP and GTP, each at 100μM (lane 2). Following UV irradiation and nuclease digestion, the proteins were analyzed by one-dimensional gel electrophoresis. From Braam, et al., (1983).

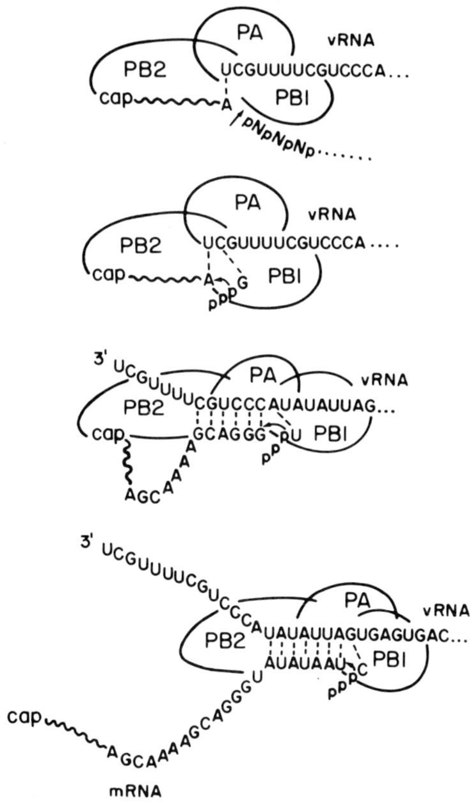

FIGURE 7. Model of the Functions, Interactions and Movements of the Three P Proteins During Capped RNA-Primed Viral mRNA Synthesis. The sequence shown is that of vRNA and mRNA chain number 5 coding for the nucleocapsid protein (Robertson, et al., 1979). From Braam, et al., (1983).

DISCUSSION

Because of our results, we can now formulate a coherent model for the functions and movements of the three influenza virus P proteins during capped RNA-primed transcription (Braam, et al., 1983). This model is shown in Figure 7. The top diagram shows the most likely arrangement of the P proteins during the generation of capped primer fragments. The P proteins are in the form of a complex which must be located at the 3' end of the virion RNA templates where transcription initiates. The PB2 protein recognizes and binds to the cap 1 structure of the primer RNA (Ulmanen, et al., 1981; Ulmanen, et al., 1983), allowing cleavage to yield a capped primer fragment. We do not yet know which protein catalyzes this cleavage. The simplest idea is that PB2, the cap-recognizing protein, spans the 5' terminal 10-13 nucleotides of the primer RNA and is capable by itself of catalyzing the cleavage reaction. Alternatively, the PB2 protein might catalyze this cleavage only when in association with one or both of the other two P proteins; or one of the other P proteins might catalyze cleavage. The PB1 protein is a possible candidate, as it is located at the appropriate position at the 3' end of the primer fragment poised to add the first G residue (second diagram). The third diagram shows that after PB1 adds the initial G residue, it moves down the elongating viral mRNA chains in a complex with PA and PB2. The PB1 protein leads the way, positioned at the growing end of the mRNA chain, where it most likley catalyzes the addition of each nucleotide to the chain. For the first 11 to 15 nucleotides of chain growth, PB2 is bound to both the growing chain and the cap, possibly causing the formation of a loop at the 5' end of the growing chain. Subsequently, PB2 dissociates from the cap, and the polymerase complex continues to move down the chain (last diagram).

ACKNOWLEDGMENTS

We thank Barbara A. Broni for expert technical assitance.

REFERENCES

1. Braam, J., Ulmanen, I. and Krug, R.M. (1983). Cell 34, 609-618.
2 Bouloy, M., Plotch, S.J. and Krug, R.M. (1978). Proc. Natl. Acad. Sci. USA 75, 4886-4890.
3. Horisberger, M.A. (1980). Virology 107, 302-305.
4. Inglis, S.C., Carroll, A.R., Lamb, R.A. and Mahy, B.W.J. (1976). Virology 74, 489-503.
5. Plotch, S.J., Bouloy, M. and Krug, R.M. (1979). Proc. Natl. Acad. Sci. USA 76, 1618-1622.
6. Plotch, S.J., Bouloy, M., Ulmanen, I. and Krug, R.M. (1981). Cell 23, 847-858.
7. Robertson, J.S. (1979). Nucl. Acids Res. 6, 3745-3757.
8. Ulmanen, I., Broni, B.A. and Krug, R.M. (1981). Proc. Natl. Acad. Sci. USA 78, 7355-7359.
9. Ulmanen, I., Broni, B.A. and Krug, R.M. (1983). J. Virol. 45, 27-35.

IN VIVO TRANSCRIPTION AND TRANSLATION OF DEFECTIVE-INTERFERING PARTICLE-SPECIFIC RNAS OF INFLUENZA VIRUS[1]

Thomas M. Chambers, Ramesh K. Akkina
and Debi P. Nayak

Jonsson Comprehensive Cancer Center and
Department of Microbiology and Immunology,
UCLA School of Medicine
Los Angeles, California 90024

Repeated high-multiplicity passage of influenza virus yields defective-interfering (DI) particles. These contain, in addition to the eight standard viral gene segments, other small RNA segments not found in the standard viral genome (see Figure 1, lanes 1 and 2). These DI-specific RNAs have been shown to be responsible for interference (for review see ref. 1).

Our laboratory and others have characterized several DI RNAs by oligonucleotide mapping and by cDNA cloning and sequencing (1). Unlike the majority of DI RNAs of vesicular stomatitis virus, which retain the 5' but not the 3' terminus of their progenitor RNA, most if not all influenza DI RNAs retain both the 5' and 3' genomic termini with relatively few base mismatches. The gene segment of origin is usually one of the three polymerase genes, from which the DI RNA is derived by single or, occasionally, multiple internal deletions. The DI RNAs we have studied to date are all of polymerase origin. However, recent studies indicate that internal deletion does occur in the other viral RNA segments during high-multiplicity passages (J. Robertson, this meeting). Since we have studied DI RNAs which have been evolutionarily selected for their capability for interference it is possible that although internally deleted RNAs may arise from other standard viral segments, they may lack the replicative advantage to become a major DI RNA species. We have repeatedly seen that all DI RNA bands are not of

[1]This work was supported by grants from the National Science Foundation (PCM 7823220) and the USPHS grants of the National Institute of Allergy and Infectious Diseases (RO1 AI 12749 and RO1 AI 16348). TMC was the recipient of a training grant (5T32CA09030) from the National Cancer Institute.

Copyright © 1984 by Academic Press, Inc.
All rights of reproduction in any form reserved.
ISBN 0-12-183501-4

equal intensity and the relative number of DI RNA molecules does not depend on their size, suggesting that the structure of a DI RNA in addition to its size may provide a replicative edge. Alternatively, strains of the virus and host cell may provide selection pressure for generating DI RNAs of specific genes.

DI RNAs are apparently generated by an aberrant replication event. Sequence studies suggest that RNA splicing is probably not involved, because of the absence of any consensus splicing sequence in the regions flanking the deletion points (1). Since most DI RNAs are of monogenic origin, it is likely that the polymerase skips to a new site without completely detaching from the template. Sequence studies have also indicated that different DI RNAs can arise independently, either from a standard viral progenitor segment or from a precursor DI RNA, suggesting a continuous evolution of DI RNAs. It is known that with repeated passage larger DI RNAs disappear and progressively smaller DI RNAs are formed (1).

Because the influenza DI RNAs retain both the 5' and 3' termini of their progenitor gene, presumably the recognition signals for initiation and termination of transcription and for 3' poly(A) addition are also preserved in the DI RNAs. Thus we have hypothesized that these DI RNAs may be capable of undergoing transcription and possible translation in infected cells, and may interfere with standard virus replication by a mechanism involving these products. Recently we have shown (2) that three influenza DI viruses of WSN origin are capable of synthesizing polyadenylated complementary RNAs for each DI RNA segment in an *in vitro* transcription system (3) (see Figure 1, lanes 5 and 6). Like transcripts of standard virus RNAs these DI cRNAs: a) contain poly(A) of heterogeneous length ranging from 60 to 350 nucleotides; b) are slightly smaller than their corresponding DI RNAs after the poly(A) has been removed; and c) require ApG or capped RNA as primer for initiation. Hybridization analysis of specific DI transcripts with their corresponding DI RNA segments indicate that they are not incomplete polymerase gene transcripts but are faithful transcripts of the DI RNAs.

Little is known concerning replication, transcription, or translation of influenza DI RNAs in vivo, and the relation of these to DI-mediated interference. The following experiments were performed to determine whether DI-specific transcription and translation products could be detected in infected cells.

TRANSCRIPTION OF DI RNAS IN STANDARD AND DI-INFECTED CELLS

All experiments reported here were done with WSN virus and MDBK cells. For *in vivo* transcription MDBK cell monolayers were infected with WSN ts$^+$ standard virus (4 pfu/cell) or coinfected with standard virus and DI-ts$^+$ (Tobita) (4 DIU/cell), then incubated in a phosphate-free medium. At various times post-infection (pi), $^{32}PO_4$ was added to the medium to a concentration of 0.3 mCi/ml. Cytoplasmic RNA was harvested by the method of Favaloro, et al. (4).The product was separated into poly(A)$^+$ and poly(A)$^-$ RNA fractions by oligo(dT) cellulose chromatography. Poly(A) tails were removed from the poly(A)$^+$ RNA by annealing to poly(dT) and digestion with RNase H. The product RNAs were electrophoresed on 3% polyacrylamide/6 M urea gels and autoradiographed (2).

Poly(A)$^+$ RNAs corresponding to vRNA and DI RNA segments were first detected in cells labeled between 2 and 4 h pi and became more prominent in cells labeled between 4 and 6 h or between 6 and 9 h pi. Figure 1, lanes 3 and 4 show the cytoplasmic poly(A)$^+$ RNAs of cells infected with standard virus or coinfected with DI and standard virus, respectively. DI-specific transcripts corresponding to each of the four smallest DI RNA segments are apparent. In lane 4, the transcript band corresponding to the DI-2 segment appears broad and sometimes as a triplet, suggesting that the DI-2 RNA represents more than one comigrating RNA species. As expected, each poly(A) transcript band migrates slightly faster than its corresponding vRNA band.

To determine whether the DI-specific poly(A)$^+$ RNAs were true complements of their corresponding DI RNAs, the ^{32}P-labeled DI transcripts and corresponding unlabeled DI RNAs were eluted from gels and specifically hybridized to each other. The hybrids were then treated with RNases A and T1, denatured and electrophoresed in a denaturing gel. DI transcripts were protected from RNase digestion to essentially their full length by their corresponding DI RNA. The DI-specific transcripts thus appeared to be true complements of the DI RNAs and not merely incomplete transcripts of standard viral segments. We have not yet shown, but we anticipate that the DI-specific poly(A)-cRNAs will possess host-derived 5' cap structures, as has been shown for the mRNAs of the standard viral segments (3).

Molar ratio analysis (Table 1) of standard and DI transcripts bands indicate that except for DI-1 each DI RNA was transcribed in molar excess over any standard viral segment in coinfected cells. Taken together the DI transcripts apparently represented about 80% of total transcript molecules.

We have also analyzed on gels the poly(A)$^-$ cytoplasmic RNA fraction from standard- and DI-infected cells. Bands corresponding to each DI RNA were visible from cells labeled from 2 to 4 h pi, and were the predominant species in cells labeled from 4 to 6 h pi. Presumably these include replicative template cRNA and progeny vRNA.

The predominance of DI-specific RNAs among both poly(A)$^+$ and poly(A)$^-$ cytoplasmic RNAs in coinfected cells suggests the preferential synthesis of DI-specific RNAs. It has been previously suggested that a mechanism exists for regulating disproportionate transcription of different influenza standard virus RNA segments, giving preference to nucleoprotein and non-structural protein transcripts early in infection, and hemagglutinin , neuraminidase and matrix protein transcripts late in infection (5), but the exact nature of this mechanism is unknown. Systems have now become available that will allow us to individually quantitate the rates of synthesis of poly(A)$^+$ cRNA, poly(A)$^-$ cRNA and vRNA for each standard and DI RNA segment in infected cells.

TRANSLATION OF STANDARD VIRAL AND DI-SPECIFIC POLYPEPTIDES

Because influenza DI-specific RNAs are transcribed into polyadenylated RNAs resembling messenger RNAs both *in vitro* and *in vivo*, it is likely that these cRNAs may be translated into DI-specific polypeptides. We have identified two DI virus preparations for which candidate DI-specific polypeptides are present in infected cell lysates (6). The DI-3 includes a prominent DI RNA species of approximately 700 nucleotides which is transcribed both *in vitro* and *in vivo*. MDBK cells coinfected with this and standard virus gave rise to an additional polypeptide (Figure 2, lane C) of about 22,000 molecular weight. A similar polypeptide was immunoprecipitated with whole WSN antiserum. The DI-7 includes two prominent DI RNA species of approximately 600 and 500 nucleotides. DI-7-coinfected cell lysates contained an additional polypeptide (Figure 2, lane E) of about 8,000 molecular weight, which was not immunoprecipitated with whole WSN antiserum.

Polypeptides similar to those of DI-3 and DI-7 were produced in an *in vitro* translation system (7) using poly(A)$^+$ cytoplasmic RNA from the DI-coinfected cells but not from standard virus-infected cells. With other DI preparations we have not observed a clear DI-specific polypeptide although all major DI RNAs are transcribed. We are

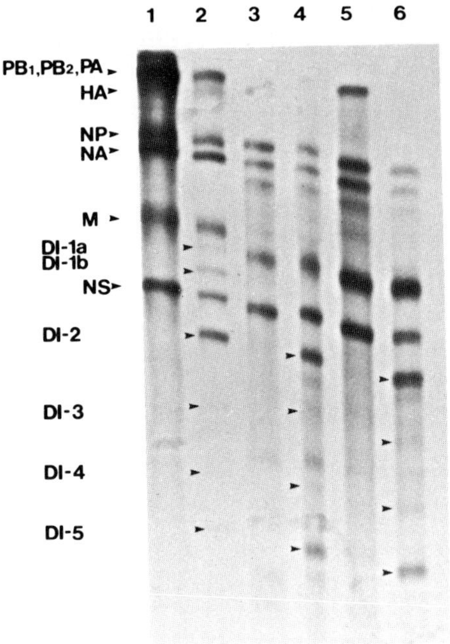

FIGURE 1

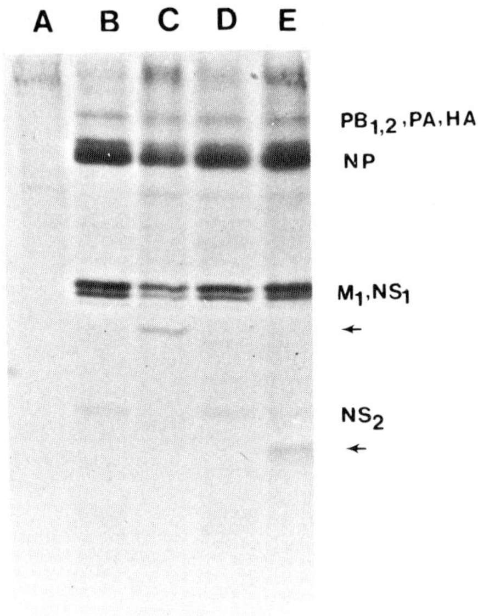

FIGURE 2

Figure 1. Comparison of the vRNA (lanes 1 and 2), poly(A)$^+$ cRNA synthesized in vivo (lanes 3 and 4), and poly(A)$^+$ cRNA synthesized in vitro (lanes 5 and 6) of standard virus (lanes 1,3, and 5) and DI-ts$^+$ (Tobita) (lanes 2,4, and 6). vRNA and in vitro poly(A)$^+$ cRNA were prepared as described in (2). In vivo poly(A)$^+$ cRNA was prepared from cytoplasms of infected cells as described in the text (4); cells were labeled with $^{32}PO_4$ from 3 h pi until harvest at 6 h pi. In vivo and in vitro poly(A)$^+$ cRNAs were selected by oligo(dT)-cellulose chromatography, and poly(A) tails removed.

Figure 2. Viral polypeptides in DI-infected cells. MDBK cells were infected with standard virus or coinfected with DI and standard virus. At 5 h after infection the cells were labeled with ^{35}S-methionine in methionine-free medium (50 uCi/ml) for one hour. Labeled cells were collected and boiled for one minute in SDS-PAGE sample buffer and subjected to electrophoresis in a 13% SDS-polyacrylamide gel containing 4 M urea (6,8). Lane A, mock-infected cells; lane B, cells infected with standard virus; lanes C, D, E: cells coinfected with standard virus and DI-3 (C), DI-4 (D), or DI-7 (E). Arrows at right indicate the positions of DI-specific polypeptides in lanes C and E.

TABLE 1
MOLAR RATIOS OF IN VIVO TRANSCRIPTS OF SPECIFIC VIRAL AND DI RNA SEGMENTS[a]

RNA	Viral segments					
	PB1 + PB2 + PA	HA	NP	NA	M	NS
DI vRNA	.09	.39	.43	.55	1.0	.95
DI cRNA	.07	.43	.34	.34	1.0	1.62
Standard vRNA	.29	.62	.58	.71	1.0	1.10
Standard cRNA	.09	1.07	.63	.56	1.0	1.54

RNA	DI Segments				
	1a + b	2	3	4	5
DI vRNA	.73	1.68	.65	.59	1.04
DI cRNA	1.46	6.19	1.81	2.80	4.17

a $^{32}PO_4$ added to infected cells at 3 h pi. RNA harvested at 6 h pi. Molar ratios normalized to M segment.

investigating to determine if these DI transcripts are not translated or if the nascent DI-specific protein is rapidly degraded.

Penn and Mahy (this meeting) have also reported the synthesis of DI-specific polypeptides in coinfected cells.

In order to understand the mechanism of DI interference with standard virus, we are analyzing the synthesis of individual viral proteins in DI-coinfected cells compared to standard-infected cells. Our preliminary data suggest: 1) That host-cell protein synthesis is greatly reduced in standard-infected cells but not in DI-coinfected cells. This correlates with the reduced cytopathic effect seen in DI infections. 2) Some DIs cause a decrease of synthesis of PB1 and PB2 proteins, while others cause a decrease of PB1 only. PA protein synthesis in these cases was almost unchanged. With other DIs the inhibition of specific polymerase proteins was not detected. This may explain why some DI preparations cause suppression of the total viral particle production in DI-coinfected cells.

SUMMARY

We have shown that influenza DI RNAs were transcribed in infected cells into both nonpolyadenylated RNA and into polyadenylated complementary RNA resembling messenger RNA. The DI-specific transcripts were synthesized in excess over the standard segment transcripts. Putative DI-specific polypeptides were also detected. Synthesis of one or more polymerase proteins was decreased in cells coinfected with some DI preparations. It is not yet known what role the transcription and translation of DI RNAs may play in the mechanism of DI-mediated interference in influenza.

REFERENCES

1. Nayak, D.P. and Sivasubramanian, N. (1983). In "Genetics of Influenza Viruses." (P.Palese and D.W. Kingsbury, eds.), p. 256. Springer-Verlag, Vienna.
2. Chanda, P., Chambers, T., and Nayak, D.P. (1983). J. Virol. 45, 55.
3. Krug, R.M. (1981). In "Current Topics in Microbiology and Immunology" (A. Shatkin, ed.), vol. 93, p. 125. Springer-Verlag, Berlin.
4. Favaloro, J., Treisman, R. and Kamen, R. (1980). In "Methods in Enzymology" (L. Grossman and K. Moldave, eds.), vol. 65, p. 718. Academic Press, New York.
5. Skehel,J.J. and Hay,.A.J. (1978). J. Gen. Virol. 39, 1.
6. Lamb, R.A. and Choppin, P.W. (1976). Virology 74, 504.
7. Etkind, P.R. and Krug, R.M. (1975). J. Virol. 16, 1464.
8. Laemmli, U.K. (1970). Nature 227, 680.

MUTANTS OBTAINED FOLLOWING UNDILUTED PASSAGE OF AN INFLUENZA A (FOWL PLAGUE) VIRUS

Christoph Scholtissek, Karin Müller, Ralph T. Schwarz, and Hans-Dieter Klenk

Institut für Virologie, Justus-Liebig-Universität Giessen, D-6300 Giessen, West Germany

INTRODUCTION

After undiluted passage, influenza viruses occasionally contain additional small RNA segments, which are derived from genome segments by extended deletions and insertions (1,2,3, 4,5). Therefore, we tried to obtain mutants following undiluted passage in the hope that in addition to point mutations, some might have small deletions. Such mutants might have properties different from those already obtained by chemical mutagenesis, since small deletions or insertions might be tolerated differently in the various genes. We have chosen the inability to multiply at 40° C as a selected marker, which can be easily monitored.

MATERIAL AND METHODS

In our studies we used the avian influenza A virus strain A/fowl plague/Rostock/34 (H7N1; FPV) and temperature-sensitive (<u>ts</u>) mutants derived from FPV by mutagenesis with 5-fluorouracil (6,7,8) as listed in Table 2. Furthermore, the influenza virus strains virus N (A/chick/Germany "N"/49 (H10N7)); PR8 (A/PR/8/34 (H1N1)); and Parrot Ulster (A/parrot/Ulster/73 (H7N1)) were used in these studies.

FPV was passaged without dilution at 33° C in chick embryo cells and plaques of different morphologies (tiny, small, large, turbid, fuzzy) were picked from each passage. Those isolates exhibiting a plaquing efficiency at 33° C which was at least by a factor of 10^5 higher than at 40° C were further examined.

The <u>ts</u> defects in isolates obtained by undiluted passage were characterized by complementation tests (rescue) with a set of standard <u>ts</u> mutants with known defects (see Table 2).

Copyright © 1984 by Academic Press, Inc.
All rights of reproduction in any form reserved.
ISBN 0-12-183501-4

FREQUENCIES OF TS MUTATIONS IN THE VARIOUS GENES

The frequencies of mutations occurring in the various genes after mutagenesis by undiluted passage at 33° C and after treatment with mutagens are compared in Table 1. In order to avoid in the calculations multiple isolations of the same mutant, we have included only such isolates of the same experiment for which mutations in different genes were detected. Indeed, only very few isolates in each experiment had similar properties. Most mutants of the first passage cannot be rescued by at least two different standard ts mutants; most of the third passage seem to be triple mutants. The highest frequencies of mutation were, according to our test system, found in RNA segments 3 (PA gene), 4 (HA gene), and 8 (NS gene). This is in contrast to ts mutants obtained using mutagens, in which the frequency of mutation follows roughly the size of the corresponding gene.

We studied the new isolates in detail by investigating further well-defined ts mutants with point mutations in RNA segment 4. As can be seen in Table 2 several of the isolates obtained by undiluted passage, e.g. ts 7/15 can be rescued by ts 227 but not by several other ts mutants with lesions in the HA gene. On the other hand, ts 1/93 cannot be rescued by ts 227 but is rescued by all other ts mutants with lesions in the HA gene. For comparison, double infection was also done with several influenza A virus strains which carry a HA non-cleavable in chick embryo cells. Generally it should not be possible to rescue mutants with a lesion in the HA gene with these influenza A virus strains (8). However, some mutants could be rescued by e.g. the parrot Ulster strain, but not by PR8.

INTRINSIC INTERFERENCE BY MUTATION

Several of our mutants which according to the results of rescue experiments should carry ts defects in two or three genes exhibited the reversion rate expected for a single point mutation (ca. 10^{-5}). One possible explanation for this observation could be that a mutant with a ts defect in one gene might interfere with the multiplication of a mutant with a ts defect in another gene at the permissive temperature, and in this way rescue to wild-type could not occur, in spite of the fact that the two mutations were located in different genes. A similar type of intrinsic interference has been shown recently to occur between FPV and swine influenza viruses (10). Therefore, we have studied two of our new mutants (ts 1/93 and ts 1/1) in more detail in the following way: Chick embryo cells were infected doubly with equal multiplicities of ts 1/93 and ts 227 at 33° C. A plaque test was

TABLE 1

DISTRIBUTION OF TS LESIONS IN MUTANTS OBTAINED EITHER BY UNDILUTED PASSAGE OR AFTER TREATMENT WITH MUTAGENS [a]

Passage Nr. (33° C)	Number of isolates	Number of mutations in segment (% of total)							Total mutations
		1 (PB2)	2 (PB1)	3 (PA)	4 (HA)	5 (NP)	6 (NA)	8 (NS)	
M*	54	20 (32)	10 (16)	8 (13)	8 (13)	8 (13)	5 (8)	3 (5)	62
1	10	2 (10)	2 (10)	5 (25)	5 (25)	1 (5)	3 (15)	2 (10)	20
2	1	1		1	1				3
3	8	2 (7.4)	2 (7.4)	6 (22.3)	8 (29.6)	2 (7.4)	1 (3.7)	6 (22.2)	27
4									
5	1				1			1	2
6	3	1 (16.6)	1 (16.6)		1 (16.6)		1 (16.6)	2 (33.3)	6
7	7	2 (13.3)	3 (20)	2 (13.3)	3 (20	1 (6.7)		4 (26.7)	15
1-7	30	8 (11)	8 (11)	14 (19.2)	19 (26)	4 (5.5)	5 (6.8)	15 (20.5)	73

[a] Ts lesions are defined by failure to rescue the ts isolate after double infection with the set of standard ts mutants as mentioned in Methods.

*M: Ts mutants were obtained by chemical mutagenesis in the laboratories in Giessen (6) and Cambridge (9).

TABLE 2

RESCUE OF TS MUTANTS OBTAINED BY UNDILUTED PASSAGE BY DEFINED TS MUTANTS WITH DEFECTS IN RNA SEGMENT 4 OR PROTOTYPE INFLUENZA VIRUS STRAINS, WHICH DO NOT PLAQUE ON CHICK EMBRYO CELLS

		Rescue to ts^+ possible with							
ts isolate	mutation in segment [a]	ts 227	ts 1	ts 482	ts 532	ts 651	virus N	PR8	Parrot Ulster
1/1	1,3,4	–	–	–	–	–	–	–	–
1/2	3,6	++[b]	++	+[b]	++	++	–	–	–
1/11	3,4	–	–	–	–	–	–	–	–
1/40	1,4,5	–	n.d.	–	–	–	–	n.d.	n.d.
1/93	3,4	–	++	++	++	++	++	++	++
2/4	1,3,4	–	–	–	–	–	–	–	–
3/3	3,4	–	++	+	++	++	+	n.d.	n.d.
3/9	1,3,4,8	–	–	–	–	–	–	–	–
3/18	3,4,5	–	++	+	+	+	–	–	++
3/46	3,4,8	–	–	–	++	+	–	–	–
5/4	4,8	–	++	–	++	+	+	n.d.	n.d.
6/4	1,4	–	–	–	–	–	–	n.d.	n.d.
7/15	2,3,8	++	++	–	+	–	+	n.d.	n.d.
7/33	3,4,8	–	+	++	++	++	++	–	++

[a]Mutation in segment 4 as defined by failure to be rescued with ts 227.

[b]++ = large clear plaques, + = smaller than wild-type plaques.

performed with the supernatant at 33° C, and plaques were picked at random. About 95% of the progeny plaques were ts 1/93 as shown by the ability to be rescued with virus N at 40° C. On the other hand, if cells were preinfected with ts 227 and superinfected 2 hours later with ts 1/93 ca. 95% of the plaque isolates at 33° C were ts 227 since they could not be rescued by virus N at 40° C. These experiments demonstrate that ts 1/93, when added simultaneously, interferes with the multiplication of ts 227 at the permissive temperature, presumably at the level of viral RNA synthesis, preventing reassortment to wild-type FPV. After preinfection with ts 227, the multiplication of ts 1/93 is blocked. These observations imply that ts 1/93 does not have a mutation in the HA gene. Indeed, the ts^+ reassortants between ts 1/93 and virus N all had replaced the RNA segment 3 of FPV, but none had RNA segment 4 of FPV replaced.

Similar experiments performed with ts 1/1 and the corresponding ts mutants revealed that ts 1/1 also carries only a single mutation, but the ts defect of ts 1/1 is located in the HA gene, and not in RNA segments 1 or 3. Thus, in these two mutants we have detected a new type of mutation causing intrinsic interference. Accordingly, the real number of ts lesions in mutants obtained by undiluted passage (as outlined in Table 1) is lower than that determined by rescue with the standard ts mutants.

ALTERATION OF GLYCOSYLATION BY MUTATION

If our assumption is correct that following undiluted passage mutations are caused also by deletions or insertions, they may be detectable by analyzing the viral RNA segments and/or proteins by polyacrylamide gel electrophoresis (PAGE) for small differences in migration rates. During examination of several of our mutants ts 1/1 was found to exhibit an unexpectedly fast migration rate of the HA_2 molecule (XHA_2 of Fig. 1) when grown at 33° C. In contrast to HA_2 of FPV wild-type, the XHA_2 cannot be labeled by ^{3}H-fucose, however, it can be labeled by ^{3}H-mannose and to a lower extent by ^{3}H-glucosamine. Further analysis of the tryptic peptides of ^{3}H-glucosamine-labeled XHA_2 as described in ref. (11) revealed that of the two oligosaccharidic chains attached to HA_2 the complex type was missing, while the mannose-rich one was still present. At 40° C the HA of ts 1/1 is only slowly cleaved, accumulating an HA with a faster migration rate when compared with the HA of FPV wild-type (Fig. 1). The molecular weight of XHA_2 as deduced from the migration rate during PAGE is smaller than would be expected if the loss of the complex oligosaccharide chain (molecular weight ca. 3000) were the only difference to HA_2. Thus, it is not excluded that the HA of ts 1/1 carries in addition a deletion in the protein backbone.

Five independent ts^+ revertants of ts 1/1 were analyzed by PAGE (see Fig. 1). Two of them synthesized an HA_2 with a normal migration rate (RIII and RV), two behaved like the mutant (RI (not shown) and RIX). One revertant (RIV) at the permissive temperature produced only the normal HA_2. and at 40° C a mixture of the altered XHA_2 and normal HA_2. The results on the ts^+ revertants suggest that (1) the loss of the oligosaccharide side chain in XHA_2 is not responsible for the ts phenotype, because two of the ts^+ revertants still synthesized XHA_2 instead of the normal HA_2. (2) The mutation leading to the ts phenotype is not located at the position where the complex oligosaccharide side chain is bound to the polypeptide. Otherwise it would be difficult to explain why RIV synthesizes at 40° C both HA_2 as well as XHA_2. Thus, the mutation in ts 1/1 should be located somewhere else in the HA molecule which causes a conformational change in such a way that the corresponding enzyme cannot transfer the oligosaccharide from the lipid-linked precursor $Glc_3Man_9(GlcNAc)_2$-PP-Dol to the correct position. During reversion, selection was for the ts^+ phenotype. Thus the reversion to ts^+ might be caused by suppressor mutations in the same gene which might or might not affect the conformation of the HA in such a way that the missing oligosaccharide side chain can be introduced again. Sequence analyses have been started on the HA genes of ts 1/1 and their ts^+ revertants to examine these possible explanations.

CONCLUSION

The new type of mutants causing intrinsic interference were so far found only after undiluted passage. It cannot be said yet whether this new type of mutation is due to a specific mechanism of mutation caused by undiluted passage, or whether such mutants occasionally occur also after chemical mutagenesis, but accumulate specifically during undiluted passage. Furthermore, we do not know whether the other new type of mutation causing a conformational change and loss of a carbohydrate side chain as found with ts 1/1 is typical for mutagenesis following undiluted passage. However, mutagenesis by multiple infection is expected not to be a rare event under natural conditions. If in a tissue an infected cell produces about 200 infectious virus particles, the cells in the immediate neighbourhood have a high probability of becoming infected with more than one infectious unit, the precondition for inducing the von Magnus phenomenon (12). Thus, mutagenesis resulting from multiple infection might contribute significantly to the high genetic variability of influenza viruses, including the deletions and/or insertions found with naturally occurring influenza virus strains, especially in the NA and HA genes (for a review see 13).

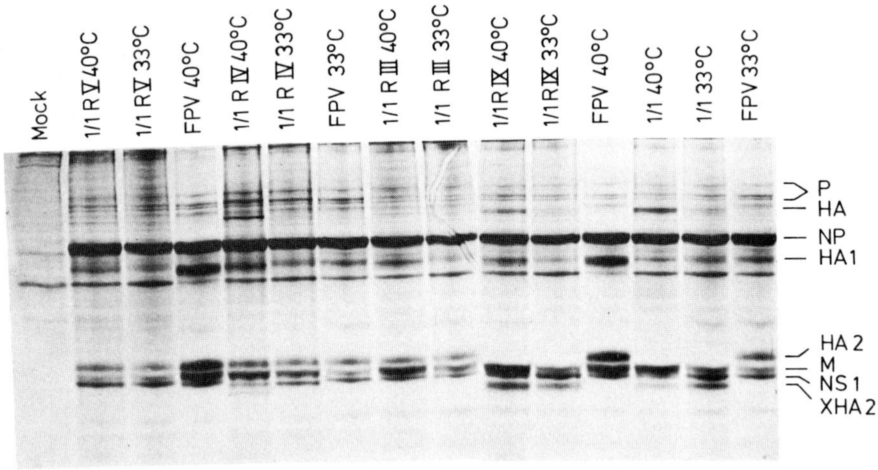

FIGURE 1. Polyacrylamide gel electrophoresis of viral proteins of *ts* 1/1, revertants (R) of *ts* 1/1, and wild-type FPV after labeling with ^{35}S-methionine.

Cells were incubated either at 33° C or 40° C. 10 µCi of ^{35}S-methionine were added 3 hr after infection. Three hr later the cells were processed. P = the three P proteins; HA = the uncleaved hemagglutinin; NP = nucleoprotein; HA_1 = the larger HA cleavage product; HA_2 = the normal smaller HA cleavage product; M = membrane protein; NS_1 = nonstructural protein 1; XHA_2 = the modified smaller HA cleavage product lacking the complex carbohydrate side chain.

ACKNOWLEDGMENTS

The work was supported by the Sonderforschungsbereich 47, Virologie, of the Deutsche Forschungsgemeinschaft.

REFERENCES

1. Davis, A. R., and Nayak, D. P. (1979). Proc. Nat. Acad. Sci. U.S.A. 76, 3092-3096.
2. Nakajima, K., Ueda, M., and Sugiura, M. (1979). J. Virol. 29, 1142-1148.
3. Davis, A. R., Hiti, A. L., and Nayak, D. P. (1980). Proc. Nat. Acad. Sci. U.S.A. 77, 215-219.
4. Moss, B. A., and Brownlee, G. G. (1981). Nucl. Acids Res. 9, 1941-1947.
5. Fields, S., and Winter, G. (1982). Cell 28, 303-313.

6. Scholtissek, C., and Bowles, A. F. (1975). Virology 67, 576-587.
7. Koennecke, I., Boschek, C. B., and Scholtissek, C. (1981). Virology 110, 16-25.
8. Scholtissek, C., Harms, E., Rohde, W., Orlich, M., and Rott, R. (1976). Virology 74, 332-344.
9. Almond, J. W., McGeoch, D., and Barry, R. D. (1979). Virology 92, 416-427.
10. Rott, R., Orlich, M., and Scholtissek, C. (1981). Arch. Virol. 69, 25-32.
11. Schwarz, R. T., Schmidt, M. F. G., Anwer, U., and Klenk, H.-D. (1977). J. Virol. 23, 217-226.
12. Barry, R. D. (1961). Virology 14, 389-397.
13. Scholtissek, C. (1983). In "Genetics of Influenza Viruses" (P. Palese and D. W. Kingsbury, eds.), pp.99-126. Springer-Verlag, Vienna.

GENETIC STABILITY OF A/ANN ARBOR/6/60 COLD-MUTANT (TEMPERATURE-SENSITIVE) LIVE INFLUENZA VACCINE STRAINS[1]

N. J. Cox and A. P. Kendal

Centers for Disease Control
Center for Infectious Diseases
Division of Viral Diseases
Atlanta, Georgia 30333

INTRODUCTION

Live, attenuated vaccines obtained by reassortment between epidemic wild-type (wt) strains and the cold-adapted (ca) A/Ann Arbor/6/60 donor strain are continually being evaluated biochemically and genetically (1-3) as well as being tested in human volunteers (for a review 1, 4-6). These ca viruses have been demonstrated to be infectious and immunogenic for children and young adults, but rarely cause unacceptable vaccine-associated side effects.

The simple and rapid technique of oligonucleotide mapping of RNA permits examination at the molecular level of the genetic stability of the ca recombinant viruses used in human volunteer studies. By comparing T_1 oligonucleotide maps of the A/Ann Arbor/6/60 mutant virus and the wild-type virus from which it was derived, we have previously identified oligonucleotide changes in RNAs 1 and 3 (polymerase genes), RNA 4 (hemagglutinin gene), and RNA 6 (nucleoprotein gene) of the mutant (3). In this study, we examined 11 ca candidate vaccine recombinant strains for the presence of these "marker oligonucleotides." We also compared the T_1 oligonucleotide maps of the RNA of 25 viruses isolated at various times post-vaccination from seronegative children or young adults to the T_1 maps of the parent vaccine strain in order to assess genetic stability during replication in humans.

RESULTS

<u>Oligonucleotide Mapping Of ca Recombinant Candidate Vaccine Strains</u>. T_1 oligonucleotide maps of total virion RNA were prepared as described previously (7) for the ca recombinant candidate live vaccine strains listed in Table 1. The genotype of each of the ca, ts recombinants has been determined by PAGE of viral RNA (3).

---

[1]This work was partially supported by NIAID contract no. 401-AI-8001.

Copyright © 1984 by Academic Press, Inc.
All rights of reproduction in any form reserved.
ISBN 0-12-183501-4

The T_1 oligonucleotide maps of these recombinants were compared with maps of the A/Ann Arbor/6/60 mutant and A/Ann Arbor/6/60 wt viruses in order to determine if the oligonucleotide changes observed in the mutant were reproducibly transferred to each recombinant. Results indicated that for each of the ca recombinants examined, the markers present in the T_1 map of the A/Ann Arbor/6/60 mutant designated B, C, and D (Figure 1A) were indeed present and oligonucleotides E, F, G, and H which were present in the A/Ann Arbor/6/60 wt RNA (Figure 1B) but absent in the mutant RNA were absent in the reassortants, as would be expected if the observed oligonucleotide changes were stable.

T_1 oligonucleotide maps of the ca recombinant AA-CR19 Cl 0 (with arrows pointing out the presence of T_1 oligonucleotide markers) is shown as an example in Figure 1C. Similar findings were observed in the other 10 ca reassortant viruses examined (Table 1).

Oligonucleotide Mapping of Reisolates of ca Recombinant Vaccine Strains Obtained From Human Vaccines. Five viruses (numbers 1537, 1641, 1645, 1726, 1727) which had been isolated from children five days post-vaccination during a study with the AA-CR 29 Cl 2 ca recombinant vaccine (5), and eight virus isolates (numbers 23, 51, 96, 100, 106, 108, 111, and 114) which had been isolated from young adults on days 2, 7, 4, 4, 4, 4, 4, and 4 respectively were examined by T_1 oligonucleotide mapping. Two of the isolates from children (1727 and 1645) showed partial loss of the ts marker in MDCK cells, but all five retained temperature-sensitivity and cold-adapted properties in primary chick kidney (pCK) cells (5). A detailed comparison of T_1 maps of total viral RNA for all 13 isolates and of the RNA from AA-CR 29 Cl 2 (Figure 1D), which had been maintained by the authors, revealed no detectable oligonucleotide changes (Figures 1E, 1F) between the viruses, even in the two isolates with "modified" ts phenotype in MDCK cells.

Likewise, oligonucleotide maps were obtained for eight virus isolates from individuals vaccinated with the cold recombinant AA-CR 37 Cl 1. Isolates 5832, 6339, 6555, 6711, and 7361 were obtained respectively, from nasal swab specimens taken from children on days 5, 7, 10, 7, and 10 post-vaccination, while isolates 11, 12, and 15 were obtained from young adults on the first day post-vaccination. Viruses retained temperature sensitivity properties in tissue culture (Robert Belshe and Brian Murphy, personal communication). Detailed comparisons of T_1 oligonucleotide maps of these five isolates and a preparation of the AA-CR 37 Cl 1 vaccine strain maintained by the authors

(Figure 1G) revealed differences only in isolate 7361, in which one oligonucleotide (N) appeared and another (M) disappeared (Figure 1H). This may be explained by a single base change and consequent charge difference in the oligonucleotide designated M. The other isolates had identical oligonucleotide maps when compared to each other and AA-CR 37 Cl 1 (Figure 1I and data not shown). By comparing the T_1 map of the isolate 7361 with T_1 maps of the individual segments of the A/Ann Arbor/6/60 mutant, it was determined that the oligonucleotide spot missing from isolate 7361 had been assigned to the nucleoprotein gene of the A/Ann Arbor/6/60 mutant, indicating that a change had occurred in the RNA coding for NP.

Finally, oligonucleotide maps were obtained for four isolates from young adults vaccinated with the AA-CR 35 Cl 2 ca recombinant vaccine. Isolates 21, 45, 49, and 71 were obtained, respectively on days 6, 8, 9, and 7 post-vaccination. A difference between the T_1 oligonucleotide pattern of the RNA of the vaccine strain maintained by the authors (Figure 1J), and the patterns of the four isolates was detected only for isolate 71 (Figure 1K). Again by comparing the T_1 map of isolate 71 with T_1 maps of individual segments of the A/Ann Arbor/6/60 mutant it was determined that the missing oligonucleotide designated 0 had been assigned to RNA 2 of the A/Ann Arbor/6/60 mutant.

DISCUSSION

By examining the oligonucleotide maps of live attenuated influenza ca recombinant vaccine strains both before use and after replication in humans, we have provided the first assessment, using molecular techniques, of the overall genetic stability of these viruses. Though T_1 mapping examines only 10 to 15 percent of the genome (8), next to sequencing this method is the most sensitive method available for detecting changes at the nucleotide level. Thus, it provides a useful method for the rapid screening of closely related viruses for an initial determination of the extent of their genetic relatedness.

In this study we demonstrated the reproducible transfer of marker oligonucleotides from the ca mutant A/Ann Arbor/6/60 parent to its ca recombinants. These findings indicate that it is feasible to require the existence of defined oligonucleotide markers as part of the identity testing of candidate live vaccine strains prepared from the ca A/Ann Arbor/6/60 strain. In order to assess the genetic stability of the ca recombinants during replication in humans, we compared the T_1 oligonucleotide maps of virus isolates recovered from vaccinated individuals to the maps of the original vaccine strain. We have shown that the 25 ca recombinant viruses analyzed exhibit a high degree of stability with minimal changes occurring in only two isolates.

Previously we have shown that mutations have occurred in all eight genes of the A/Ann Arbor/6/60 mutant (3) during adaption to growth at 25° (9). The presence of multiple mutations in the genome of the mutant virus provides a theoretical basis for postulating a lower rate of reversion to virulence for the ca mutant than has been observed for single-step ts mutants with lesions in one or two genes. The present study supplements other laboratory data indicating that the ca A/Ann Arbor/6/60 virus has genetic properties appropriate for production of live attenuated vaccines.

TABLE 1

ANALYSIS OF OLIGONUCLEOTIDE MARKERS IN THE T_1 OLIGONUCLEOTIDE MAPS OF A/ANN ARBOR/6/60 WILD TYPE, A/ANN ARBOR/6/60 MUTANT, AND SELECTED COLD-ADAPTED RECOMBINANT VIRUSES.

		Oligonucleotide markers							
Viruses analyzed	Wild-type parent[a]	A[b]	B	C	D	E	F	G	H
A/Ann Arbor/6/60 wt		–[c]	–	–	–	+[d]	+	+	+
A/Ann Arbor/6/60 ca		+	+	+	+	–	–	–	–
AA-CR 6 Clone 0	A/Queensland/6/72	–	+	+	+	–	–	–	–
AA-CR 13 Clone 9	A/Dunedin/4/73	–	+	+	+	–	–	–	–
AA-CR 18 Clone 7	A/Scotland/840/74	–	+	+	+	–	–	–	–
AA-CR 19 Clone 0	A/Victoria/3/75	–	+	+	+	–	–	–	–
AA-CR 22 Clone 1	A/Victoria/3/75	–	+	+	+	–	–	–	–
AA-CR 29 Clone 2	A/Alaska/6/77	–	+	+	+	–	–	–	–
AA-CR 31 Clone 3	A/Alaska/6/77	–	+	+	+	–	–	–	–
AA-CR 31 Clone 10	A/Alaska/6/77	–	+	+	+	–	–	–	–
AA-CR 35 Clone 2	A/Hong Kong/123/77	–	+	+	+	–	–	–	–
AA-CR 37 Clone 1	A/California/10/78	–	+	+	+	–	–	–	–
AA-CR 44 Clone 6	A/Beijing/2/79	–	+	+	+	–	–	–	–

[a]Wild type epidemic strain from which recombinant derived surface glyco-proteins.

[b]This marker for RNA 4 (3) is present only in the cold-adapted A/Ann Arbor/6/60 mutant because all the recombinants possessed HA from the wild-type parent.

[c]Indicates that the oligonucleotide marker is not present in the virus analyzed.

[d]Indicates that the oligonucleotide marker is present in the virus analyzed.

FIGURE 1. Oligonucleotide maps of the total virion RNA of wt and ca A/Ann Arbor/6/60 viruses and several ca recombinants. (A) A/Ann Arbor/6/60 mutant (B) A/Ann Arbor/6/60 wt (C) AA-CR 19 Cl 0 (D) AA-CR 29 Cl 2 (E) Isolate 1645 (F) Isolate 106 (G) AA-CR 37 Cl 1 (H) Isolate 7361 (I) Isolate 15 (J) AA-CR 35 Cl 2 (K) Isolate 71 (L) Isolate 21. Arrows point out oligonucleotide spot differences between mutant and wt RNAs (ABCD in Figure 1A and EFGH in Figure 1B) as well as the presence of marker oligonucleotides in ca recombinants. Additional arrows point out differences between parent vaccine strains and isolates 7361 (H) and 71 (K). The positions of dye markers bromophenol blue and xylene cyanol FF are shown by the upper and lower X's.

ACKNOWLEDGEMENTS

We thank Drs. Peter Wright, Robert Belshe, and Brian Murphy for providing the virus isolates obtained from vaccinees and Ms. Renee Black for excellent technical assistance. Cold-adapted recombinant viruses were originally prepared and selected in collaboration with Dr. H. F. Maassab.

REFERENCES

1. Kendal, A. P., Maassab, H. F., Alexandrova, G. I., and Ghendon, Y. Z. (1981). Antiviral Res. 1, 339.
2. Cox, N. J., Maassab, H. F., and Kendal, A. P. (1979). Virology 97, 190.
3. Cox, N. J., Konnecke, I., Kendal, A. P., and Maassab, H. F. (1981). "Genetic Variation Among Influenza Viruses" (D. Nayak and C. F. Fox, eds.) p. 639 Academic Press, New York.
4. Cisneros, L. A., Rennels, M. B., Miller, E. H., Black, R. E., Levine, M. M., Betts, R. F., Douglas, R. G., Jr., Maassab, H. F., Cox, N. J., and Kendal, A. P. (1981). Infect. Immun. 32, 693.
5. Wright, P. F., Okabe, N., McKee, K. T., Jr., Maassab, H. F., Karzon, D. T. (1982). J. Infect. Dis. 146, 71.
6. Murphy, B. R., Nelson, D. L., Wright, P. F., Tierney, E. L., Phelan, M. A., and Chanock, R. M. (1982). Infect. Immun. 36, 1102.
7. Cox, N. J., and Kendal, A. P. (1983). J. Infect. Dis. Accepted for publication.
8. Young, J. F., Desselberger, U., and Palese, P. (1979) Cell 18, 73.
9. Maassab, H. F. (1967). Nature (London) 213, 612.

Gene Expression, Protein Synthesis, and Protein Modification

SUBGENOMIC RNA OF ARENAVIRUS PICHINDE AND ITS IMPLICATION IN REGULATION OF VIRAL GENE EXPRESSION IN PRODUCTIVE INFECTION AND PERSISTENCE[1]

Wai-Choi Leung,[a] Arlene Ramsingh,[a2]
Guozhong Jing,[a] Kam Mong,[a]
Ashok K. Taneja,[b] and Robert S. Hodges[b]

Department of Medicine,[a] Department of Biochemistry and Medical Research Council Group in Protein Structure and Function,[b]
University of Alberta,
Edmonton, Alberta, Canada.

GENE MAPPING OF PICHINDE VIRUS

The arenaviruses readily induce persistent infection in their natural host. Viral persistence can also be established in cell culture in vitro and certain features of the virus gene expression in vitro mimic those observed in vivo.(1,2,3,4,) As a concerted effort to understand virus persistence, we have strived to define the genetic composition of the virus and the arrangement of genes on the viral genome.(5) Past efforts had demonstrated that Pichinde virus, a member of the Arenaviridae, contained a segmented genome with two RNA species (L and S RNA). The viral RNAs were of negative polarity and an RNA transcriptase were shown to associate with the viral nucleocapsid. Although it has not been demonstrated, it is reasonable to expect an RNA replicase function encoded by the virus genome.(6,7,8,9)

Virus Gene Products. There were three primary gene products in Pichinde virus; a L protein (~200,000 M.W.), NP protein (64,000 M.W.) and a GPC protein (79,000 M.W.).

---

[1]This work was supported by the Medical Research Council of Canada and the Alberta Heritage Foundation for Medical Research.

[2]Present address: Department of Human Genetics, Yale University School of Medicine, New Haven, Conn., U.S.A.

Copyright © 1984 by Academic Press, Inc.
All rights of reproduction in any form reserved.
ISBN 0-12-183501-4

The GPC was first derived from a nonglycosylated precursor of 42,000 M.W. and was later glycosylated to the 79,000 M.W. protein which was found associated only with the infected cells. The GPC was then cleaved into 2 glycoproteins GP-1 and GP-2, presumably on the cell surface, and also incorporated into the envelope of mature virion. Distinct tryptic peptide maps have been found for each polypeptide, hence suggesting that they were distinct primary gene products.(10)

The virus specificity of these 3 polypeptides were established by their appearance in virus infected cells, in mature virion as well as in subviral components. They were recognized by immune serum against Pichinde virus. In the case of NP and GPC, the virus specificity were further strengthened by the similar, but not identical tryptic peptide maps found in the polypeptides of Pichinde virus and another closely related arenavirus, Munchique.(15) This finding suggested strongly against the possibility that these two polypeptides were contaminating host cell proteins. As will be discussed later, the NP can even be related to the virus RNA at nucleotide sequence level.

The less well defined polypeptide, at present, is the L protein. It only constituted about 2-3% of the total ^{35}S-methionine labeled polypeptide in purified virion preparation and also in 100,000Xg supernatant in infected cell lysate. However, our recent findings indicated an abundance of this L protein was associated with a particulate structure of size larger than 100S. This particulate-associated L protein could be recovered by immunoprecipitation of a low speed (about 15,000Xg) supernatant. It will facilitate the characterization of this polypeptide.

Although there were reports about additional virus associated polypeptides found in virion or in virus infected cells, the virus specificity of these polypeptides were not present in molar amounts which leads to difficulties in assigning meaningful biological functions to them. However, using the present biochemical and immunological technologies, we cannot exclude the possibility of additional viral gene products specified by virus. More conclusive results will await the completion of total nucleotide sequence of both the L and S RNA. The presence of additional open reading frame will suggest the possibility of new primary gene products.

From the biological point of view, the 3 proteins associated with Pichinde virus will be able to supply the essential viral functions sufficient for virus replication and also as basic components of the virion structure. The NP will complex with viral RNAs to form the ribonucleoprotein in the nucleocapsid, whereas the derivatives of GPC, the GP-1 and GP-2 will form the surface spikes on the virion envelope. By deduction, the L protein is likely to exhibit an RNA transcriptase and replicase function, which is analagous to findings in other negative strand RNA viruses.

Fine Mapping of Viral Proteins on the Genome. The viral RNA of Pichinde virus had been carefully measured by gel electrophoresis under denaturing conditions to be 2.73×10^6 M.W. (8.5Kb) for the L RNA and 1.28×10^6M.W. (4.0 Kb) for the S RNA.[11] The demonstration of the presence of L protein immediately suggested the assignment of the L protein to the L RNA, since the S RNA was too small to code for such a large protein.[5] The NP and GPC were confirmed to reside on the S RNA by the analyses with a recombinant virus derived from Pichinde and Munchique parents. It was found that the NP and GPC segregated genetically together with the S RNA, thus indicating their gene locations.[5,12,13]

We reported here the further assignment of the NP on the 3' terminal portion of Pichinde viral RNA. Viral RNA sequence of 120 base long has been obtained from the 3' terminus of the S RNA of Pichinde virus.[14] Analysis of the nucleotide sequence revealed a first translation initiation codon (AUG) was present at the nucleotide residue 84-86, if the codon was deduced from the complementary RNA strand. A protein sequence could then be deduced as follows: met ser asp asn ile pro ser phe arg trp val gln. We then synthesized an eight amino acid peptide with the sequence ser asp asn ile pro ser phe arg using Merrifield's solid phase method with a Beckman amino acid synthesizer. The peptide was coupled to bovine serum albumin and was used as antigen to immunize rabbits. Antisera from immunized rabbits were able to immune precipitate the NP, but not GPC and L proteins from the virus infected cells. No immune precipitation of NP was found in mock infected cells. This result indicated that the anti-peptide sera recognized NP, and suggested that the NP gene resided on the 3' portion of the S RNA. By deduction, the GPC resided on the 5' portion of the S RNA. This observation was confirmed independently by nucleotide sequence analysis of the Pichinde virus S RNA. A 62,000 M.W. protein with basic charges (a putative NP) was found at the 3' end of the S RNA, and another protein with a hydrophobic NH_2-termini sequence (a putative signal sequence of GPC) followed immediately. No additional reading frame was found at the NP protein gene locus of the S RNA. (Auerpin and Bishop, personal communication).

Besides fine mapping the NP and GPC genes on the S RNA, the above experiment indicated that the NP gene did utilize the first initiation codon for translation.[15] More importantly, since the amino acid sequence was deduced from the complementary RNA sequence, it represented the most conclusive evidence for the negative polarity of the Pichinde genomic RNA.

GENOME ORGANIZATION OF PICHINDE VIRUS

The search for primary gene products and the assignment of gene loci on the viral RNA conveyed a picture of a virus of very simple genetic constituents. It consisted of 2 viral RNA and 3 viral proteins. The limited sequence data available of the S RNA also suggested a very simple arrangement of genes on the RNA. However, Pichinde virus, as well as other arenavirus, were able to exercise a wide range of gene expression observed in the replication of virus and also in persistent infection. We then investigated the organization of viral genome as a possible means of gene regulation.

A previous collaborative study has indicated that Pichinde virus was able to attenuate the infectivity in an infected cell culture during the initial infective stage.[16] The infected cell produced extracellular virus at a logarithm rate for the first 2 to 3 days of infection. Then the rate of virus production was reduced abruptly for two logs during the following days. This change in extracellular virus titer was likely due to a concomittant 2 log decrease of virus infectious centers in the cell culture. It suggested that more than 95% of the cells in culture had shut off virus production while the remaining 1 to 5% of the cells still producing virus contributed to the reduced level of virus released into the culture medium. However, almost all of the cells in culture still synthesized cytoplasmic antigen as detected by immuno-fluorescence. The cytoplasmic antigen represents the presence of NP. On the other hand, no significant amount of surface antigen (the GP1) was detected on these cells which were not releasing infectious virion. (Leung, unpublished observation). The cells exhibited no apparent cytopathic effect and replicated at a rate similar to that of the mock infected cells. It represented the initial step which the Pichinde was able to attenuate the virus infectivity in cell culture by reducing the synthesis of glycoprotein, the reduction of infectious center but continuously producing NP as cytoplasmic antigen.

Since these infected cells will readily develop into persistence, we then focused our attention on the molecular events which elicit these exchanges in viral gene expression. Two smaller RNA species (3.1Kb and 1.9Kb) were observed to arise concomittantly with the changes of viral antigen expression and persisted throughout the experiments. The nature of these smaller RNAs were investigated as follows. Because of the lack of hybridization specificity in using cDNA probes transcribed from viral RNA isolated from agarose gel, we decided to obtain recombinant cDNA clones and used them as hybridization probes. Oligonucleotides derived from digests of calf thymus DNA were used to prime cDNA synthesis randomly along the viral RNA. The cDNAs were cloned into the Pst I restriction endonuclease site of pBR322. The recombinant clones were characterized and grouped according to the pattern of hybridization on RNA blots. It was found that the majority of the recombinant cDNA clones hybridized predominantly to either L or S RNA, thus confirming the previous observation that the L or S RNA contained mainly unique sequences.[7] It also provided an operational criteria to separate the recombinant clones into two groups - the L cDNA clones and the S cDNA clones. However, some S cDNA clones showed minor degrees of cross-hybridization with the L cDNA clones and vice versa, suggesting that minor degrees of sequence homology did exist between the viral L and S RNA. The 3.1Kb RNA cross-hybridized with selected members of the S cDNA and L cDNA clones. It suggested a virus specificity for this RNA species but did not yield further information on the nature of this RNA. However, the 1.9Kb RNA hybridized with only a subset of the S cDNA clones. The latter were found to localize on the 3' half of the S RNA. It indicated that the 1.9Kb RNA derived from the 3' half of the S RNA. The immediate question of whether it was a monocistronic mRNA or a subgenomic RNA was examined by the following experiments. Strand specific hybridization probe was generated by subcloning the cDNA clones and obtained cDNA clones with inserts at opposite orientations at the Pst 1 site. One specific strand was then labelled by the sequential action of the exonucleolytic and polymerization activity of T_4 DNA polymerase and used as strand specific hybridization probe. The S RNA and the 1.9Kb RNA was found to hybridize with only one strand specific hybridization probe but not with the other probe. It indicated that they shared the same polarity and the 1.9Kb RNA was a subgenomic RNA derived from the 3' half of the S RNA. Since it derived from the gene locus of the NP protein and its size was sufficient to code for the NP protein. It suggested the appearance of the 1.9Kb RNA represented an amphification of the NP gene.

HYPOTHESIS OF SUBGENOMIC RNA AMPHIFICATION

It has not escaped our attention that the appearance of a 1.9Kb subgenomic S RNA immediately suggested a mechanism where the virus with a simple genetic constituent will be able to regulate their gene expression. The S RNA coded for and expressed the NP and GPC genes whereas the 1.9Kb RNA expressed only the NP. Appearance of the 1.9Kb RNA will result in a net increase in the synthesis of NP. One could then visualize the ratio between the S RNA and the subgenomic 1.9Kb RNA will determine the ratio of NP to GPC protein. A high ratio will ensure the balanced production of NP and GPC proteins which in turn will provide for cytoplasmic antigen, surface antigen and the formation of infectious virus. Conversely, a low ratio will result in the production of a considerable amount of NP but a very small amount of GPC. The infected cells will exhibit cytoplasmic antigen but no significant amount of surface antigen and infectious virus will be produced.

Besides offering a molecular mechanism to explain the changes in Pichinde virus gene expression during productive infection in cell culture, it is conceivable that the amphification of the 1.9Kb RNA and the concommittant reduction of the S RNA would also serve to explain the perpetuation of cytoplasmic antigen and the reduction of surface antigen and extracellular virus commonly observed in long term persistent cultures as well as in experimental animals chronically infected with arenaviruses.(1,2,3,4,17)

For the 1.9Kb RNA to persist and express the NP in long term culture or in experimental animals, a transcriptase and replicase function has to be provided for by the L RNA or its equivalent.(5) At present the nature of the 3.1Kb RNA is very much unknown but the time kinetics when it arises together with the 1.9Kb RNA would suggest a possible role as the transcriptase and replicase.

The main thrust of this hypothesis is through the amphification of a subgenomic RNA, the virus is able to selectively express a subset of its viral genes. Therefore the hypothesis does not call for the generation of subgenomic RNA at the exact gene size for NP. Alternatively, subgenomic RNA for arenaviruses could code for the NP and also part of the GPC sequences which will not form functional glyco-proteins. It will still be able to selectively express NP but not GPC.

In order for the 1.9Kb to amphify over the S RNA, the former must have a selective advantage during RNA replication. This feature might be analogous to that of the defective interfering particles commonly observed in other RNA viruses.[18] However, except with few exceptions, the defective interfering particles serve to inhibit virus replication and repress overall viral gene expression. It is the emphasis on the selective regulation of gene expression of the subgenomic RNA which distinguishes itself from the defective interfering particles.

The subgenomic RNA described here and the defective interfering particles described previously in arenaviruses by various laboratories might not be mutually exclusive. Each observation might serve to describe a facet of the phenomenon in the complicated pattern of gene expression of productive infection and persistence in arenavirus. Additional controlling mechanisms, such as multifunctional proteins serving as positive or negative gene regulators, processing of glycoproteins, variant viruses and temperature sensitive mutants, etc. will come into play.[19-22] Undoubtedly, the unraveling of how a virus of extremely simple genetic constituents is able to exercise complicated patterns of gene expression promises to be an exciting and fruitful research experience in the coming years.

REFERENCES

1. Rawls, W.E. and Leung, W-C. (1979). In "Comprehensive Virology" (H. Fraenkel-Conrat and R.R. Wagner, eds.), Vol. 14, p. 157. Plenum Publishing Corp.
2. Pedersen, I.R. (1979). In "Advances in Virus Research." Vol. 24, p. 227. Academic Press.
3. Howard, C.R. and Simpson, D.I. (1980). J. Gen. Virol. 51:1.
4. Rawls, W.E., Chan, M.A. and Gee, S.R. (1981). Can. J. Microbiol., 27, 568.
5. Leung, W-C., Harnish, D., Ramsingh, A., Dimock, K., and Rawls, W.E. (1981). In "The Replication of Negative Strand Viruses." (D.H.L. Bishop and R.W. Compans, eds.) p. 51. Elsevier North Holland, Inc.
6. Leung, W-C., Ghosh, H.P. and Rawls, W.E. (1977). J. Virol. 22, 235.
7. Leung, W-C., Leung, M.F.K.L. and Rawls, W.E. (1979). J. Virol. 30, 98.
8. Leung, W-C., Ramsingh, A., Dimock, K., Rawls, W., Petrovich, J. and Leung, M. (1981). J. Virol. 37, 48.

9. Leung, W-C., and Rawls, W.E. (1977). Virology 81, 174.
10. Harnish, D.G., Leung, W-C., and Rawls, W.E. (1981). J. Virol. 38, 840.
11. Ramsingh, A., Dimock, K., Rawls, W. and Leung, W-C. (1980). Intervirology 14, 31.
12. Vezza, A., Cash, P., Jahrling, P., Eddy, G., and Bishop, D. (1980). Virology 106, 250.
13. Harnish, D., Dimock, K., Bishop, D., and Rawls, W., J. Virol. 46, 638.
14. Auerpin, D., Dimock, K., Cash, P., Rawls, W., Leung, W-C., and Bishop, D. (1982). Virology 116:363.
15. Kozak, M. (1983). Microbiol. Rev. 47, 1.
16. Dimock, K., Harnish, D., Sisson, G., Leung, W-C., and Rawls, W.E. (1982). J. Virol. 43, 273.
17. Oldstone, M.B. and Buchmeier, M.J. (1982). Nature 300, 360.
18. Huang, A.S., and Baltimore, D. (1977). In "Comprehensive Virology", (H. Fraenkel-Conrat and R.R. Wagner, eds.) Vol. 10, p. 73. Plenum Press, New York.
19. Dutko, F.J. and Pfau, C.J. (1978). J. Gen. Virol. 38, 195.
20. Gimenz, H.B. and Compans, R.W. 91980). Virology, 107, 229.
21. Popescu, M. and Lehmann-Grube, F. (1977). Virology 77, 78.
22. Hotchin, J. (1962). Nature (London), New Biol., 241, 270.
23. Help, G.I. and Coto, C.E. (1980). Medicina (B. Aires), 40, 531.
24. Jacobson, S. and Pfau, C.J. (1980). Nature, 283, 311.

BIOCHEMICAL CHARACTERIZATION OF HANTAAN VIRUS

Connie S. Schmaljohn and Joel M. Dalrymple

Department of Viral Biology, Virology Division
United States Army Medical Research Institute
of Infectious Diseases
Fort Detrick, Frederick, MD 21701

Hantaan virus is the prototype of a group of viruses implicated as etiologic agents of clinically similar diseases which have collectively been termed "Hemorrhagic Fever with Renal Syndrome" (HFRS). Three well recognized diseases in the syndrome include: Korean hemorrhagic fever, nephropathia epidemica, and epidemic hemorrhagic fever, various forms of which are endemic throughout Korea, Scandinavia, Europe, and China. Hantaan virus has no documented serological relationship to viruses in any characterized family but shows broad cross-reactivity with a number of viruses recently isolated from rodents both in countries known to have the disease syndrome as well as in the United States and other non-endemic areas (1, 2, 3). The recent isolation and cell culture adaptation of Hantaan virus strain 76-118 and the subsequent optimization of propagation and purification methods (4, 5, 6, 7), has permitted biological and molecular examination of virions. Physical characteristics of Hantaan virions as well as a tripartite, single stranded RNA genome and virion associated polymerase (6, 7), were supportive of previous morphological observations that Hantaan virus was similar to the Bunyaviridae (8, 9,). The 3' terminal sequence of each of the Hantaan virion RNA segments, however, indicated that Hantaan virus could not be classified within any of the four existing Bunyaviridae genera and led us to propose that a fifth genus may be required to accommodate the serologically and moleculary unique group of viruses related to Hantaan virus (7). To confirm that Hantaan virus is representative of a group of viruses that are similar to, but unique from other Bunyaviridae, we have compared the RNAs of Hantaan virus to those of other serologically related viruses, and expanded the biochemical characterization of Hantaan virus with a basic description of viral proteins.

Copyright © 1984 by Academic Press, Inc.
All rights of reproduction in any form reserved.
ISBN 0-12-183501-4

VIRUSES

Origin of Virus Strains. Hantaan virus, strain 76-118, was isolated in Korea from lungs of the Korean striped field mouse, *Apodemus agrarius corea* (4). The Lee strain of Hantaan virus was recovered from the blood of a Korean hemorrhagic fever patient by passage in seronegative *Apodemus* (P. W. Lee, and H. W. Lee, personal communication). Girard point and Tchoupitoulas viruses were isolated from wharf rats (*Rattus norvegicus*) in Philadelphia and New Orleans, respectively (2, 3). All viruses were routinely propagated and plaque assayed in a cloned line of Vero cells (Vero E6; ATCC C1008). None of the strains produced detectable cytopathology. Growth characteristics of the virus strains were quite similar. Maximum virus infectivity titers of approximately 10^7 pfu/ml were achieved eight to ten days post infection. Korean isolates could be differentiated from U. S. rat isolates by plaque reduction neutralization assays utilizing specific antisera.

VIRION RNA

Agarose Gel Electrophoresis of Virion RNA. Although we previously reported molecular weight estimates for the large (L), medium (M) and small (S) segments of Hantaan virus to be 2.7, 1.2, and 0.6 x 10^6 daltons respectively based upon relative migration in methyl-mercury agarose gels, we have found that such estimates are extremely dependent upon the denaturing conditions employed. We have used three common RNA denaturants, glyoxal (10), methyl-mercury (11), and acid urea (12), in conjunction with agarose gel electrophoresis, to compare the RNAs of Hantaan virus strain 76-118 and two of the serologically related rodent viruses (Figure 1). In the three gel systems examined, molecular weight estimates relative to host cell 28 and 18S ribosomal RNAs varied from 2.2-3.6 x 10^6 for L, 1.4-1.9 x 10^6 for M, and 0.6-0.75 x 10^6 for S. Corresponding genome segments of each of the three viruses appeared to have very similar electrophoretic migration rates. Glyoxal denaturation revealed the only obvious apparent molecular weight difference, in that the S segment of Hantaan virus appeared to migrate slightly ahead of the S segment of Tchoupitoulas virus. The molecular weight estimates for the genome segments of these three rodent viruses are consistent with their classification in the family Bunyaviridae and provide evidence of molecular similarity between the prototype Hantaan virus and serologically related viruses isolated from widely diverse geographical areas.

FIGURE 1. Agarose gel electrophoresis of virion RNA. Denaturation of Hantaan 76-118 (H), Tchoupitoulas (Tch), and Girard Point (GP) virion RNAs and host cell ribosomal 28 and 18S RNA species was with glyoxal, methylmercury, or acid urea (10, 11, 12). Electrophoresis in 1.2% agarose was performed at 100 V.

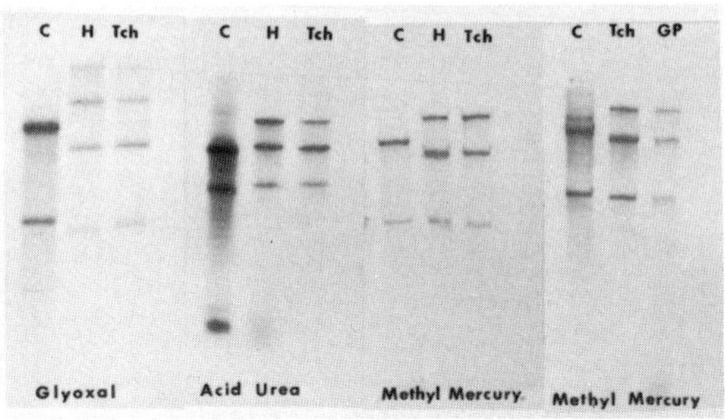

Oligonucleotide Map Comparison of the Individual Genome Segments of Two Strains of Hantaan Virus. Hantaan virus, strain 76-118, has been implicated as the etiologic agent of Korean hemorrhagic fever largely due to isolation of this virus from rodents in an endemic disease area and serological reactivity using convalescent human sera (4). Consequently, it was of interest to compare this rodent virus to the human isolate (Lee strain). Two-dimensional, electrophoresis of ribonuclease T1 resistant fragments (oligonucleotide maps) has been demonstrated to be a sensitive and reliable means for comparison of Bunyaviridae RNAs (13). We have used this method to compare the individual virion genome segments of the 76-118 and Lee strains (Figure 2). The oligonucleotide maps of corresponding genome segments of these two viruses showed considerable diversity and indicated that at least two variants of Hantaan virus exist within a defined endemic disease area, either or both of which may be involved in classical Korean hemorrhagic fever.

Terminal 3' Sequence Analysis of Virion RNA. The conserved 3'terminal nucleotide sequence of the three genome segments of Hantaan strain 76-118 (3'AUCAUCAUCUG) was different from those of other viruses in the four recognized genera of the Bunyaviridae and suggested that Hantaan and serologically related viruses should be classified in a new and separate genus (7). We have now compared the 3' terminal nucleotide sequences of the human isolate (Lee strain) to the 76-118 virus strain. Chemical sequence analysis of the L, M, and S RNA species of Lee virus demonstrated that each segment possessed 3' terminal sequences of at least 14 residues which were identical to those at the corresponding 3' termini of

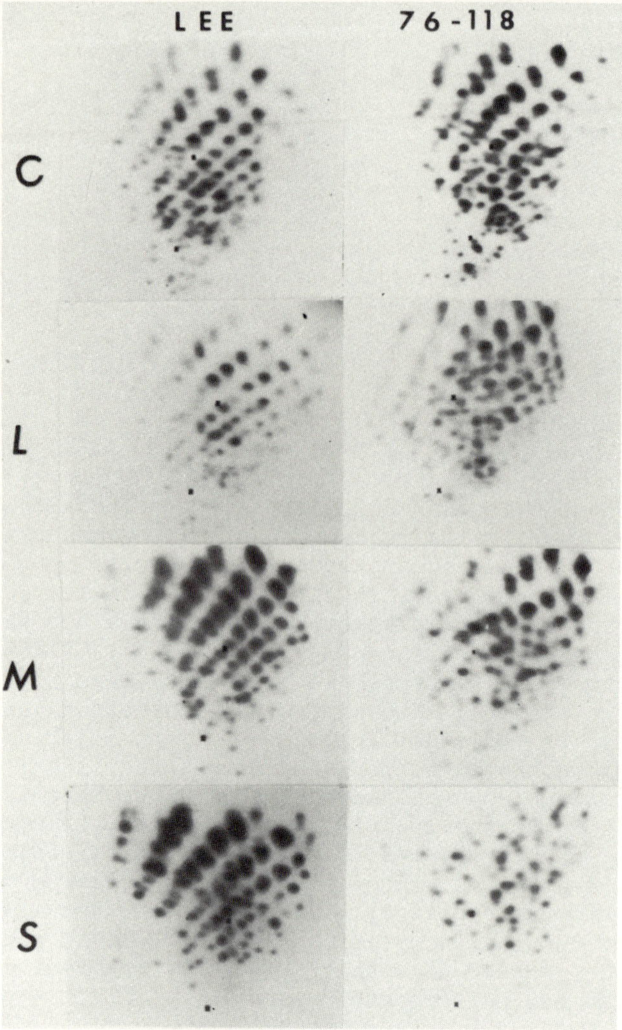

FIGURE 2. Oligonucleotide map comparison of the large (L), medium (M), and small (S) genome RNA segments and composites of all three segments (C) of Hantaan virus strains 76-118 and Lee. Individual genome segments were resolved by methyl-mercury agarose gel electrophoresis, digested with T1 ribonuclease, and radiolabeled at their 5' termini with gamma-ATP^{32} prior to electrophoresis (14). Electrophoretic conditions were as previously described (15).

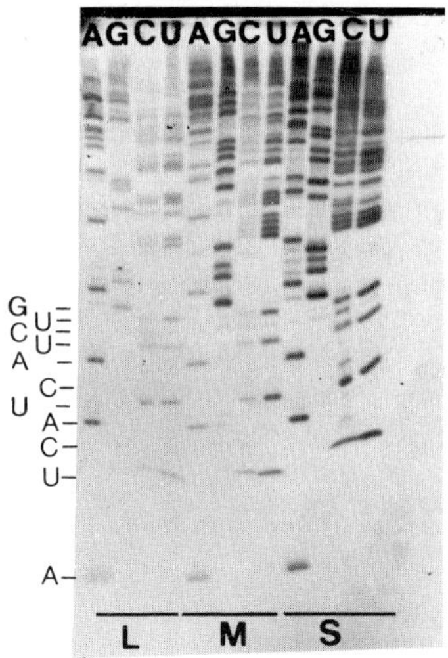

FIGURE 3. Sequence analysis of the 3' termini of the L, M, and S, genome segments of the Lee strain of Hantaan virus. Chemical cleavage of the genome segments radiolabeled at their 3' termini with cytidine-bis-phosphate (^{32}P-pCp) and RNA ligase and subsequent resolution of fragments by acrylamide gel electrophoresis was as described by Peattie (16).

the Hantaan virus genome segments (Figure 3). In preliminary studies, this same sequence was observed on the 3' termini of the three Tchoupitoulas virion RNA species (H. Regnary and C. Schmaljohn, unpublished observations). Although other strains have not yet been examined, it appears that the 3' terminal sequence of virion RNAs will prove to be useful for classification of Hantaan-like viruses. These data also serve to support our earlier proposal to establish a new genus of Bunyaviridae to include Hantaan and related viruses.

VIRAL PROTEINS

Polypeptides Associated with Purified Virions. Virions in the family Bunyaviridae possess three major structural polypeptides: a nucleocapsid protein and two envelope glycoproteins. The nucleocapsid proteins of viruses in three of the Bunyaviridae genera (Bunyavirus, Uukuvirus, and Phlebovirus) are 20-30,000 daltons in size, and members of the fourth genus (Nairovirus) have nucleocapsid proteins with estimated molecular weights of 48-60,000 daltons. The envelope proteins of the Bunyaviridae occur either as one large and one small protein (Bunyavirus and Nairovirus) or as two proteins with similar molecular weights (Phlebovirus and Uukuvirus). An additional large protein, presumably the polymerase protein, has also been reported in association with Bunyaviridae (17). We have examined the structural protein profile of Hantaan virions and compared it to those of viruses in the four recognized Bunyaviridae genera.

Hantaan virions (strain 76-118) radiolabeled with ^{35}S-methionine were disrupted with non-ionic detergent and nucleocapsids separated from envelope proteins by density sedimentation in cesium chloride gradients (6). Two peaks of radioactivity and antigenicity were observed at densities of approximately 1.26 and 1.12 (data not shown). Polypeptides contained in these fractions as well as intact virion proteins were examined by polyacrylamide gel electrophoresis (Figure 4A). Three major polypeptides were observed: a nucleocapsid protein with a molecular weight of about 52,000 daltons and two detergent soluble and presumed envelope proteins with apparent molecular weights of 55 and 65,000 daltons. A fourth, high molecular weight polypeptide was also evident in the virion and nucleocapsid preparations, and while we have previously demonstrated transcriptase activity to be associated with Hantaan virions, we have no conclusive evidence that the polymerase is represented by this polypeptide (7). The nucleocapsid protein of Hantaan virions, therefore, appears similar to that of viruses in the *Nairovirus* genus while the envelope proteins appear more similar in size to those of the *Phlebovirus* and *Uukuvirus* genera.

Immune Precipitation of Intracellular Hantaan Virus Proteins. Examination of the kinetics of synthesis of intracellular Hantaan virus specific proteins revealed that at a high multiplicity of infection (MOI=10), viral proteins could first be detected at 6 hours post infection, quantitatively increased to 24 hours and subsequently remained constant up to 52 hours post-infection (data not shown). Immune precipitation of proteins radiolabeled for one, two, three or four hours during the earliest period of maximum incorporation (24-28 hours post infection) revealed no polypeptides other than those described for virions (Figure 4B). The high molecular weight protein previously observed in association with virions could not be detected in these cell extracts. This could either be a result of low concentrations present, suboptimal radiolabeling times, antisera containing limited antibody to this protein or alternatively, that this large protein is not virus specified.

In vitro Translation of Virion Protein. Hantaan virus messenger RNA (mRNA) was isolated by nonionic detergent disruption of infected cells and cesium chloride centrifugation of the cell lysates (18). Material which pelleted through the gradients (containing mRNA) was recovered and examined for the ability to program a rabbit reticulocyte lysate *in vitro* translation system (IVT). An

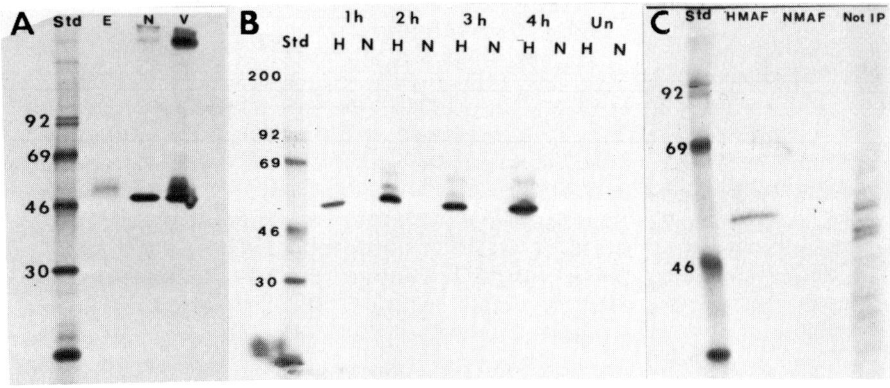

Figure 4. Polyacrylamide gel electrophoresis of Hantaan virus polypeptides radiolabeled with ^{35}S methionine. A. Virion structural proteins (V); nucleocapsid associated polypeptides (N); envelope polypeptides (E). B. Hantaan virus infected cell lysates and uninfected Vero E6 cell lysates (Un). Immune precipitation was performed using either Hantaan virus hyperimmune mouse ascitic fluid (H) or normal mouse ascitic fluid (N) and protein A sepharose. C. Reticulocyte lysate IVT (Bethesda Research Lab.). Electrophoresis was as previously described (6).

immune precipitable polypeptide with an approximate molecular weight of 52,000, presumably the Hantaan virus nucleocapsid protein, was synthesized under these conditions. (Figure 4C). No polypeptides were detected by immune precipitation of IVT lysates programmed with uninfected Vero E6 cell RNA or from Hantaan virion RNAs (data not shown).

SUMMARY

The data presented here support the earlier proposal that Hantaan and serologically related viruses be included in a new genus in the virus family Bunyaviridae. Oligonucleotide map analysis of virion RNAs served to differentiate virus strains while 3' terminal nucleotide sequence analysis demonstrated molecular homogeneity. The protein profile of Hantaan virus similarly demonstrated consistency, but not identity with the other Bunyaviridae genera. Continued biochemical characterization of isolates from diverse hosts and geographic regions will be essential for definition of the scope and disease potential of this group of viruses.

REFERENCES

1. Kitamura T., Morita, C., Komatsu, T., Sugiyama, K., Arikawa, J., Shiga, S., Takeda, H., Akao, Y., Imaizuma, K., Oya, A., Hashimoto, N., Urasawa, S. (1983). Japan. J. Med. Sci. Biol. 36, 17.
2. LeDuc, J. W., Smith, G. A., Bagley, L. R., Hasty, S. E., and Johnson, K. M. (1982). New Eng. J. Med. 307, 624.
3. Tsai, T. F., Bauer, S. P., Sasso, D. R., McCormick, J. B., Bradford, H., Caraway, C. T., McFarland, L. M., Medrand, O. and Soulie, G. (1982). New Engl. J. Med. 307, 623.
4. Lee, H. W., Lee, P. W., and Johnson, K. M. (1978). J. Infect. Dis. 137, 298.
5. French, G. R., Foulke, R. S., Brand, O. A., Eddy, G. A., Lee, H. W., and Lee, P. W. (1981). Science 211, 1046.
6. Schmaljohn, C. S., Hasty, S. E., Harrison, S. A., and Dalrymple, J. M. (1983). J. Infect. Dis. in press.
7. Schmaljohn, C. S., and Dalrymple, J. M. (1983). Virology, submitted.
8. McCormick, J. G., Palmer, E. L., Sasso, D. R., and Kiley, M. P. (1982). Lancet 1, 765.
9. White, J. D., Shirey, F. G., French, G. R., Huggins, J. W., and Brand, O. M. (1982). Lancet 1, 768.
10. McMaster, G. K., and Carmichael, G. C. (1977). Proc. Natl. Acad. Sci. 74, 4835.
11. Bailey, J. M., and Davidson, N. (1976). Anal. Biochem. 70, 75.
12. Lehrach, H., Diamond, D., Wozney, J. M. and Boedtker, H. (1977). Biochem. 21, 4743.
13. Bishop, D. H. L., and Shope, R. W. (1979). In "Comprehensive Virology" vol. 14. (H. Fraenkel-Conrat and R. R. Wagner eds.) Plenum Press, New York.
14. Pederson, F. S., and Haseltine, W. A. (1980). Methods in Enzymol. 65, 680.
15. DeWachter, R., and Fiers, W. (1972). Anal. Biochem. 49, 184.
16. Peattie, D. A. (1979). Proc. Natl. Acad. Sci. 76, 1760.
17. Bishop, D. H. L., Calisher, C. H., Casals, J., Chumakov, M. P., Gaidamovich, S. Ya., Hannoun, C., Lvov, D. K., Marshall, I. D., Oker-Blom, N., Pettersson, R. F., Porterfield, J. S., Russell, P. K., Shope, R. E., and Westaway, E. G. (1980). Intervirology 14, 125.
18. Leppert, M. L., Rittenhouse, L., Summers, D. F. and Kolakofsky, D. (1979). Cell 18, 735.

EUKARYOTIC EXPRESSION OF CLONED cDNA CODING FOR INFLUENZA VIRAL GLYCOPROTEINS USING AN SV40 VECTOR: USE OF RECOMBINANT DNA MUTANTS TO STUDY STRUCTURE-FUNCTION RELATIONSHIPS[1]

Timothy J. Bos, Nancy L. McQueen, Alan R. Davis[2] and Debi P. Nayak

Jonsson Comprehensive Cancer Center and
Department of Microbiology and Immunology
UCLA School of Medicine
Los Angeles, CA 90024

Influenza virus has been widely used to study the biosynthesis, sorting, distribution and orientation of membrane proteins (1,2). Influenza has 3 membrane associated proteins; hemagglutinin (HA), neuraminidase (NA) and matrix protein (M). HA and NA are integral membrane proteins located on the outer envelope whereas M protein is associated with the inner lining of the viral membrane. Both HA and NA play important roles in viral epidemiology, viral pathogenesis, and in maturation and release of viral particles (3). HA and NA have been studied extensively at the molecular level (4-8). Their 3 dimensional structure (9,10), antigenic epitopes, receptor binding sites, glycosylation points and subunit organization have been defined (11,12).

HA and NA, although both integral membrane proteins, have quite different characteristics. HA conforms to the "classical" type of integral membrane protein in that it is anchored by hydrophobic residues near the COOH-terminus and it contains an NH_2-terminal hydrophobic "signal" peptide not present in the mature protein (see ref. 4). NA, however, contains only one hydrophobic region which is located at the NH_2-terminus. This region is not cleaved from the mature protein but remains embeded in the lipid bilayer of the cellular or viral membrane (see ref. 4). Thus, HA and NA

[1]This work was supported by NIH research grants Al-16348, AI-12749, National Science Foundation grant PCM 78-23220, National Institute of Allergy and Infectious Diseases R01 AI2749 and R01 AI16348. TJB and NLM were recipients of USPSH National Research Awards GM07104 and GM7185.

[2]Present address: Wyeth Laboratories, Inc., Radnor, PA 19108.

Copyright © 1984 by Academic Press, Inc.
All rights of reproduction in any form reserved.
ISBN 0-12-183501-4

are structurally quite different and yet both are transported to the same domain of the viral and cellular membrane.

Because influenza contains these two types of integral membrane proteins it is an ideal candidate for the study of the mechanism by which integral membrane proteins are transported from their site of synthesis, via the rough endoplasmic reticulum (RER) and golgi apparatus, to their final destination in eukaryotic cells. Cotranslational and post-translational modifications along with specific structural features of a protein are presumably responsible for guiding the protein through the elaborate cellular machineries to its final destination.

As a first step in defining the mechanism by which proteins are transported, we have separately cloned cDNA copies of HA and NA into an SV40 late replacement vector containing the early and late promoters as well as the early genes. Both HA and NA are expressed as functionally active proteins (4,6,7). In this report we describe several modifications of specific structural domains of HA and NA and the consequent effect on function and transport.

SV40 AS AN EXPRESSION VECTOR

The methods used to express HA and NA in CV1P cells (4) using SV40 as a vector have been described (4,7). Briefly, the cDNA is cloned into the late region of SV40 between Hpa II and Bam HI, replacing the late SV40 genes. CV1P cells are transfected with the recombinant virus and SVsal.32; a helper virus defective in the early genes (4). Stock viral lysates are prepared and used to infect CV1 cells for expression analysis.

Expression of Hemagglutinin. The HA gene from A/WSN/33 was shown, by indirect immunofluorescence and immunoprecipitation with WSN antibody of ^{35}S labeled proteins, to be expressed on the surface of CV1 cells, to be fully glycosylated and to be functionally active (hemadsorption) (Table 1) (2,6). Furthermore the expressed HA is located on the apical surface of polarized primary African green monkey cells (2) as in influenza infected cells. In order to elucidate some of the structural features involved in biogenesis, transport, and sorting, we have constructed and expressed chimeric cDNA clones (7) in which DNA encoding the NH_2-terminus of HA (GHA) or the COOH-terminus of HA (HAG) have been replaced with VSV G (Figure 1). VSV G, a glycoprotein functionally similar to influenza HA, was chosen because influenza and VSV bud from different domains of the plasma membrane in polarized epithelial cells. Therefore HA and G most likely contain different structural properties responsible for directing them to their final locations.

Table 1 summarizes the properties of the expressed proteins. Both GHA and HAG are synthesized, glycosylated, and transported to the RER. Surface immunofluorescence studies show that the chimeric proteins do not insert into the plasma membrane (7). Studies using endoglycosidase H, which cleaves high mannose carbohydrates but not complex carbohydrates, indicate that although glycosylated, these chimeric proteins are not processed through the Golgi apparatus (7). Furthermore, these proteins are not biologically active as determined by hemadsorption and hemagglutination.

Several conclusions can be drawn from these results. 1)The NH_2-terminal region of VSV G can act as a signal for translocation of the GHA chimera, however, further processing is blocked. 2)Lack of complete processing is not due solely to the foreign NH_2-terminus because replacement of the COOH-terminus (HAG) leaving the NH_2-terminus intact, shows the same results.

It is likely that an incorrect conformation of these chimeric proteins is responsible for blocking transport.

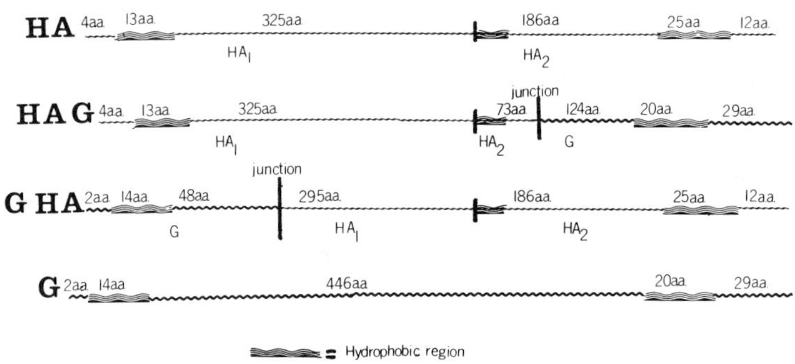

Figure 1. Schematic representation of the primary structures of HA, G and the chimeric proteins HAG and GHA. Amino acids from influenza HA and VSV G are denoted by wavy lines of light and heavy intensity respectively. Large vertical bars denote the junction sites of the chimeric proteins. Small vertical bar denotes the junction between HA1 and HA2.

TABLE 1

Properties of Expressed Influenza Proteins[1]

	VIRAL		EXPRESSED HA			EXPRESSED NA		
	HA	NA	HA	GHA	HAG	SNC	SN10	SN26
Synthesis	+	+	+	+	+	+	+	-
Glycosylation	+	+	+	+	+	+	+	-
Translocation across RER	+	+	+	+	+	+	+	-
Complex Carbohydrate[2]	+	+	+	-	-	+	+	-
Transport to Surface	+	+	+	-	-	+	+	-
Biological Activity								
Hemadsorption	+	*[3]	+	-	-	*	*	*
Enzyme Activity	*	+	*	*	*	+	+	-
Polarized Distribution	+	**[4]	+	**	**	**	**	**

[1]References 4 and 7 for data and methods.

[2]Complex carbohydrate determined by treatment with endoglycosidase H.

[3] * indicates not applicable.

[4] ** indicates not done.

<u>Expression of Neuraminidase</u>. The NA of A/WSN/33 was expressed in a similar manner as HA (4). The expression vector is shown in Figure 2A. Complete NA (SNC) was expressed and readily detected by immunofluorescence and immunoprecipitation of ^{35}S labeled proteins with monoclonal antibodies (4). SNC was found to be of the same molecular weight as authentic NA, glycosylated, transported to the cell surface and to exhibit enzyme activity by cleaving the sialic acid off of α 2,3-sialylactitol (4) (Table 1). Because the amino terminus presumably acts both as a signal and anchor, we have made two deletions in this domain to determine the effect on expression and function. SN10 and SN26 are fusion proteins containing NH_2-terminal amino acids derived from the SV40 agnogene and linker sequences followed by NA starting at amino acids 11 and 27 respectively (Figure 2B). A comparison of the properties of these mutants is shown in Table 1. It is clear that replacement of the first 10 amino acids of NA has no effect on expression, glycosylation or transport whereas further deletion to amino acid 26 abolishes expression completely.

It has been recently shown that the first 12 amino acids of all type A N1 and N2 strains are strictly conserved (13). The functions of these conserved residues is not clear, however, our data indicate that strict conservation is not required for synthesis, glycosylation or enzyme activity (SN10). The conserved residues may, however, play an important role in sorting, directional transport or interactions with other viral proteins such as M in the viral membrane. SN26 is missing all but 9 of its NH_2-terminal hydrophobic amino acids. It is not clear why no detectable amounts of protein were produced. It is possible that the 9 amino acid hydrophobic stretch is not large enough to act as a signal. The protein would then be made in the cytoplasm where it is presumably degraded. It appears that either

A

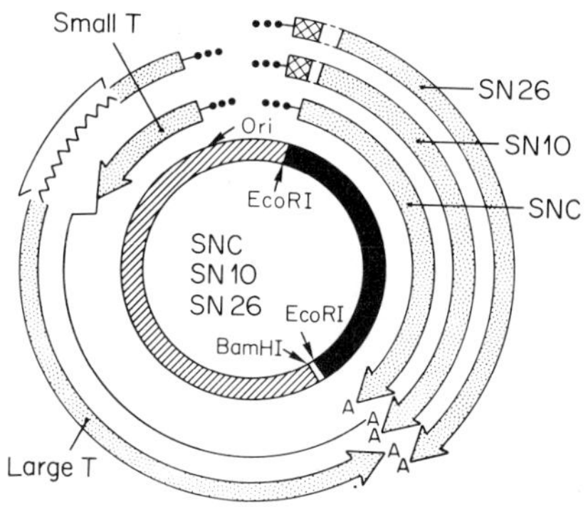

B

SNC NH_2 - M N P N Q K I I T I G S I C M V V G I I S L I L Q I G N I I

SN10 NH_2 - M V L R P D P N S G S I C M V V G I I S L I L Q I G N I I

SN26 NH_2 - M V L R P D P N S G N I I

FIGURE 2. A) Schematic representation of SV40-NA recombinant DNA that express influenza viral proteins. Solid bar, the cloned NA DNA segments; hatched bar, SV40 DNA sequences; the mRNA and protein products are shown outside the circle. Cross hatched areas represent the agnoprotein leader. Wavy lines indicate regions that are excised from the RNA. B) Predicted amino acid sequence encoded by SNC, SN10, and SN26 using one letter amino acid code. Underlined amino acids are derived from the SV40 agnoprotein leader; double underlined amino acids come from linker sequences. All are identical from amino acid 27 (gly) of NA to the COOH-terminus.

lack of a critical sequence between amino acid 10 and 26 or lack of a long hydrophobic stretch is responsible for inhibiting transport across the RER.

It is evident that this SV40 expression system can be useful in mapping the structural domains of a protein associated with specific functions. In this report we have used this system to study the structural properties of the two major types of integral membrane proteins. Experiments are in progress to further characterize, by sequence specific alterations, the roles of the NH_2- and COOH-terminal regions of HA as well as the NH_2-terminal region of NA.

REFERENCES

1. Boulan, E.R. and Pendergast, M. (1980). Cell 20, 45-54.
2. Roth, M.G., Compans, R.W., Giusti, L., Davis, A.R., Nayak, D.P., Gething, M.H, and Sambrook, K.J. (1983) Cell 33, 435-443.
3. Bucher, D. and Palese, P. (1975). In "The Influenza Viruses and Influenza". (Kilbourne, E.D., ed.) pp. 83-123. Academic Press, New York.
4. Davis, A.R., Bos, T.J. and Nayak, D.P. (1983) Proc, Natl. Acad. Sci. USA 80, 3976-3980.
5. Gething, M.J. and Sambrook, J. (1982) Nature 300, 598-603.
6. Hartman, J.R., Nayak, D.P. and Fareed, G.C. (1982) Proc. Natl. Acad. Sci. USA 79, 233-237.
7. McQueen, N.L., Nayak, D.P., Jones, L. and Compans, R.W. (1983) Proc.Natl. Acad. Sci. USA Submitted.
8. Sveda, M.M., Markoff, L.J., and Lai, C.J. (1982) Cell 30, 649-656.
9. Wilson, I.A., SKehel, J.J., and Wiley, D.C. (1981) Nature 289, 366-373.
10. Varghese, J.N., Laver, W.G., and Colman, P.M. (1983) Nature 303, 35-40.
11. Wiley, D.C., Wilson, I.A., and Skehel, J.J. (1981) Nature 289, 373-378.
12. Colman, P.M., Varghese, J.N. and Laver, W.G. (1983) Nature 303, 41-44.
13. Blok, J. and Air, G.M. (1980) Virology 107, 50-60.

STUDIES ON THE SECOND PROTEIN ENCODED BY THE NEURAMINIDASE GENE OF INFLUENZA B VIRUS[1]

Michael W. Shaw, Otto Haller[2], Robert A. Lamb[3], and Purnell W. Choppin

The Rockefeller University
New York, New York 10021

Studies on the genomes of types A and B influenza viruses revealed similar coding strategies, suggesting a common ancestor for the two virus types (1). Analogous proteins are encoded by the corresponding genome segments of the two virus types, and there are significant homologies at the nucleotide and amino acid levels (2-5). However, the nucleotide sequence of genome segment 6, which encodes the neuraminidase glycoprotein [NA], revealed that influenza B viruses employ a coding arrangement that is not seen in influenza A viruses (5,6). Unlike the influenza A virus NA gene (7-10), and all other influenza A and B virus genes with known sequences (rev. in 1), the influenza B virus NA is translated from a reading frame initiated at the second AUG codon (5). The first AUG codon, which is separated from the second by four nucleotides, begins an open reading frame from which a 100 amino acid polypeptide designated NB is translated (6). NB is a glycoprotein with a molecular weight of ~18,000. The synthesis of NB in vitro in a wheat germ system and in infected cells has been demonstrated, and NA and NB have been shown to be translated from overlapping reading frames on a bicistronic mRNA (6).

We describe here the kinetics of the synthesis of NB in infected cells, and the antibody response to it in mice. Results of computer analyses of predicted structural features of NB will be discussed.

---

[1]Supported by Research Grants AI-05600 and AI-20201 from the NIAID and PCM 80-13464 from the NSF. M.W.S. was an American Cancer Society Postdoctoral Fellow and R.A.L. is an Established Investigator of the American Heart Association.

[2]Present address: Institute for Immunology and Virology, University of Zurich, Switzerland.

[3]Present address: Department of Biochemistry, Molecular and Cell Biology, Northwestern University, Evanston, Illinois.

Copyright © 1984 by Academic Press, Inc.
All rights of reproduction in any form reserved.
ISBN 0-12-183501-4

SYNTHESIS OF NB IN INFECTED CELLS

Previous studies on the synthesis of influenza B virus polypeptides in infected cells failed to reveal NB (2), however, the deduced amino acid sequence of NB suggested that this was due to the labeled amino acids used. As shown in Fig. 1A, the use of an abundant residue, cysteine, allows detection of NB in cells infected with 3 different influenza B virus strains. The strain-specific variation in the electrophoretic mobility of NB confirms the viral origin of the protein. Similar differences in the mobilities of the polypeptides synthesized in vitro have been observed [not shown].

As shown previously, the NB and NA glycoproteins are translated from a single bicistronic mRNA transcribed from genome RNA segment 6 (6). Thus NB and NA should appear simultaneously in infected cells. This is shown in Fig. 1B.

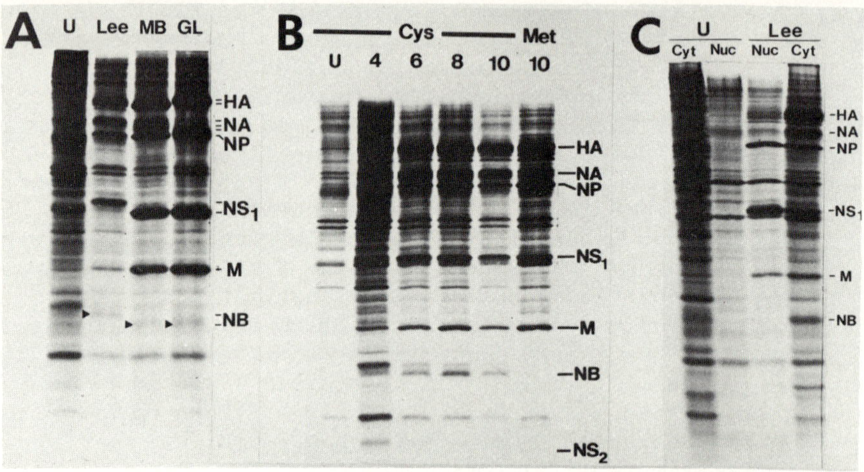

FIGURE 1. [A] Synthesis of polypeptide NB in HeLa cells infected with 3 different influenza B virus strains. 10 hr after infection, cells were labeled with cysteine for 2 hr. Lanes: Lee, B/Lee/40; MB, B/MB/50; GL, B/GL/1760/54; U, uninfected cells. [B] Time-course of NB synthesis in HeLa cells infected with B/Lee virus. At the indicated hours after infection, cells were labeled with [^{35}S]cysteine or methionine for 30 min. [C] Intracellular localization of NB. MDCK cells were labeled with cysteine from 8 to 10 hr after infection with B/Lee virus and fractionated as described elsewhere (11). Lanes: Cyt, cytoplasmic supernatant; Nuc, nuclear pellet; Lee, B/Lee virus; U, uninfected cells.

Gels in all figures were 20% polyacrylamide.

Both polypeptides are readily detected by 4 hours after infection in cells labeled with cysteine and NB becomes a prominent viral protein. To investigate the localization of NB, nuclear and cytoplasmic fractions were obtained by detergent disruption and low-speed centrifugation (11). The NB polypeptide was found in the cytoplasm; none was evident in the nuclear pellet [Fig. 1C].

The stability of NB in the infected cell was examined in pulse-chase experiments. Unlike the virion glycoproteins [HA and NA], NB disappeared from infected cells soon after synthesis, with little or none evident after a 3 hr chase [Fig. 2A]. To determine whether the loss of NB is a result of degradation or of the release from the cell after synthesis, medium from infected cells was fractionated as shown in Fig. 2B. HKCC cells infected with the B/GL/1760/54 [B/GL] strain, a system

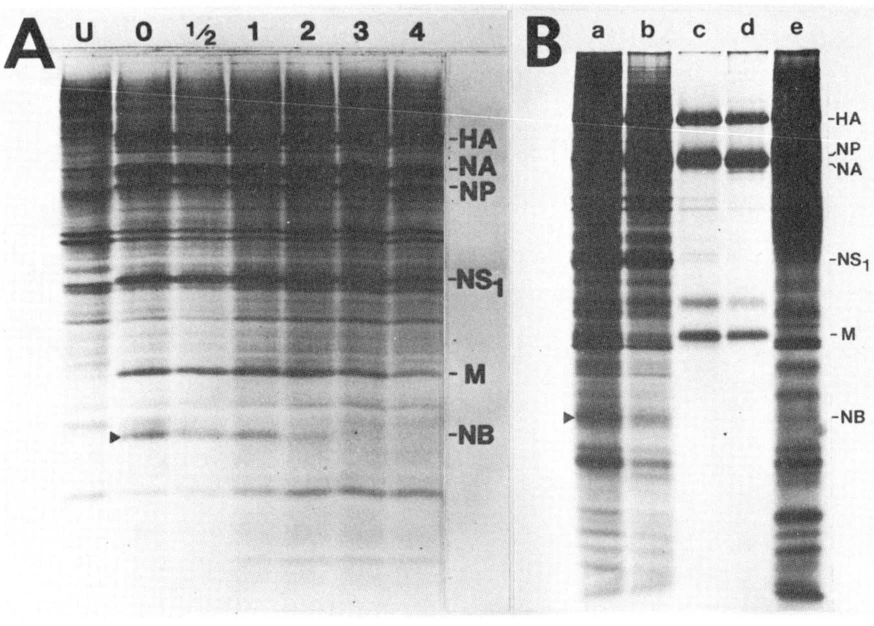

FIGURE 2. [A] Pulse-chase experiments with NB. 8 hr after infection with B/Lee virus, HeLa cells were labeled with cysteine for 30 min and then incubated with unlabeled medium for the indicated hours before harvest. [B] NB in culture medium fractions. HKCC cells were labeled with cysteine from 4 to 24 hr after infection with B/GL virus. Lanes: a, attached cells; b, 2000 g pellet; c, 10,000 g pellet; d, 400,000 g pellet; e, 400,000 g supernatant.

that produces high titers of infectious virus (12), were labeled continuously from 4 to 24 hours after infection, and the cells and medium were fractionated by differential centrifugation. NB was found primarily in attached cells and the 2000 g pellet, which consisted of detatched cells and cell debris. NB was also detected in the 400,000 g supernatant fraction; however, the amount of NB found in all fractions was less than would be expected to be synthesized during the 20 hr labeling period. The pulse-chase experiments and medium analysis suggest that, although a small amount of NB is released into the medium, most of it may be degraded after synthesis. Preliminary studies with mouse antiserum against NB suggest that some 12-15,000 MW polypeptides detected in infected cells and medium may be breakdown products of NB [results not shown]. If so, the discrete bands seen in gels of the immunoprecipitates may represent specific cleavage products of NB.

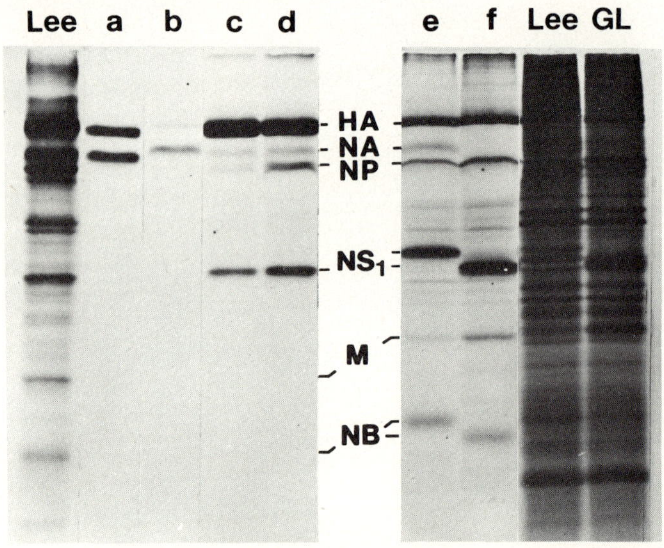

FIGURE 3. Immune mouse serum precipitates of cysteine-labeled infected HeLa cell extracts. Lanes: Lee, B/Lee-infected cells; GL, B/GL-infected cells; a, serum 10 d after primary infection with B/Lee; b, serum 10 d after primary infection with B/GL; c, serum 14 d after secondary infection with B/Lee; d, serum 14 d after secondary infection with B/GL; e and f, serum from mice infected twice with B/GL and boosted i.p. with homogenized B/GL-infected lungs in FCA. Lanes a to e, immunoprecipitates of B/Lee-infected cells; lane f, B/GL-infected cells.

ANTIBODIES TO NB IN SERUM FROM INFECTED MICE

The observation that NB is glycosylated in infected cells and that it can be found in the medium as well as in cells suggested that animals infected with influenza B virus might mount an antibody response to NB. Mice of the A2G strain (13,14) were inoculated intranasally with B/Lee or B/GL virus and serum samples taken at intervals after infection. Viral proteins to which antibodies were produced were identified by immunoprecipitation of infected cell extracts (15) as shown in Fig. 3. Serum taken after a primary infection with B/Lee or B/GL [lanes a and b, respectively] had activity directed primarily against the HA and NA glycoproteins and the nucleocapsid protein, NP. Reinfection resulted in the appearance of anti-NS1 antibodies [lanes c and d]. To enhance any low level antibody responses not detected in the convalescent sera, the mice were boosted by the intraperitoneal injection of homogenized lung tissue from acutely infected mice. In these mice, antibody responses to NB comparable to those to the other viral proteins were detected with the serum precipitating both the B/Lee and B/GL NB polypeptides [lanes e and f]. Thus, NB synthesized in infected mice is antigenic. Failure to detect anti-NB activity in convalescent mouse sera may indicate that the amount of NB presented to the immune system was insufficient to ellicit an IgG response [immunoprecipitates selected by protein A binding detect only IgG]. The total antibody response in infected mice, including IgM, is being examined.

STRUCTURAL FEATURES OF NB

The deduced amino acid sequence of B/Lee NB (6) permits predictions concerning its hydrophobicity and possible secondary structure. Using conformational parameters and bend frequencies determined for each amino acid, Chou and Fasman (16) derived probabilities for a given amino acid participating in helical, sheet, turn, or coiled secondary structures as influenced by the neighboring residues in the sequence. As shown in Fig. 4, top, according to the predictions, NB could have a predominately beta-sheet conformation [residues 1-60 and 79-94] with 7 regions of beta-turn permissible. Formation of a beta-sheet conformation requires the interaction of at least 2 polypeptide chains. Therefore, the regions of NB predicted to favor a beta-sheet conformation would not be, of themselves, capable of that secondary structure. The presence of beta-turn and random-coil regions might allow the NB polypeptide chain to fold back on itself to form beta-sheets entirely from intramolecular interactions or, alternatively, monomers of NB could interact intermolecularly to form a multimeric configuration.

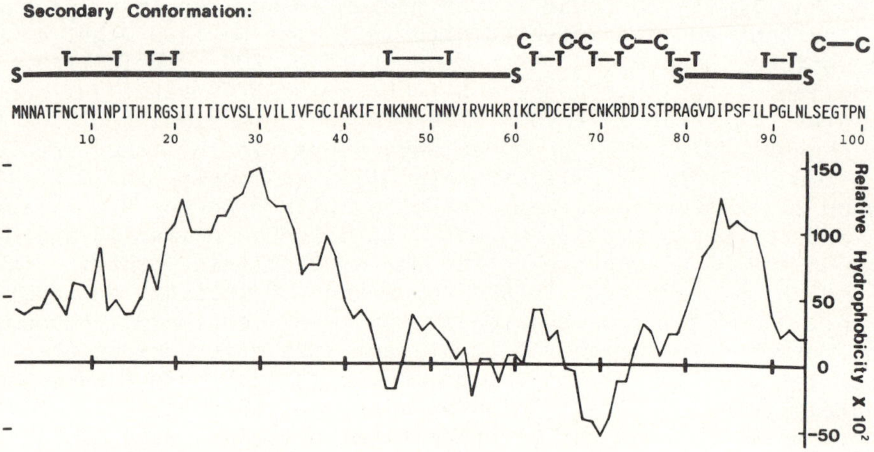

FIGURE 4. Predicted structural features of the B/Lee NB polypeptide. [Top] Chou-Fasman projection of secondary conformation. S=beta-sheet, T=beta-turn, C=random coil. [Middle] Deduced amino acid sequence of NB. [Bottom] Relative hydrophobicity of NB. Values for each residue n represent the average for residues n through n+6 at neutral pH.

Three of the four potential glycosylation sites are within these beta-sheet, beta-turn regions which include 85% of NB. According to Chou-Fasman predictions, the remainder of NB would be random coil with no areas of alpha helix evident.

Fig. 4, bottom, summarizes computer studies on the relative hydrophobicity of the NB amino acid sequence. Because the values are averaged over 7 residues, the values of the C-terminal 6 residues are not listed. The whole molecule is significantly hydrophobic [overall average=46 where a hydrophobicity constant > 50 is capable of stable insertion in a lipid bilayer (17)], and there are 2 regions of high hydrophobicity at amino acids 8-40 and 80-92. The locations of 3 of the 4 potential glycosylation sites relative to these regions suggest that not all of the sites would be glycosylated if the polypeptide is bound by one of these regions to a membrane. The hydrophobicity analysis also reveals clusters of hydrophilic residues in the middle of the sequence, which give NB an overall positive charge [+6 at neutral pH]. The B/Lee NB sequence includes an odd number of cysteine residues, 7, thus some residues could participate in intermolecular bonding.

The amino acid sequence suggests that NB is a relatively

"sticky" protein that could have an affinity for diverse microenvironments. The effects of glycosylation or other potential post-translational modifications of NB are difficult to assess without knowing its function. However, the NB amino acid sequence and its probable secondary conformation suggest that NB could be capable of extensive interactions in the host, both intra- and extracellularly. The possibility that NB might play a role in some of the observed biological differences between influenza A and B viruses [e.g., possible greater "toxicity" of influenza B virus, lack of an animal reservoir for influenza B virus, and the absence of recombination between the virus types] has been discussed previously (6), however, there is as yet no evidence regarding the function of NB.

CONCLUSIONS

The finding of a new influenza B virus glycoprotein indicates that there are aspects of the replication of the virus or its interaction with host that were previously unsuspected. NB is synthesized in significant amounts in infected cells, appearing simultaneously with NA, reflecting the fact that both are translated from a single bicistronic mRNA. Unlike NA, NB has been seen to disappear from infected cells within 3 hours after synthesis. The only other virus-specific polypeptide seen to disappear in pulse-chase experiments was NS2 [results not shown; NS2 labels poorly with cysteine as used here (6)]. The fate of NB after synthesis is uncertain. After its disappearance from infected cells, only small amounts of intact NB can be detected in the medium. In the absence of monospecific antibody to NB it has not yet been possible to determine its exact location or to assay for the presence of cleavage products that might result from proteolytic processing.

The NB glycoprotein is apparently unique to influenza B viruses; influenza A virus neuraminidase genes examined thus far lack a second open reading frame (5-10). The use of appropriate precursors to label virus-specific polypeptides in infected cells has revealed that NB is one of the major virus-coded proteins synthesized. On the basis of such studies, and the number of labeled residues present in the NB and NA polypeptides as deduced from the nucleotide sequences (5), it can be estimated that, on a molar basis, more NB than NA is synthesized in infected cells at any given time. The NB reading frame is conserved in the influenza B strains examined, from the original 1940 isolate to a 1983 isolate [results not shown] in spite of the fact that a single point mutation could eliminate NB without affecting NA. NB represents a previously unrecognized consequence of influenza B virus infection; its possible role in viral replication or pathogenesis remains to be elucidated.

ACKNOWLEDGEMENTS

We thank Dr. Bruce W. Erickson for expert advice and helpful discussion on the computer analyses and Andrew Koopman for excellent technical assistance.

REFERENCES

1. Lamb, R.A. and Choppin, P.W. (1983). Ann. Rev. Biochem. 52, 462.
2. Briedis, D.J., Lamb, R.A., and Choppin, P.W. (1981). Virology 112, 417.
3. Briedis, D.J. and Lamb, R.A. (1982). J. Virol. 42, 186.
4. Krystal, M., Elliot, R.M., Benz, E.W., Jr., Young, J.F., and Palese, P. (1982). Proc. Natl. Acad. Sci., U.S.A. 79, 4800.
5. Shaw, M.W., Lamb, R.A., Erickson, B.W., Briedis, D.J., and Choppin, P.W. (1982). Proc. Natl. Acad. Sci., U.S.A. 79, 6817.
6. Shaw, M.W., Choppin, P.W., and Lamb, R.A. (1983). Proc. Natl. Acad. Sci., U.S.A. 80, 4879.
7. Fields, S., Winter, G., and Brownlee, G.G. (1981). Nature (London) 290, 213.
8. Hiti, A.L. and Nayak, D.P. (1982). J. Virol. 41, 730.
9. Markoff, L. and Lai, C.-J. (1982). Virology 119, 288.
10. Blok, J. and Air, G. (1982). Virology 121, 211.
11. Briedis, D.J., Conti, G., Munn, E.A., and Mahy, B.W.J. (1981). Virology 111, 154.
12. Choppin, P.W., Lazarowitz, S.G., and Goldberg, A.R. (1975). In "Negative Strand Viruses" (B.W.J. Mahy and R.D. Barry, eds.) Vol.1 p.105. Academic Press, London.
13. Lindenmann, J., Lance, C.A., and Hobson, D. (1963). J. Immunol. 90, 942.
14. Haller, O. (1981). Cur. Top. Microbiol. Immunol. 92, 25.
15. Lamb, R.A. , Etkind, P.R., and Choppin, P.W. (1978). Virology 91, 60.
16. Chou, P.Y. and Fasman, G.D. (1978). Ann. Rev. Biochem. 47, 251.
17. Pliska, V. and Fauchere, J.-L. (1979). In "Peptides. Structure and Biological Function" (E. Gross and J. Meierhofer, eds.) p.249. Pierce Chem. Co., Rockford, Il.

FUNCTIONAL EXPRESSION AND MUTATIONAL ANALYSIS OF INFLUENZA VIRUS SURFACE GLYCOPROTEINS

Ching-Juh Lai, Lewis Markoff, Kenji Sekikawa
and Joanna Hansen

Molecular Viral Biology Section
Laboratory of Infectious Diseases
National Institute of Allergy and Infectious Diseases
Bethesda, MD 20205

We investigated the hydrophobic terminal sequences for functional expression of influenza virus hemagglutinin (HA) and neuraminidase (NA) in African green monkey kidney cells. A simple in-phase deletion was made in HA DNA resulting in loss of internal amino acids from the signal peptide. The signal deletion recombinant produced HA that was not glycosylated but stably accumulated in the cytoplasm. Similarly, a series of deletions were also introduced into NA DNA coding for the N-terminal sequence. These mutants produced either glycosylated or unglycosylated NA depending on the extent of deletion. These studies delineated the peptide sequence that is involved in translocation across the membrane, glycosylation and surface expression of NA. Results from site-specific mutagenesis revealed that many mutations in the signal sequence of HA had no effect on functional activity. However two mutants failed to accumulate HA at the cell surface. Sequence analysis of one mutant showed that among multiple substitutions an alteration of amino acid at the signal cleavage had occurred and such mutation was presumably responsible for the defective phenotype. Analysis of the other mutant showed two mutations were involved and the defect seemed to depend on both mutations in combination suggesting a conformational alternation of the signal sequences. These intracellular HAs contained only the endoglycosidase H sensitive carbohydrate component whereas the endo H resistant sugar moiety normally present in the wild type HA was not found. These results indicate that both mutant HAs, regardless whether the signal peptide had been cleaved or not, sustained a defect that resulted in the failure to acquire modified, endo H resistant carbohydrate components.

Copyright © 1984 by Academic Press, Inc.
All rights of reproduction in any form reserved.
ISBN 0-12-183501-4

INTRODUCTION

Influenza hemagglutinin (HA) and neuraminidase (NA) constitute the two surface antigens of the viral particle. HA is crucially involved in determining host immunity presumably because HA is responsible for attachment of infectious viral particles to the cell receptors and subsequent penetration and uncoating by fusion with intracelular membranes (1-4). On the other hand, NA is an enzyme that plays a probable role in promoting release of mature viral particles from the cell surface (5). The synthesis of both viral antigens undergoes a series of post-translational processings including glycosylation, intracellular transport and final accumulation at the outer membrane. Earlier we showed that the HA polypeptide synthesized from cloned influenza DNA in primate cells is functionally active (6). This provides a valuable tool to investigate amino acid sequences that are responsible for various properties of the viral antigen. In this communication we study the requirement of the amino-terminal hydrophobic sequences, so-called signal peptide, cell surface expression of the polypeptide. Results from characterization of functional abnormalities of HA and NA containing altered signal peptide sequences are presented.

RESULTS AND DISCUSSION

<u>(1) Functional Defects of Influenza Surface Antigens Lacking the Signal Sequences</u>. We wish to elucidate the functional role that the signal peptide plays in mediating the transport and modification of the polypeptide. A simple in-phase deletion was made in the HA DNA resulting in loss of 11 internal amino acids from the 16 amino-acid signal peptide. The deletion HA DNA in an SV40 vector was used to infect African green monkey kidney cells (Figure 1) Mutant HA was not detected on the cell surface but stably-accumulated in the cytoplasm. Polypeptide analysis from the infected cell lysate showed that mutant HA was not glycosylated despite the presence of all normal glycosylation sites. The mutant HA failed to agglutinate red blood cells, probably due to a defect directly or indirectly associated with the lack of carbohydrate side chains.

In order to identify the signal sequences and their involvement in NA functional expression, we also constructed deletion in NA DNA resulting in removal of amino acids from acids from the 29 amino acid hydrophobic sequence. Two NA deletion mutants were analyzed: one (dlK) retained 22 amino acids and the other (dlZ) retained 6 amino acids of the entire hydrophobic sequences (Figure 1). Both mutant NA's

```
                           MboII
                             ↓
wt HA    ATG AAG ACT ATC ATT GCT TTG AGC TAC ATT TTC TGT CTG GTT CTC GGC CAA GAC....
         TAC TTC TGA TAG TAA CGA AAC TCG ATG TAA AAG ACA GAC CAA GAG CCG GTT CTG
         Met Lys Thr Ile Ile Ala Leu Ser Tyr Ile Phe Cys Leu Val Leu Gly Gln Asp

dl HA    ATG AAG ACT ATC  (           33 bp deletion            ) GGC CAA GAC....
         TAC TTC TGA TAG                                           CCG GTT CTG
         Met Lys Thr Ile                                           Gly Gln Asp

wt NA    ATG AAT CCA AAT CAA AAG ATA ATA ACA ATT GGC TCT GTC TCT CTC ACC ATT GCA
         TAC TTA GGT TTA GTT TTC TAT TAT TGT TAA CCG AGA CAG AGA GAG TGG TAA CGT
         Met Asn Pro Asn Gln Lys Ile Ile Thr Ile Gly Ser Val Ser Leu Thr Ile Ala

         ACA ATA TGC TTC CTC ATG CAG ATT GCC ATC CAG GTA ACT ACT GTA ACA TTG CAT....
         TGT TAT ACG AAG GAG TAC GTC TAA CGG TAG GTC CAT TGA TGA CAT TGT AAC GTA
         Thr Ile Cys Phe Leu Met Gln Ile Ala Ile Gln Val Thr Thr Val Thr Leu His

dl K     ATG AAT CCA AAT CAA AAG ATA ATA ACA ATT GGC TCT GTC TCT CTC ACC ATT GCA
(NA)     TAC TTA GGT TTA GTT TTC TAT TAT TGT TAA CCG AGA CAG AGA GAG TGG TAA CGT
         Met Asn Pro Asn Gln Lys Ile Ile Thr Ile Gly Ser Val Ser Leu Thr Ile Ala

                                                -174 bp
                                                   ▼
         ACA ATA TCC TTC CTC ATG CAG ATT GCC ATC TGG TCA AAG CCG....
         TGT TAT AGG AAG GAG TAC GTC TAA CGG TAG ACC AGT TTC GGC
         Thr Ile Cys Phe Leu Met Gln Ile Ala Ile Trp Ser Lys Pro

                                                       -282 bp
                                                          ▼
dl Z     ATG AAT CCA AAT CAA AAG ATA ATA ACA ATT GGC TCA ATT CGG CTT TCT....
(NA)     TAC TTA GGT TTA GTT TTC TAT TAT TGT TAA CCG AGT TAA GCC GAA AGA
         Met Asn Pro Asn Gln Lys Ile Ile Thr Ile Gly Ser Ile Arg Leu Ser
```

Figure 1. Amino-terminal hydrophobic sequences of HA, NA and their derived deletion mutants (data from references 7 and 8).

produced from their respective SV40 recombinants were not detected on the cell surface by a live cell fluorescence assay. Analysis of lysates from infected cells showed that similar to wild type NA, dlK NA was glycosylated since unglycosylated dlK NA of predicted molecular size was found in the presence of tunicamycin. Mutant NA from dlZ, on the other hand, was not glycosylated and additon of tunicamycin had no effect on the synthesis and molecular size of the polypeptide product. These results indicate that the hydrophobic sequence near the N-terminus of NA plays a role similar to that does the transient signal peptide of HA.

(2) Hemagglutinins Containing Point Mutations in the Signal Sequence of Influenza Virus HA. The availability of both the wt and dl cloned pHA-SV40 DNA enabled us to generate amino acid substitution in the signal sequence using the method of local mutagenesis. Heteroduplex DNA was prepared between the wt and dl linear pHA-SV40 cleaved separately by a single-cut enzyme. Sodium bisulfite treatment of the heteroduplex DNA was performed according to the method of Shortle and Nathans (9). Several clones with full-length HA inserts were selected for sequence analysis to determine amino acid

alterations that had occurred. As shown in Figure 2, many base substitutions were silent mutations, but others resulted in amino acid substitutions. Mutants 3, 14, and 44 sustained one amino acid change, while mutants 7, 28, and 5 acquired two or more amino acid substitutions. One of these, the Gly to Ser change of mutant 28 occurred at the presumptive cleavage site. No charged amino acids were introduced into this region, thus the overall hydrophobicity of the signal peptide was maintained among these isolates.

In order to examine the functional activity of these HA proteins, we prepared circular HA-SV40 DNA from each mutant for infection of AGMK cells. Expression of intracellular HA was detected by indirect immunofluorescence and the results showed that all mutants produced an HA immunologically indistinguishable from wt HA. A guinea pig erythrocyte binding assay and a "live" cell immunofluorescence staining were performed to examine if all these mutant HAs were expressed on the cell surface. Mutants 3, 14, 8 and 5 produced HA proteins detectible on the cell surface by both assays. On On the other hand, mutants 7 and 28 were concistently negative indicating that these two mutants did not accumulate HA HA at the cell surface (Figure 2). Mutant 7 sustained two amino acid substitutions, each of which was found, respectively, in mutants 3 and 14 which were apparently normal for cell surface expression. It is possible that the two substitutions together sufficiently altered the conformation of the

																			Glycosylation	Detection at Cell Surface
wt-HA	ATG TAC 1 Met-	AAG TTC 2 Lys-	ACT TGA 3 Thr-	ATC TAG 4 Ile-	ATT TAA 5 Ile-	GCT CGA 6 Ala-	TTG AAC 7 Leu-	AGC TCG 8 Ser-	TAC ATG 9 Tyr-	ATT TAA 10 Ile-	TTC AAG 11 Phe-	TGT ACA 12 Cys-	CTG GAC 13 Leu-	GTT CAA 14 Val-	CTC GAG 15 Leu-	GGC CCG 16 Gly-	CAA GTT Gln-	GAC CTG Asp-	+	+
dl-HA	Met-	Lys-	Thr-	Ile-(					11 amino acids deleted						)	-Gly-	Gln-	Asp-	-	-
3	—	—	—	—	—	—	—	AGT Ser	TAT Tyr	—	TTT Phe	—	TTG Leu	—	TTT Phe	—	—	—	+	+
14	—	—	—	—	—	GTT Val	—	—	—	—	—	—	—	—	—	—	—	—	+	+
44	—	—	—	—	—	—	TTA Leu	—	—	—	—	TAT Tyr	—	—	—	—	—	—	+	+
7	—	—	—	—	—	GTT Val	—	—	TAT Tyr	—	—	—	TTG Leu	—	TTT Phe	—	—	—	+*	-
28	—	—	—	—	—	ACT Thr	TTA Leu	AAC Asn	—	—	—	TAT Tyr	CTA Leu	ATT Ile	—	AGC Ser	—	—	+*	-
5	—	—	—	—	—	ACT Thr	TTA Leu	AAC Asn	—	—	—	TAT Tyr	—	ATT Ile	—	—	—	—	+	+

Figure 2. Mutational analysis of hemagglutinins containing amino acid substitutions in the signal sequence. The amino terminal sequences that are cleaved (arrow) from the wild type hemagglutinin are presented. Altered HAs from a deletion mutant and several substitution mutants were analyzed as described. (*) indicates altered glycosylation was found.

signal peptide thus rendering a defect in transport. Sequence analysis showed that except for the Gly to Ser change, other substitutions in mutant 28 were all found in mutant 5 which exhibited the normal surface expression phenotype. These data suggest the Gly Ser alteration at the putative cleavage site may be responsible for the observed abnormality of mutant 28.

(3) Glycosylation Analysis of Mutant Hemagglutinins. In order to test the hypothesis that mutant hemagglutinins had undergone altered intracellular processing, endoglycosidase H (endo H) was employed to characterize the carbohydrate components that were contained in mutant HAs. As shown in Figure 3, the wt HA (70-75Kd) after endo H digestion yielded two labelled bands reduced in molecular size: one migrated at 56 Kd, the predicted value for the unglycosylated HA; the other migrated at 65 Kd representing the HA component that retained endo H-resistant carbohydrates normally present in wt HA. Analysis of mutant 5 HA that exhibited an apparently normal phenotype revealed a glycosylation pattern similar to that of wild type HA. Endo H digestion of HA's from mutants 7 and 28, on the other hand, showed that the 70-75 Kd HA was each converted entirely to the 56 Kd form corresponding to the unglycosylated HA. This finding indicates that both mutant 7 and 28 HA's contained only the mannose-rich sugar moiety that was added at the first stage of glycosylation and further modifications of the carbohydrates had not occurred. A comparison of electrophoretic mobility revealed that after endo H digestion the product of mutant 28 HA migrated slower than the product of mutant 7 did indicating the molecular size of mutant 28 HA is larger than that of mutant 7 HA. These results are consistent with the notion that the signal peptide was not cleaved from mutant 28 HA. On the other hand hand, the signal peptide cleavage appeared to be normal for mutant 7 HA. In any event these data showed that both defective mutant HAs did not acquire the endo H-resistant component. We speculate that both mutant HAs fail to proceed to the Golgi complex where further glycosylation and modifications take place. One interpretation for the finding is that conformational alteration in the signal peptide of mutant 7 HA gives rise to incorrect folding of the mature polypeptide thus preventing the normal transport. Alternatively, the signal peptide is involved in the molecular sorting directing the intracellular polypeptide transport to various locations. Since mutant 28 appeared to be cleavage defective the mature HA would retain two permanent hydrophobic regions. It is possible that both regions interact with intracellular membranes and this additional interaction

at the amino-terminus may prevent normal transport to the cell surface. Taken together, these results suggest that the signal peptide may be directly or indirectly involved in programming the transport process through post-translational modification by glycosylation.

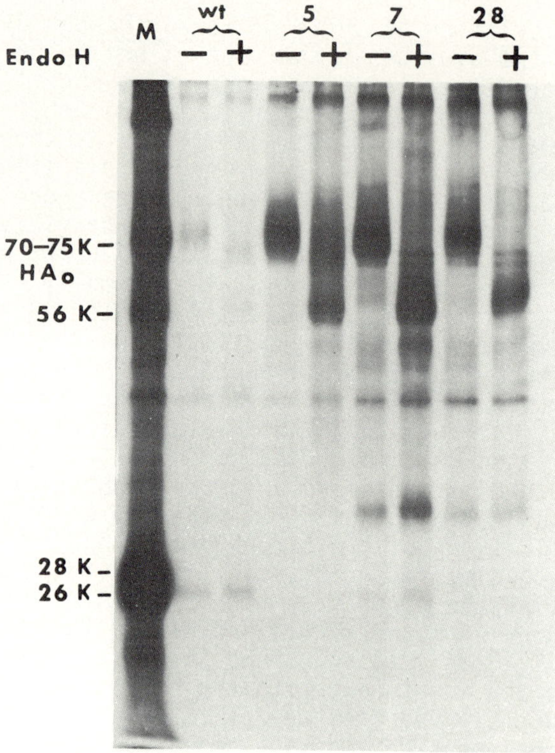

Figure 3. Digestion of hemagglutinins with endoglycosidase H. Radio-labelled HA from mutant and wild type HA-SV40 infected cells was isolated by immunoprecipitation. Digestion with endo H was carried out and the digests analyzed by electrophoresis on SDS-polyacrylamide gel. M indicates the molecular size markers of influenza viral proteins.

ACKNOWLEGEMENTS

The authors wish to thank Jo Ann Berndt, Salome Kruger and Esther Huang for their excellent technical assistance. We are also grateful to Jean Worsley and Linda Jordan for preparing the manuscript.

REFERENCES

1. Klenk, H.-D., Rott, R., Orlich, M. and Blodorn, J. (1975). Virology 68, 426.
2. Lazarowitz, S.G. and Choppin, P.W. (1975) Virology 68, 440.
3. Matlin, K.S., Reggio, H., Helenious, A., and Simons, K. (1981). J. Cell Biol. 91, 601.
4. Maeda, T., Kawasaki, K. and Ohnishi, S.-I. (1981). Proc. Nat. Acad. Sci. USA 78, 4133.
5. Palese, P., Toshita, K., Ueda, M., and Compans, R.W. (1974). Virology 61, 397.
6. Sveda, M.M. and Lai, C.-J. (1981). Proc. Nat. Acad. Sci. USA 78, 5488.
7. Sekikawa, K. and Lai, C.-J. (1983). Proc. Natl. Acad. Sci. USA, 80, 3563.
8. Markoff, L., Lin, B.-C., Sveda, M. and Lai, C.-J. (1983). Mol. Cell. Biol. in press.
9. Shortle, D. and Nathans, D. (1978). Proc. Natl. Acad. Sci. USA 75, 2170.

THE NONSTRUCTURAL GENE SEGMENT OF INFLUENZA A VIRUS: EXPRESSION OF NS1 PROTEIN IN MAMMALIAN CELLS; ANALYSIS OF A DELETION MUTANT

Mark Krystal, Susumu Nakada, Deborah A. Buonagurio, Dan C. DeBorde*, H.F. Maassab* and Peter Palese

Mount Sinai School of Medicine of CUNY, N.Y., N.Y. 10029

*School of Public Health
University of Michigan, Ann Arbor, MI 48109

INTRODUCTION

Many different procedures have been used for the study of the function of specific viral proteins. A classical and highly successful approach concerns the isolation of mutant viruses with lesions in the gene coding for the protein of interest. Alternatively, the development of viral vectors which can be used for the expression of proteins coded for by cloned DNAs has proven useful for the study of viral protein function. In this way it is possible to specifically alter the DNA clone to create defined protein mutations, and this mutant protein can then be expressed in mammalian cells. We are attempting to use both of these approaches to study the function(s) of the nonstructural proteins of influenza virus. We report here the analysis of a host range mutant of influenza virus with a deletion in the NS gene segment and we describe the successful expression of NS1 protein from cloned cDNA in CV-1 cells using an SV40 viral vector.

EXPRESSION OF NS1 IN MAMMALIAN CELLS

RNA segment 8 of influenza A virus codes for two overlapping polypeptides (NS1 and NS2) which are synthesized during viral infection and are not present in mature virions (1-3). The functions of these proteins in the infected cell have not been elucidated, but it is known that the NS1 protein will migrate to the nucleoli (4-7) or nucleoplasm soon after infection (8). A number of influenza viral proteins, including the hemagglutinin (HA), neuraminidase (NA), nucleoprotein (NP) and matrix (M) protein, have been successfully expressed in mammalian cells by viral vectors containing cloned influenza viral cDNA copies (9-18). The most common vector system employed involves

Copyright © 1984 by Academic Press, Inc.
All rights of reproduction in any form reserved.
ISBN 0-12-183501-4

SV40 recombinants which contain foreign genes inserted into either the early or late region of the viral DNA. We have used the SV40 late replacement vector SVL_2 (19) to express the influenza A/PR/8/34 viral NS1 protein in CV-1 cells.

The pAPR801 plasmid (20) was digested with EcoRl to liberate the full length NS insert. The EcoRl sticky ends were end-filled and the synthetic BglII linker AGATCT was ligated to the DNA fragment. After BglII digestion, the fragment was inserted into the BamHI site in the expression vector $pSVL_2$. Three chimeric constructions with respect to

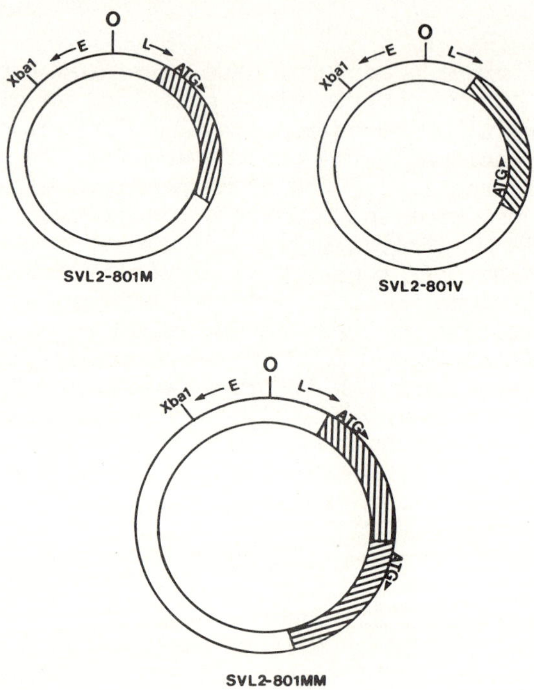

FIGURE 1. Structure of SV40-NS recombinants. The striped region represents the A/PR/8/34 NS clone. The ATG indicates the position of the first met codon of the NS1 protein. The SVL_2 vector has a deletion in the late region spanning nucleotides 282-2469. For propagation in E. coli, a pBR322 derivative is inserted at the indicated XbaI site.

the orientation and number of NS inserts were isolated (fig. 1). Two of the constructions have a single copy of the NS insert, one with the clone in the mRNA sense with respect to the SV40 promoter (pSVL$_2$-80lm) and the other in the vRNA sense (pSVL$_2$-80lv). The third chimeric plasmid has two tandem copies of the NS clone, both in the mRNA orientation with respect to the SV40 promoter (pSVL$_2$-80lmm).

Each plasmid was digested to remove the pBR322 sequences and religated to produce circular SV40-NS recombinant molecules (fig. 1) (21). Each DNA molecule was then co-transfected into CV-1 cells along with a helper SV40 recombinant. This recombinant, pSV-r-INS-7 (21,22) is an SV40 vector which has the insulin gene replacing portions of the early region of SV40. After a two week incubation period, cells were freeze thawed three times and the supernatant was used as the viral stock.

CV-1 cells were infected with the viral stock preparations and labelled three days post infection. Figure 2 shows that both the SVL$_2$-80lm and SVL$_2$-80lmm (lanes 4 and 5) viral stocks produce in infected cells a protein which migrates identically to the NS1 protein produced in A/PR/8/34 virus infected cells (lane 6). In contrast, the SVL$_2$-80lv stock does not produce the presumptive NS1 protein. To confirm that this band is indeed NS1, cell extracts were immunoprecipitated with polyclonal antisera specific for NS1 protein (8). After immunoprecipitation, bands corresponding to NS1 protein were seen only with the SVL$_2$-80lm and SVL$_2$-80lmm infected cells (data not shown). As of yet we have not been able to observe the synthesis of NS2 protein in cells infected with the SVL$_2$-NS viral stocks. We are now attempting to produce polyclonal antisera against E. coli synthesized NS2 protein which could be used to detect small amounts of protein in immunofluorescence, immunoprecipitation or RIA assays.

Immunofluorescence studies were undertaken to determine whether the NS1 protein produced in infected cells by the SVL$_2$-80lm and SVL$_2$-80lmm viral stocks is localized within the nucleus (4-8). In influenza A/PR/8/34 virus infected cells, the NS1 protein was shown to be located in the nucleoplasm throughout the infectious cycle (8). This is in contrast to influenza A/Bel/42 virus infected cells. Early in infection the NS1 protein is localized in the nucleoplasm and then becomes associated with nucleoli at later times during infection (8). The immunofluorescence data in figure 3 illustrates that the CV-1 cells infected with the SVL$_2$-80lm viral stock produced NS1 protein which was nucleoplasm associated. This nucleoplasmic association was observed two and three days post infection of the viral stocks. The SVL_2-80lmm viral stock exhibits an identical immunofluorescence pattern while the SVL_2-80lv stock did not show any positive immunofluorescence (data not shown).

ANALYSIS OF A DELETION MUTANT

Several groups have studied NS1 function by deriving viral mutants with temperature sensitive (ts) lesions in

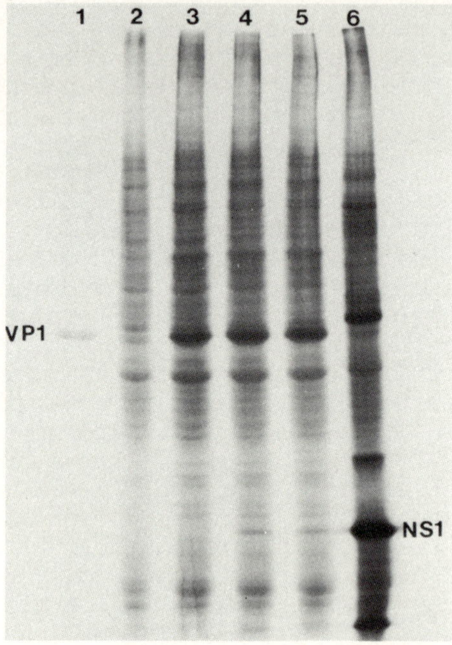

Figure 2. Expression of NS1 protein in CV-1 cells. Linear 7-14% SDS-PAGE of infected CV-1 cells labelled for 3 hours with ^{35}S-met at 2 days p.i. Lane 1, SV40 infected CV-1 cells; lane 2, mock infected cells; lane 3, SVL_2-80lv infected CV-1 cells; lane 4, SVL_2-80lm infected CV-1 cells; lane 5, SVL_2-80lmm infected CV-1 cells; lane 6, A/PR/8/34 infected CV-1 cells labelled for 1 hour at 6 hours p.i. The VP1 protein of SV40 virus and influenza viral NS1 are indicated.

RNA 8 (23-28). Recently, Maassab and DeBorde (29) isolated a NS mutant virus which displays a restricted host range (hr) phenotype that is not temperature dependent. This mutant virus, CR43-3, will form plaques on primary chick kidney cells (PCK) but will not form plaques in Madin-Darby canine kidney cells (MDCK). CR43-3 was accidentally generated in a recombination experiment following co-infection in PCK cells with the A/ALA/6/77 virus and the cold-adapted (ca) and temperature-sensitive (ts) A/AA/6/60 virus (29,30). Genotypic analysis of the CR43-3 virus and the parental strains revealed that segment 8 of the mutant virus migrated faster than either parent on polyacrylamide-agarose gels (29). Maassab and DeBorde also found that the host range phenotype of the virus is associated with this apparently smaller NS RNA. In order to determine the origin of this RNA in the mutant virus and its precise defect, we have cloned and sequenced cDNA copies of segment 8 from CR43-3 virus and the parental A/ALA/6/77 virus.

Double-stranded cDNAs from all three viruses (A/AA/6/60, A/ALA/6/77 and CR43-3) were synthesized from viral RNAs using reverse transcriptase and synthetic dodecanucleotide primers as described (20). In figure 4, single (ss) and double strand (ds) cDNA copies of the A/AA/6/60 and CR43-3 viruses are examined by polyacrylamide gel electrophoresis before and

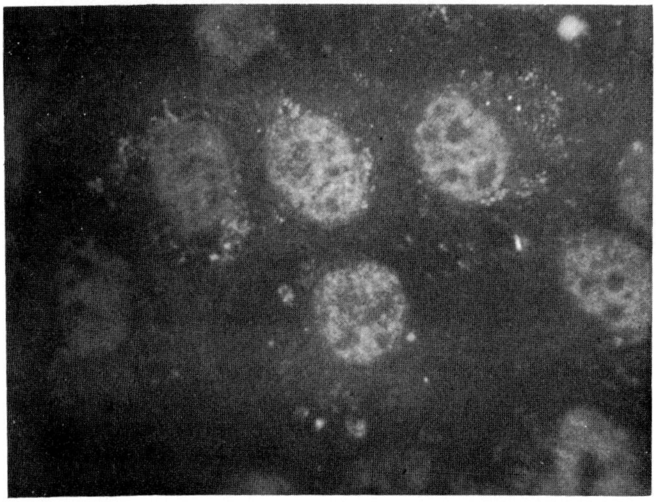

FIGURE 3. Immunofluorescence of SVL_2-80lm infected CV-1 cells two days p.i. Polyclonal rabbit antisera to E. coli produced NS1 protein was used as described (8).

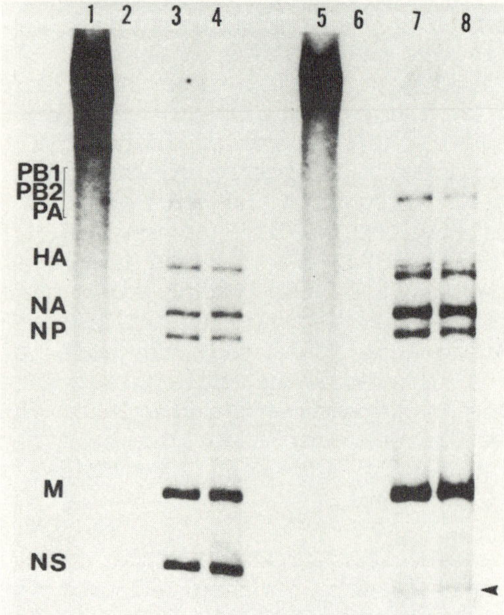

FIGURE 4. Electrophoresis of the reverse-transcribed cDNA copies of A/AA/6/60 and CR43-3 viral genes in a 5% polyacrylamide gel. Lane 1, A/AA/6/60 ss cDNA; lane 2, A/AA/6/60 ss cDNA following S1 nuclease treatment; lane 3, A/AA/6/60 ds cDNA; lane 4, A/AA/6/60 ds cDNA following S1 nuclease treatment; lane 5, CR43-3 ss cDNA; lane 6, CR43-3 ss cDNA following S1 nuclease treatment; lane 7, CR43-3 ds cDNA; lane 8, CR43-3 ds cDNA following S1 nuclease treatment. It should be noted that an extra band is present in lanes 7 and 8 which migrates slightly slower than the HA segment. The origin of this band is unknown.

after S1 nuclease treatment. Lanes 4 and 8 show the ds cDNA of A/AA/6/60 and CR43-3 viruses after S1 treatment. In each case, bands are present which correspond to full length cDNA copies of all eight RNAs. The cDNA molecule corresponding to the NS RNA from CR43-3 virus is migrating faster than the corresponding segment from A/AA/6/60 virus (lane 4). Size estimates suggest that this CR43-3 segment is approximately 40 bases shorter than the A/AA/6/60 NS segment. As expected, the NS RNA of A/ALA/6/77 virus migrates identically to the A/AA/6/60 NS RNA (not shown).

Double stranded cDNA copies of segment 8 from CR43-3 and A/ALA/6/77 viruses were cloned into pBR322 and sequenced

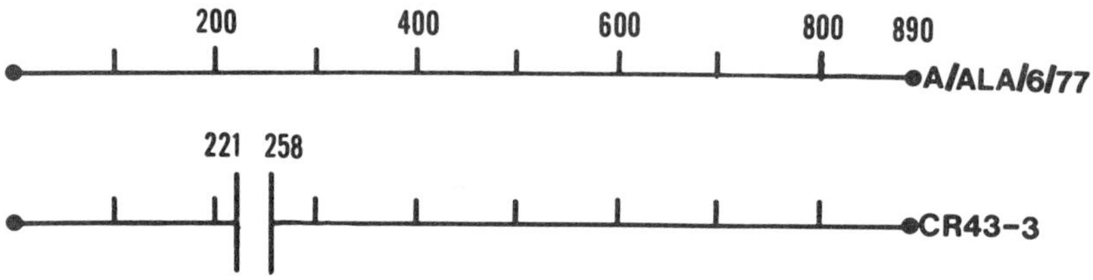

FIGURE 5. Graphic representation of RNA segment 8 of influenza A/ALA/6/77 and CR43-3 virus. The parent A/ALA/6/77 NS1 protein is 237 amino acids in length, while the deleted CR43-3 NS1 polypeptide is 225 amino acids long. The NS2 polypeptide in both viruses contains 121 amino acids.

as described (31). The complete sequences will be presented elsewhere (31). The sequences of the two RNA segments are identical except for an internal deletion of 36 nucleotides in the CR43-3 segment (fig. 5). Therefore, the CR43-3 NS RNA probably originated through a deletion from the A/ALA/6/77 segment. We also compared the CR43-3 sequence to partial sequence data available for 55 bases of segment 8 from the wild type A/AA/6/60 virus (32). In this region, six bases differed from those in the CR43-3 and A/ALA/6/77 viral sequences. This confirms that the A/ALA/6/77 virus is the parent which provided the NS segment to the CR43-3 virus.

The deletion in the CR43-3 NS cDNA corresponds to nucleotides 222-257 of the A/ALA/6/77 gene (fig. 5). The presence and specific location of the deletion in the virion RNA itself was confirmed by direct RNA sequencing. This was accomplished by using the dideoxy method and a restriction fragment as a primer on the vRNA template of CR43-3 virus (data not shown). This deletion results in an NS1 protein which is 12 amino acids shorter (deletion of amino acids 66-77 of A/ALA/6/77 virus) than the parental protein. It is likely that this deletion is responsible for the altered host range phenotype of the mutant. The coding sequence of the NS2 protein is not affected by the deletion of bases 222-257 in the RNA (1, 2).

Protein synthesis of the mutant virus was examined in the permissive PCK and the nonpermissive MDCK cell lines. Figure 6 shows that in the nonpermissive MDCK line (lane 2), the mutant NS1 protein is synthesized in amounts equal to

FIGURE 6. Linear gradient SDS-PAGE of ^{35}S-met labelled CR43-3 proteins synthesized in permissive and nonpermissive cells. Cells were infected with CR43-3 virus and labelled for 1 hour at 8 hours p.i. (31). Lane 1, CR43-3 infected PCK cells; lane 2, CR43-3 infected MDCK cells. The increase in NS2 protein observed in MDCK cells relative to PCK cells is also seen in A/ALA/6/77 infected cells (not shown).

those found in the PCK cell line (lane 1). However, the level of M1 protein synthesis is markedly reduced. Densitometer tracings indicate an approximate 50% reduction in the amount of M1 protein in the MDCK cell line relative to PCK cells. Finally, immunofluorescence studies have shown that the nuclear localization of the altered NS1 protein is unchanged in the nonpermissive MDCK cells relative to PCK cells (data not shown). Therefore, the defect in the CR43-3 NS1 protein does not affect this biological property which is associated with the NS1 protein.

SUMMARY AND DISCUSSION

We have used SV40 viral vectors to direct the expression of the NS1 protein of influenza A/PR/8/34 virus from a cDNA copy of the vRNA segment. As in influenza virus infected cells, the NS1 protein of the SV40 recombinant virus infected cells migrates to the nucleoplasm (8). This illustrates that the nuclear localization of the NS1 protein is most likely a characteristic property of the protein itself rather than a consequence of different interacting viral proteins. Since in this system there is a DNA involved in the expression of the NS1 protein, the regions of the protein responsible for the migration to the nucleus can now be studied using mutational changes. For example, deletions or site specific mutagenesis in the cloned virus specific DNA may alter the tropism of the protein within the cell. This may help in identifying the possible presence of a "signal peptide" which directs the migration of a protein to the nucleus. As other functional assays for the NS1 protein are developed, the approach of expressing the protein via SV40 recombinants may prove very useful for determining which regions of the protein are responsible for different functions.

The eukaryotic expression system may also allow for the isolation of new influenza virus mutants through complementation with influenza virus proteins expressed by recombinant vectors. In the future, it may even be possible to "rescue" the transcribed cDNA and incorporate, via recombination, a complete RNA segment into the viral genome of an influenza virus. This procedure would allow for the construction of influenza viruses with defined genomes for use in vaccine or structure/function studies.

We have also analyzed a host range mutant of influenza virus which has a deletion of RNA segment 8. The 36 nucleotide deletion occurs in a region coding for the amino terminal part of the NS1 polypeptide. The internal deletion of 12 amino acids most likely changes the protein in such a way that replication of the virus is prevented in MDCK cells. It should also be noted that in the nonpermissive cell line M1 protein synthesis is markedly reduced. This result is similar to that seen when ts mutants known to possess a defect in segment 8 are used to infect cells at nonpermissive temperatures (23-25, 33). We do not fully understand the mechanism by which the NS1 protein influences the M1 synthesis, but it is clear that the defect associated with the altered CR43-3 NS1 can be bypassed through host cell factors in the PCK cells. To our knowledge, CR43-3 virus is the first conditional lethal mutant of influenza virus with a defined deletion in a gene coding for a nonsurface protein.

ACKNOWLEDGEMENTS

We wish to thank G. Khoury for providing us with the $pSVL_2$ vector and P. Sharp for his gift of the pSV-r-INS-7 vector. P.P. is a recipient of an I.T. Hirschl Career Research Award and M.K. is the recipient of an American Lung Association Training Fellowship. This work was supported by grants from the National Institutes of Health (AI-11823 and AI-18998) and by contract NO1 AI-72521 from the National Institute of Allergy and Infectious Diseases, Development and Applications Branch.

REFERENCES

1. Inglis, S.C., Barrett, H.T., Brown, C.M. and Almond, J.W. (1982). Proc. Natl. Acad. Sci. USA 76, 3790.
2. Lamb, R.A. and Choppin, P.W. (1979). Proc. Natl. Acad. Sci. USA 76, 4908.
3. Lamb, R.A. (1983). In: "Genetics of Influenza Viruses" (P. Palese and D.W. Kingsbury, eds.), p. 21. Springer Verlag, Vienna.
4. Dimmock, N.J. (1969). Virology 39, 224.
5. Taylor, J.M., Hampson, A.W., Leyton, J.E. and White, D.O. (1970). Virology 42, 744.
6. Krug, R.M. and Soeiro, R. (1975). Virology 64, 378.
7. Shaw, M.W., Lamon, E.W. and Compans, R.W. (1981). Infection and Immunity 34, 1065.
8. Young, J.F., Desselberger, U., Palese, P., Ferguson, B., Shatzman, A.R. and Rosenberg, M. (1983). Proc. Natl. Acad. Sci. USA 80 (in press).
9. Gething, M.J. and Sambrook, J. (1981) Nature 293, 620.
10. Hartman, J.R., Nayak, D.P. and Fareed, G.C. (1982). Proc. Natl. Acad. Sci. USA 79, 233.
11. Sveda, M.M. and Lai, C.-J. (1981). Proc. Natl. Acad. Sci. USA 78, 5488.
12. Sveda, M.M., Markoff, L.J. and Lai, C.-J. (1982). Cell 30, 649.
13. Lamb, R.A. and Lai, C.-J. (1982). Virology 123, 237.
14. Lin, B.-C. and Lai, C.-J. (1983). J. Virol. 45, 434.
15. Davis, A.R., Bos, T.J. and Nayak, D.P. (1983). Proc. Natl. Acad. Sci. USA 80, 3976.
16. Paoletti, E. and Panicali, D. (1983). Proc. Natl. Acad. Sci. USA 80 (in press).

17. Gething, M.J. and Sambrook, J. (1983). In: "Genetics of Influenza Viruses" (P. Palese and D.W. Kingsbury, eds.) p. 169. Springer Verlag, Vienna.
18. Roth, M.G., Compans, R.W., Giusti, L., Davis, A.R., Nayak, D.P., Gething, M.-J. and Sambrook, J. (1983). Cell 33, 435.
19. Gruss, P., Rosenthal, N., Koenig, M., Ellis, R.W., Shih, T.Y., Scolnick, E.M. and Khoury, G. (1982). In: "Eukaryotic Viral Vectors (Y. Gluzman, ed.) p. 13. Cold Spring Harbor Laboratory, Cold Spring Harbor, N.Y.
20. Baez, M., Taussig, R., Zazra, J.J., Young, J.F., Palese, P., Reisfeld, A. and Skalka, A. (1980). Nucl. Acids Res. 8, 5845.
21. Laski, F.A., Alzner-DeWeerd, B., RajBhandary, U.L. and Sharp, P.A. (1982). Nucl. Acids Res. 10, 4607.
22. Laski, F.A., Belagaje, R., RajBhandary, U.L. and Sharp, P.A. (1982). Proc. Natl. Acad. Sci. USA 79, 5813.
23. Wolstenholme, A.J., Barrett, T., Nichol, S.T. and Mahy, B.W.J. (1980). J. Virol. 35, 1.
24. Almond, J.W., McGeoch, D. and Barry, R. (1977). Virology 81, 62.
25. Koennecke, I., Boschek, C.B. and Scholtissek, C. (1981). Virology 110, 16.
26. Shimizu, K., Mullinix, M.G., Chanock, R.M. and Murphy, B.R. (1982). Virology 117, 38.
27. Shimizu, K., Mullinix, M.G., Chanock, R.M. and Murphy, B.R. (1982). Virology 117, 45.
28. Shimizu, K., Mullinix, M.G., Chanock, R.M. and Murphy, B.R. (1983) Virology 124, 35.
29. Maassab, H.F. and DeBorde, D.C. (1983) Virology (in press).
30. Maassab, H.F., Smitka, C.W., Donabedian, A.M., Monto, Monto, A.S., Cox, N.J. and Kendal, A.P. (1981). In: "The Replication of Negative Strand Viruses" (D.H.L. Bishop and R.W. Compans, eds.) p. 395. Elsevier/ North Holland, New York.
31. Buonagurio, D.A., Krystal, M., Palese, P., DeBorde, D.C. and Maassab, H.F. (1983). J. Virol. (in press).
32. Parvin, J.D., Young, J.F. and Palese, P. (1983) Virology 128, 512.
33. Robertson, J.S., Robertson, E., Roditi, I., Almond, J.W. and Inglis, S.C. (1983) Virology 126, 391.

PRODUCTION OF A SUBUNIT VACCINE FOR RIFT VALLEY FEVER VIRUS

A. F. Purchio,[1] M. S. Collett,[1] M. Parker,[2]
C. Schmaljohn,[2] and J. Dalrymple[2]

[1]Molecular Genetics, Inc.
10320 Bren Road East
Minnetonka, Minnesota 55343

[2]Dept. of Viral Biology, Virology Division
U.S. Army Medical Research Inst. of
Infectious Diseases
Fort Detrick, Frederick, Maryland 21701

Rift Valley fever virus (RVFV) is a member of the phlebotomus fever serogroup in the Bunyaviridae family of viruses. The genome of RVFV consists of three RNA segments of negative polarity termed L(m.w. = $2.7 \cdot 10^6$), M(m.w. = $1.7 \cdot 10^6$) and S($0.6 \cdot 10^6$)(1). By analogy with other Bunyaviruses (2), the M RNA species is presumed to contain the genetic information coding for the two viral glycoproteins G_1(m.w. = $65 \cdot 10^3$) and G_2(m.w. = $55 \cdot 10^3$)(1). Since the determinants recognized by neutralizing antibodies reside on the viral glycoproteins (3), we attempted to produce, in bacteria, proteins related to the glycoproteins of RVFV which may then serve as a subunit vaccine.

The general procedure used to clone this RNA is outlined in figure 1. RVFV M RNA was prepared from nucleocapsids and tailed with adenosine residues using E. coli poly (A) polymerase. The RNA was then chromatographed on oligo [dT]-cellulose to select for those M RNA molecules which were polyadenylated at their 3'-end. This RNA fraction was then used as a template for cDNA synthesis by reverse transcriptase using oligo (dT) as primer. After alkali treatment to hydrolyze the RVFV RNA, second strand synthesis was performed with the Klenow fragment of DNA polymerase I. The double stranded cDNA was then cleaved with mung bean nuclease and tailed with [dC] residues using terminal deoxnucleotidyl

Copyright © 1984 by Academic Press, Inc.
All rights of reproduction in any form reserved.
ISBN 0-12-183501-4

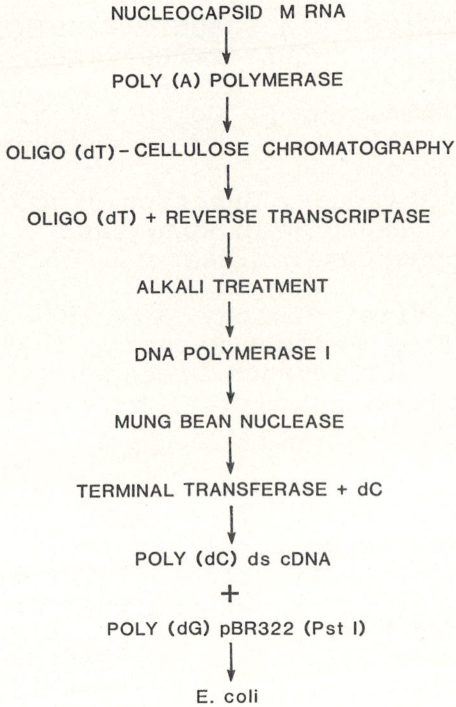

FIGURE 1. Cloning of RVFV M RNA. RVFV M RNA was extracted from purified M-nucleocapsids as described (4). The RNA was tailed with poly (A) polymerase (BRL) in a total volume of 0.3 ml containing 50 mM Tris-HCl (pH 7.9), 10 mM $MgCl_2$, 2.5 mM $MnCl_2$, 250 mM NaCl, 0.1 mM ATP, 29 units of enzyme and 0.6 mg RNA. The reaction was incubated for 10 min at 37^oC, extracted with phenol:-chloroform:isoamylalcohol (24:24:1) and chromatographed on oligo [dT]-cellulose. The poly (A)-containing RNA was made into double stranded cDNA, tailed with dC residues, hybridized to pBR322 tailed with dG residues at its unique PstI site and transformed into competent _E. coli_.

transferase. This cDNA was annealed to pBR322 which had been cut with PstI and tailed with (dG)-residues and the mixture was used to transform

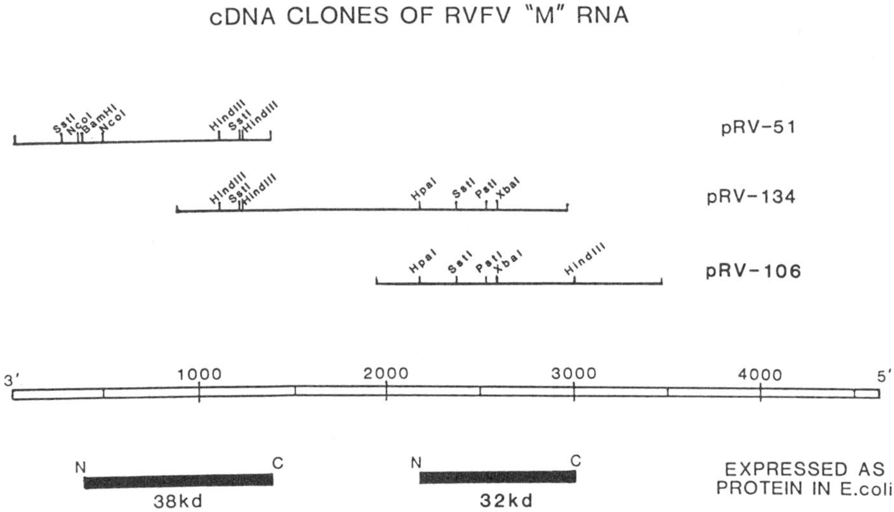

FIGURE 2. Partial restriction map of RVFV M RNA-specific clones. Plasmids containing RVFV M RNA specific sequences were cleaved with various restriction enzymes and aligned with the virion M RNA. DNA sequencing of several regions revealed open reading frames only in one direction, from left to right in the above figure. Shaded areas indicate regions currently expressed as protein in E. coli.

competent E. coli

Insertion of foreign DNA sequences into the PstI site of pBR322 as described above results in the inactivation of the gene coding for ampicillin resistance. Therefore colonies were first screened for their $Ap^{S}Tc^{R}$ phenotype. Bacteria harboring DNA sequences complementary to RVFV M RNA were identified by colony hybridization using $[^{32}P]$-cDNA made against agarose gel-purified RVFV M-RNA as probe. Approximiately 170 positive colonies were identified.

From these 170 clones, several plasmids were isolated and those containing large inserts were chosen for further study. Figure 2 shows a restriction map of three of the RVFV 'M' specific inserts. They comprise about 4100 base pairs representing approximately 90% of the RVFV M RNA (estimated in our hands to be 4700 nucleotides in length).

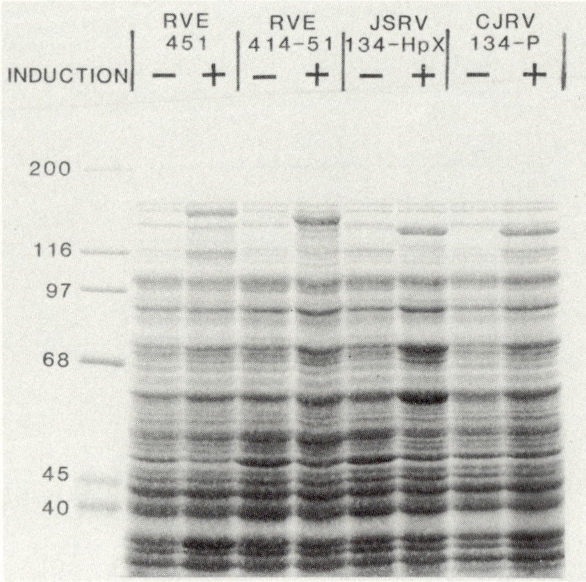

FIGURE 3. Induction of RVFV fusion proteins. E. coli, strain NF1829 containing various expression plasmids were grown in L broth in the absence or presence of inducer (1% lactose). Overnight cultures were harvested, lysed directly in SDS-gel sample buffer by boiling for 5 min., and aliquots subjected to SDS-polyacrylamide gel electrophoresis (5). Bacteria harboring these plasmids synthesize a high molecular weight polypeptide only upon induction of the lac promoter. All proteins are of the size predicted from data relating to the size of the inserted RVFV DNA sequences and the β-galactosidase gene. Molecular weight markers are presented in kilodaltons.

In order to determine which regions of the DNA could code for protein, DNA sequencing was performed on selected regions. In all cases, open reading frames were found in only one direction, going from left to right in figure 2 (i.e., 3' → 5' on the virion M RNA). Portions of these RVFV genetic sequences containing open reading frames were manipulated so as to express RVFV-related

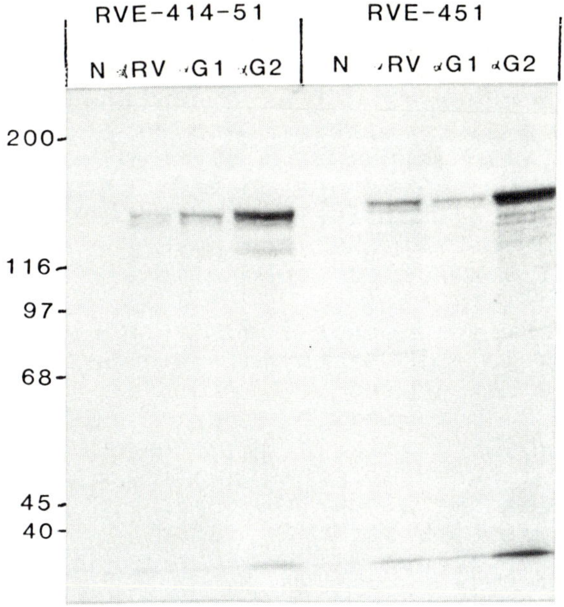

FIGURE 4. Immune precipitation of RVFV fusion proteins using anti RVFV serum. Cultures (1 ml) of pRVE451 and pRVE414-51 were induced to produce their RVFV/β-galactosidase fusion proteins by the addition of IPTG (1 mM). After 30 min. of growth at 37°C, [^{35}S]-methionine (100 μCi/ml; NEN, Boston, MA) was added for a brief period (5 min.). The bacteria were then centrifuged from the culture medium, resuspended in 50 μl of 10 mM Tris-HCl, pH 7.4, 10 mM EDTA, and 50 μl of 10 mg/ml lysozyme was added. After 15 min. at 4°, 0.9 ml of RIPA buffer (150 mM NaCl, 10 mM Tris-HCl, pH 7.4, 1% Triton X100, 1% deoxycholate, 0.1% SDS) was added and the mixture briefly sonicated. Cell debris was removed by centifugation (16,000 xg for 10 min.), and the resulting cell-free lysate was used for immunoprecipitation as described (6). Four sera were employed in this experiment: normal guinea pig serum (N), guinea pig serum from an animal hyperimmunized with whole RVFV (αRV), guinea pig serum from an animal injected with partially purified RVFV glycoprotein (αG1), and guinea pig serum from an animal injected with partially purified RVFV glycoprotein

polypeptides (shaded areas, figure 2). These regions were expressed as fusion proteins consisting of 23 amino acids of the cro gene product of bacteriophage lambda at the amino end, the RVFV sequences, and 116,000 daltons of a modified E. coli β-galactosidase gene at the carboxy end. The expression of these proteins was under control of a lac operon and all were inducible with either lactose or IPTG. Several such fusion proteins analyzed by SDS-polyacrylamide gel electrophoresis are shown in figure 3.

To determine whether the RVFV sequences expressed in E. coli contained antigenic determinants recognized by antibodies against authentic RVFV proteins, bacterial fusion proteins were radiolabelled with [^{35}S]-methionine and analyzed by immunoprecipitation followed by SDS-polyacrylamide gel electrophoresis and fluorography. Four sera were used: normal guinea pig serum (N); guinea pig serum from an animal hyperimmunized with whole RVFV (αRV); guinea pig serum from an animal injected with partially purified RVFV glycoprotein G_1 (αG_1); and guinea pig serum from an animal injected with partially purified RVFV glycoprotein G_2 (αG_2). Figure 4 shows two fusion proteins that were specifically precipitated with antisera raised against authentic RVFV proteins indicating that these bacterially-produced proteins do possess RVFV-specific antigenic determinants.

In order to determine if the RVFV bacterial fusion proteins were able to elicit an RVFV-specific immune response, various formulations containing the fusion proteins were prepared and injected into laboratory mice. After an appropriate period of time, sera were obtained from the injected animals and evaluated for antibodies directed against authentic RVFV proteins. An [^{35}S]-methionine-labeled lysate of RVFV-infected Vero cells was prepared for use in an immunoprecipitation experiment. Portions of this cleared lysate were immunoprecipitated with

---

G2 (αG2). Immune precipitates were fractionated on a 7.5% polyacrylamide-SDS gel (5) and prepared for fluorography as described (7). Molecular weight markers are presented in kilodaltons.

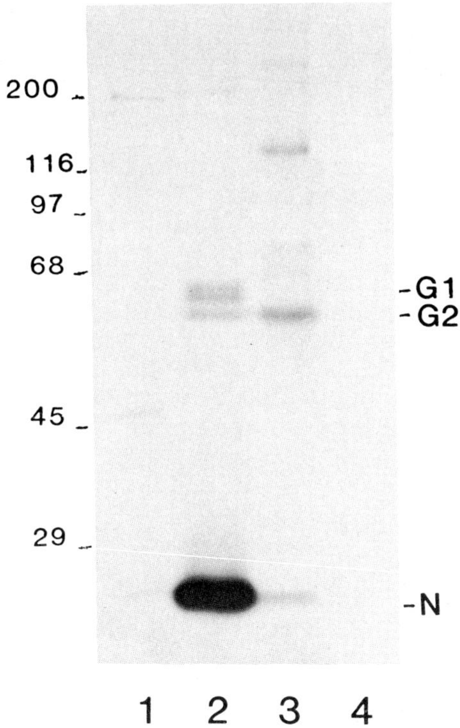

FIGURE 5. Immune precipitation of authentic RVFV proteins using sera from animals immunized with RVFV fusion proteins. Various formulations of fusion protein pRVE414-51 were injected intramuscularly into mice (1 mg fusion protein/mouse). A second injection of one-half this amount was performed 21 days after the first injection. Animals were bled 7 days later and sera prepared. This sera was tested for its ability to immune precipitate authentic RVFV proteins from an [^{35}S]-labelled lysate of RVFV-infected Vero cells. Immune precipitates were fractionated on a 10% SDS-polyacrylamide gel and fluorographed. Track 1, normal guinea pig serum; track 2, anti RVFV guinea pig serum; track 3, anti pRVE414-51 fusion protein mouse sera; track 4, normal mouse sera.

various sera and analyzed by SDS-polyacrylamide gel electrophoresis, and fluorography. A repesentative example of results obtained from such an analysis of sera generated by the RVFV fusion protein immunogens is provided in Figure 5. Serum obtained from an animal immunized with the pRVE414-51 fusion protein appeared to specifically immunoprecipitate a protein from RVFV-infected cell lysates that co-migrated with authentic RVFV glycoprotein G2. This suggests that the RVFV fusion proteins are able to act as immunogens.

The ultimate test concerning the usefulness of these fusion proteins in subunit vaccines will be in their ability to elicit an immune response capable of protecting animals against a subsequent RVFV challenge. These experiments are now in progress.

REFERENCES

1. Cash, P., Robeson, G., Erlich, R. J. and Bishop, D. H. L. (1981). In Contributions to Epidemiology and Biostatistics, Vol. 3, Rift Valley Fever, p. 1. ed. Suartz, Klingberg and Goldblum. S. Karger A G, P.O. Box CH-4009, Basel, Switzerland.
2. Gentsch, J. R. and Bishop, D. H. L. (1979). J. Virol. 30:767.
3. Gonzalez-Scarano, F., Shope, R. E., Calisher, C. E., Nathanson, N. (1982). Virology 120, 42.
4. Petterson, R. F., Hewlett, M. J., Baltimore, D., Coffin, J. (1977). Cell 11:51.
5. Laemmli, U.K. (1970). Nature 227:680.
6. Brugge, J. S. and Erikson, R. L. (1977). Nature 269:346.
7. Chamberlain, J. P. (1979). Anal: Biochem. 98:132.

PERSISTENT EXPRESSION OF INFLUENZA VIRUS NUCLEOPROTEIN IN RECOMBINANT DNA TRANSFECTED MOUSE CELLS

Bor-Chian Lin[1], Ching-Juh Lai[1], and Ming-Fan Law[2]

[1]Molecular Viral Biology Section
Laboratory of Infectious Diseases, NIAID
Bethesda, MD 20205

[2]Laboratory of Pathology
National Cancer Institute
Bethesda, MD 20205

A cloning strategy has been devised to construct recombinants between cloned influenza NP DNA (coding for the nucleoprotein) and a bovine papilloma virus DNA vector. Such NP-BPV recombinant DNA was used for transfection of a mouse cell line in order to establish persistent expression of influenza nucleoprotein. Recombinant DNA transfected cells were morphologically transformed and produced NP polypeptide predominantly detected in the nucleus during early passages. Subcloning of transformants yielded a heterogeneous cell population and more cytoplasmic distribution was apparent as regards NP production. These data suggest that persistent expression of NP was accompanied by alteration of cell phenotype. Sucrose gradient sedimentation analysis showed the NP product was detected at low molecular weight fractions suggesting that little or no encapsidation of RNA has occurred. Analysis of cell lysates from such transformants showed that a full-length NP was immunoprecipitated with specific antisera and the NP product exhibited functional characteristics namely, phosphorylation, similar to that found during influenza virus infection.

INTRODUCTION

Like other nucleocapsid protein from the virus group that contains a negative stranded RNA genome, the influenza virus nucleoprotein (NP) is involved in encapsidation of RNA that is important for replication (7). Mutants of influenza virus defective in the NP gene segment have been isolated and their physiologic defects well characterized (10). However, many important questions concerning the mechanism of RNA binding and functional involvement in transcription and viral

Copyright © 1984 by Academic Press, Inc.
All rights of reproduction in any form reserved.
ISBN 0-12-183501-4

maturation remain to be elucidated. Previously, we constructed recombinants between cloned full-length NP DNA and an SV40 vector and showed that the NP product was detected in the nucleus (5) of recombinant DNA-infected cells. Because of the SV40 cytolytic effect, such transient expression limits further analysis of the viral polypeptide. To circumvent this difficulty, we employed the recently developed BPV DNA vector system to obtain persistent expression of NP DNA in mouse cells. Such cells should provide a source of properly modified influenza virus polypeptide for functional analysis. This paper describes the construction of NP-BPV recombinant DNA and isolation of such recombinant DNA transformants that produce influenza virus NP.

RESULTS AND DISCUSSION

(1) Construction of NP-BPV DNA Recombinants. In order to provide transcripion initiation and termination signals for cloned NP DNA in the BPV DNA vector mouse cell system, we employed a recently derived expression vector that contains the promoter sequence of the cloned mouse metallothionein (MT) gene and the poly A-addition signal for SV40 early RNA transcription (4). Briefly, the procedure for construction of NP-BPV recombinant DNA is as follows: The Bam HI cleaved full-length NP DNA was inserted into the Bgl II site located between the MT promoter and the SV40 polyA addition site of cloned vector DNA containing the pML2 moiety (a derivative of pBR322). The recombinant DNA containing the NP DNA inserted in the "sense" transcription orientation was selected and partially digested to open either of the two remaining Bam HI sites: one located within the NP gene and the other located downstream of the SV40 polyA addition sequence. Such partial digest was used for insertion with Bam HI cleaved full-length BPV DNA. The NP-BPV recombinants that contained BPV DNA inserted at the Bam HI site located downstream of the SV40 polyA site were selected. Figure 1 shows the final construct of NP-BPV DNA in which the transcription orientation of the BPV insert is opposite to the transcription direction of the NP gene. Such recombinant DNA was prepared and used for transfection.

Figure 1. Physical map of pNP-BPV recombinant DNA. The sources: NP DNA, 1.57kb, encodes the full length nucleoprotein of influenza virus; SV40, 0.85kb carries the polyadenylation site of early transcription; BPV, 7.95kb, contains a full length of bovine papilloma virus type 1 DNA; pML2, 2.63kb is a derivative of pBR322; MT, 1.90kb, (between the Eco RI and Bgl II sites) contains promoter sequences The transcription direction of mRNA for NP and for the large "T" antigen of BPV is indicated by an arrow.

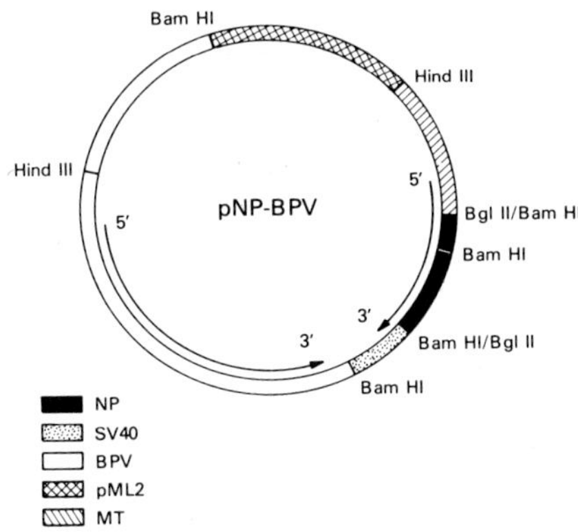

(2) Transformation of Mouse Cells with NP-BPV DNA. Mouse cell line (C127) was used for transfection with NP-BPV recombinant DNA in order to obtain stable cell transformants (2) that would also produce the encoded influenza nucleoprotein. Transformation of such a cell line offered several advantages including the high frequency of transformation and ease of detecting transformed cells that are morphologically distinguishable (1). Several foci of morphologically transformed cells were selected and passaged continuously for further analysis. An immunofluorescence assay was performed to examine the production of influenza nucleoprotein in transformed cells. Cells from one of the transformants showed positive nuclear immunofluorescence in greater than 90% of the population during early passages (Figure 2). The amount of NP product appeared to vary among cells suggesting that these cells were heterogeneous for expression of the plasmid gene. One interpretation is that these transformed cells replicate and harbor different copy numbers of NP-BPV DNA (4). Plasmid DNA extracted from these transformants showed a pattern identical to the original construct after digestion with restriction enzymes Bam HI and Hind III indicating that NP-BPV DNA appeared to be stably maintained. These results suggest that persistent expression of NP was accompanied by alteration of cell phenotype as regards the NP polypeptide import into the nucleus.

Figure 2. Detection of influenza virus nucleoprotein in NP-BPV recombinant DNA transformed mouse cells by indirect immunofluorescence. The NP-BPV recombinant DNA transformed cells were fixed with cold acetone, added with NP specific antibodies and stained with a fluorescein conjugate of goat anti-mouse serum (5).

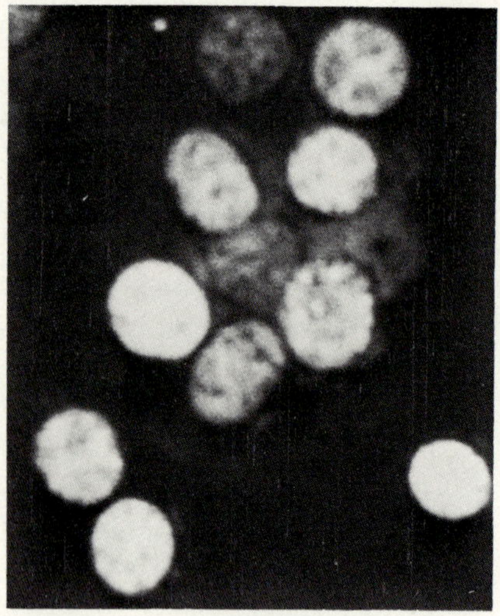

(3) Analysis of Nucleoprotein by Sucrose Gradient Sedimentation. An important functional characteristic of the NP produced during influenza infection is the formation of RNA-nucleoprotein complexes. The persistent synthesis of NP in mouse cells facilitated the analysis of functional properties, such as RNA binding activity of the polypeptide product. Sedimentation analysis in sucrose gradients was performed on lysates prepared from the radio-labelled cells and gradient fractions immunoprecipitated with NP specific antiserum. The sedimentation profile of NP produced during influenza infection in African green monkey kidney cells was used as a control (Figure 3). Labelled NP in the nucleoplasm as well as the cytoplasm sedimented at higher molecular weight fractions consistent with the earlier finding of Krug and Etkind (4) that essentially all NP Product formed complexes with RNA. Similar analysis of the NP product from NP-BPV DNA transformed cells showed a pattern different from that of influenza infection. Most, if not all of labelled NP was detected in low molecular weight fractions, suggesting that the NP product remained in the free form and was not associated with RNAs. Association of nucleocapsid protein with RNA molecules may require recognition of specific terminal sequences, as they are present in all viral RNA segments. This lack of RNA binding was also observed with the N protein of vesicular stomatitis virus (VSV) synthesized from N-SV40 recombinant DNA infected cells (10).

(4) Phosphorylation of Influenza Virus Nucleoprotein. The nucleocapsid protein of influenza virus is phosphorylated and such phosphorylation may be involved in virus maturation, since evidence indicates that in abortively infected HeLa cells the nucleocapsid protein is not phosphorylated (8). It

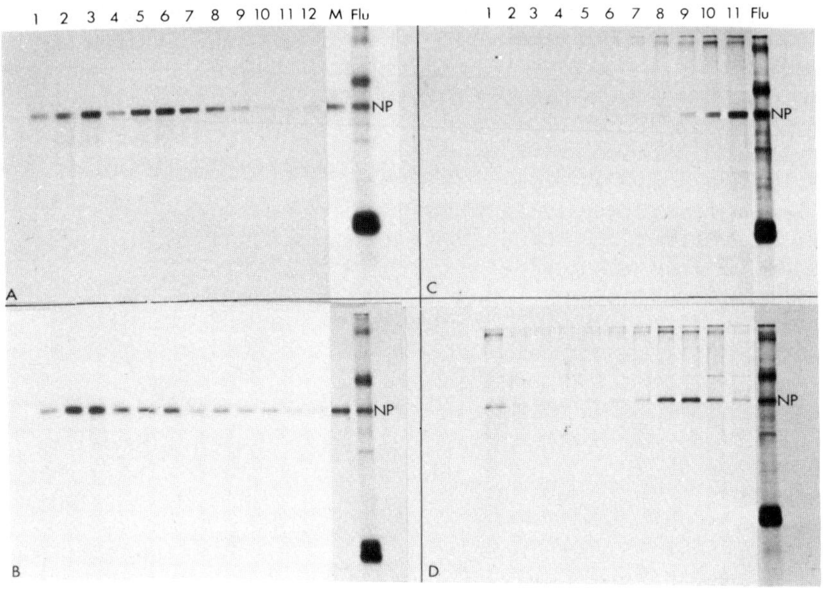

Figure 3. Sucrose gradient sedimentation analysis of influenza nucleoprotein. Recombinant NP-BPV DNA transformed mouse cells and influenza virus-infected AGMK cells were labelled with ^{35}S-methionine. Cell lysates were prepared and separated into cytoplasmic and nuclear fractions according to the method described by Penman (6). Each labelled preparation was sedimented in sucrose gradient (15-30%) and gradient fractions collected for immunoprecipitation of NP. Panels (A) and (B) show the cytoplasmic fraction and the nuclear fraction from influenza virus infected cells. Panels (C) and (D) show the cytoplasmic and nuclear fraction from cells persistently synthesizing NP, respectively.

is not clear whether phosphorylation of NP requires a ribonucleoprotein complex structure that is produced during influenza infection or phosphorylation occurs on NP in the absence of other virus infections. In order to answer these questions, we examined phosphorylation of the NP product isolated from NP-BPV recombinant DNA transformed mouse cells. Radio-labelled lysates from transformed cells and from influenza virus infected AGMK cells were immunoprecipitated with NP specific antibodies and separated on SDS-polyacrylamide gels. As can be seen from figure 4, a radio-phosphate labelled band similar in size to the ^{35}S-labelled NP polypeptide was produced during influenza virus infection. Such modification by phosphorylation did

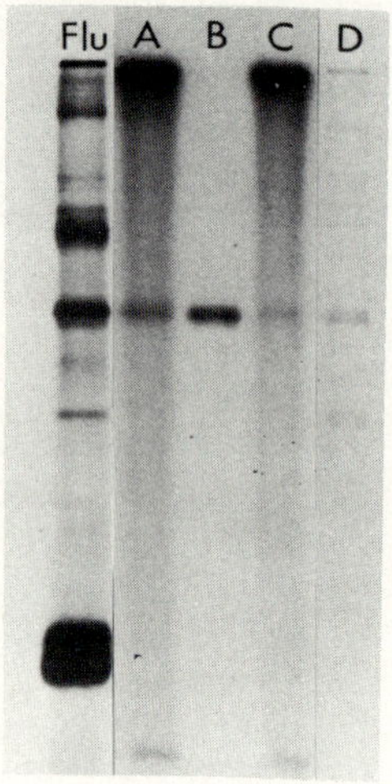

Figure 4. Polyacrylamide gel analysis of phosphorylated nucleoprotein. Radio-labelled lysates were immunoprecipitated with NP specific antibodies and analyzed on SDS-polyacrylamide gels. Lanes A and B show ^{32}P-labelled and ^{35}S-methionine-labelled bands after immunoprecipitation from lysates of influenza virus infected cells. Lanes C and D show ^{32}P-labelled and ^{35}S-methionine labelled NP produced from recombinant DNA transformed cells, respectively. "Flu" indicates the molecular size markers from influenza virus infected cells.

not apparently alter the gel mobility of the NP polypeptide. Similarly, NP-BPV recombinant DNA transformed mouse cells also produced a ^{32}P-labelled band that migrated to the same position as did the ^{35}S-labelled nucleoprotein. This finding indicates that the influenza virus nucleocapsid protein from NP-BPV DNA transformed mouse cells was phosphorylated in the absence of other influenza viral functions.

REFERENCES

1. Dvoretzky, I., Shober, R., Chattopadhyay, S.K., and Lowy, D.R. (1980). Virology 103,369.
2. Graham, F.L., and van der Eb, A.J. (1973). Virology 52,456.
3. Krug, R.M. and Etkind, P.R. (1973). Virology 56,334.
4. Law, M.F., Byrne, J.C. and Howley, P.M. (1983). Mol. Cell. Biol. in press.
5. Lin, B.-C. and Lai, C.-J. (1983). J. Virol. 45, 434.
6. Penman, S. (1966). J. Mol. Biol. 17, 117.
7. Scholtissek, C. (1978). Curr. Top. Microbiol. Immunol. 80, 139.
8. Schrom, M. and Coliguiri, L.A. (1981). ICN-UCLA Symposium on Molecular and Cellular Biology 21, 272.
9. Sprague, J., Condra, J.H., Arnheiter, H., and Lazzarini, R.A. (1983). J. Virol. 45, 773.
10. Thierry, F., Spring, S.B. and Chanock, R.M. (1980). Virology 101, 484.

EXPRESSION OF INFLUENZA VIRUS SUBGENOMIC VIRION RNAs IN INFECTED CELLS [1]

C.R.Penn and B.W.J.Mahy

Division of Virology,
Department of Pathology, University of Cambridge,
Laboratories Block, Addenbrookes Hospital,
Hills Road, Cambridge. CB2 2QQ.

When passaged at high multiplicities of infection influenza virus generates a large proportion of non-infectious virus, some of which interferes with virus replication (1). These preparations of defective interfering (DI) virus contain additional subgenomic virion RNA segments, many of which appear to be derived by internal deletion of RNA segments 1-3 (2, 3). Many of these internally deleted vRNA segments retain an open reading frame (3) and may therefore be expressed in infected cells to give polypeptides structurally related to the major segment product. We have isolated a number of temperature sensitive (ts) mutants of influenza A/FPV/Rostock/34 which induce the synthesis of new small polypeptides in infected cells, and these polypeptides appear to be the product of subgenomic (DI-type) virion RNA segments.

---

[1]This work was supported by Medical Research Council programme grant PG 979/140.

Copyright © 1984 by Academic Press, Inc.
All rights of reproduction in any form reserved.
ISBN 0-12-183501-4

ISOLATION OF TS MUTANTS THAT INDUCE THE SYNTHESIS OF NOVEL POLYPEPTIDES

Six isolates of a ts mutant defective in RNA segment 1 were obtained by recombination between a double mutant of FPV Rostock, ts C47d (defective in segments 1 and 8), and the wild-type virus. Each isolate was the product of a separate recombination experiment. Several of these ts mutant isolates induced the synthesis of additional small polypeptides in infected cells (arrowed in figure 1). One such polypeptide was seen in cells infected with tsA, C and D (designated SpA, SpC and SpD respectively) and two were seen in cells infected with tsE.

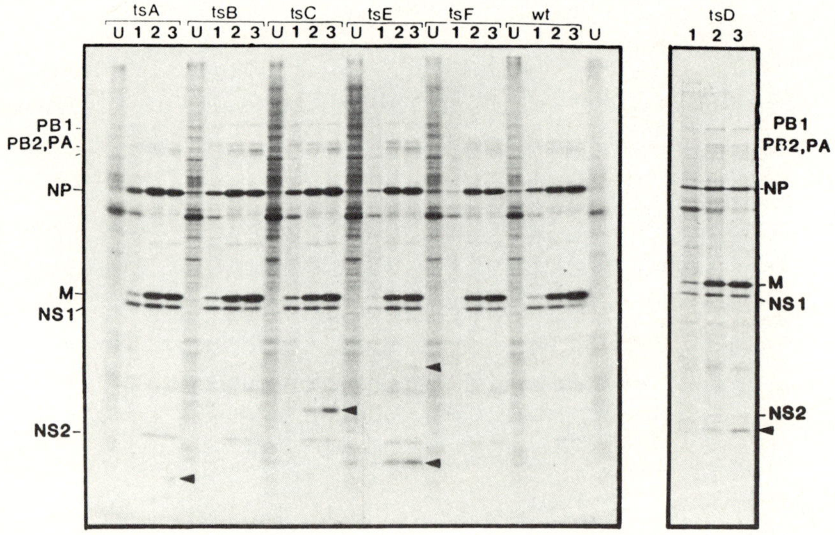

Figure 1. Polypeptides synthesised in infected cells. Chick embryo fibroblast monolayers infected at 34°C with ts mutants as shown were pulse-labelled with ^{35}S-methionine for 30 minutes at 2 (1), 4 (2) or 6 (3) hours after infection. Cell lysates were then analysed by electrophoresis in a 16% polyacrylamide gel.

Although each of these additional polypeptides had a different mobility in SDS polyacrylamide gel electrophoresis, at least three (SpA, SpC, SpD) were structurally related as shown by tryptic peptide mapping (not shown). A comparison of the tryptic peptide maps of these three polypeptides with those of the known virus-specific polypeptides (not shown) indicated that SpA, SpC and SpD were structurally related to either PB2 or PA (encoded by RNA segments 1 and 3).

TRANSLATION IN VITRO OF RNA FROM tsC-INFECTED CELLS

RNA was extracted from CEF cells infected with tsC and translated in the wheat germ cell free translation system (4). Polypeptide SpC was found to be translated *in vitro* in

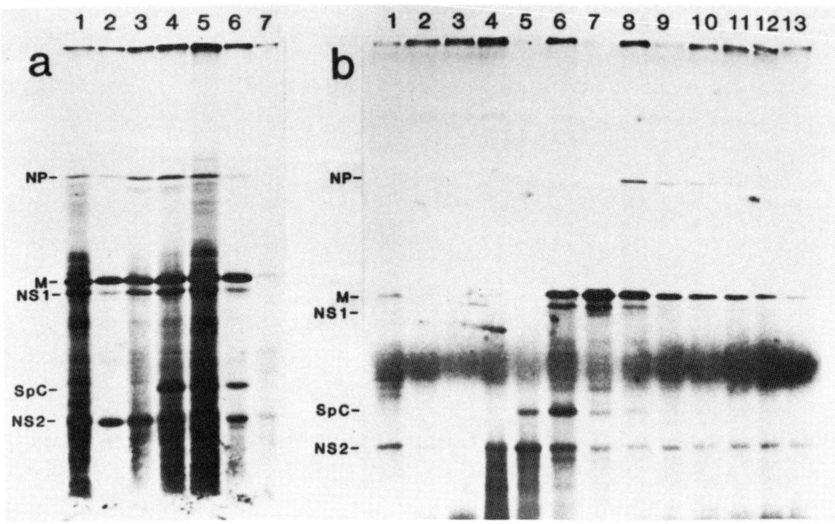

Figure 2. *In vitro* translation of RNA from infected cells. **a:** Reactions contained either no RNA (7), total cell RNA (1,4), polyA- RNA (3,5) or polyA+ RNA (2,6) from wild-type (1,2,3) or tsC infected cells (4,5,6). **b:** total cell RNA from tsC infected cells was fractionated on a 5-20% sucrose gradient and then translated *in vitro*. Fractions are from top (1) to bottom (13).

reactions containing either total cell RNA or RNA selected on oligo-dT cellulose (figure 2a). When the RNA was fractionated by centrifugation in a sucrose gradient, SpC was translated from RNA in fractions that indicated that the SpC mRNA was intermediate in size between those encoding NS1 and NS2 (figure 2b). SpC is therefore encoded by a small polyadenylated mRNA and is not the result of premature termination of transcription or degradation of a larger polypeptide.

IDENTIFICATION OF tsC-SPECIFIC SUBGENOMIC RNAs

mRNAs from infected cells were identified by oligo-dT primed reverse transcription of RNA in the presence of an (α^{32}P)-labelled deoxynucleoside triphosphate, followed by selection of cDNAs by hybridisation to plasmid DNA containing FPV Rostock-specific cDNA inserts (5). Eluted transcripts were analysed by polyacrylamide gel electrophoresis. Virion RNA segments were similarly characterised except that the RNA was labelled with (^{32}P)-pCp and RNA ligase.

When cDNA to RNA from tsC-infected cells was selected with segment 1, 3 and 8-specific DNA , mRNAs corresponding to transcripts of these three segments were identified (figure 3, lanes 4,6,8). Additional subgenomic mRNAs were also selected by the segment 1 and 3-specific DNA, for which corresponding vRNA molecules were found in tsC virions (figure 3, lanes 3-6). This contrasts with the result found for segment 8, where a major subgenomic mRNA was identified (presumed to be that for NS2) for which there was no corresponding vRNA (figure 3, lanes 7 and 8). The most prominent of the segment 1 or 3 specific subgenomic mRNAs is that derived from segment 3, and which is slightly larger than the NS2 mRNA. The size of this RNA, as a candidate for the SpC mRNA, agrees well with the result from gradient fractionation of RNA (figure 2b). Similar segment 1 and 3 specific subgenomic mRNAs were found in tsA-infected cells, and corresponding vRNA molecules were found in tsA virions; no such subgenomic RNAs were found in wild-type virions or in wild-type infected cells (not shown).

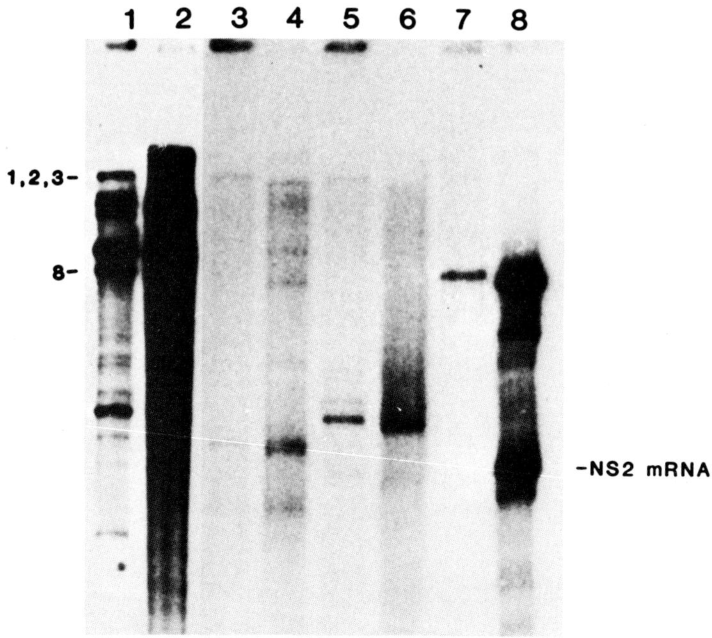

Figure 3. tsC-specific subgenomic RNAs. Virion RNA (1,3,5,7) and reverse transcripts of mRNA (2,4,6,8) selected with segment 1 (3,4), segment 3 (5,6) and segment 8 (7,8) specific DNA. Electrophoresis was in a 7.5% polyacrylamide gel.

CONCLUSION

Isolates of a segment 1 ts mutant of influenza A/FPV/Rostock/34 were found to contain additional subgenomic vRNA segments related to the genomic RNA segments 1 and 3. These vRNAs are transcribed in infected cells to give small polyadenylated mRNAs that in some cases appear to be translated. The function of the resulting polypeptides (if any) is not known, but such polypeptides may be responsible for interference in DI virus preparations as this virus also contains subgenomic vRNAs.

ACKNOWLEDGEMENTS

The plasmids containing FPV/Rostock-specific DNA were plasmids ISI 584, ISI 693 and ISI 86 prepared by I. Roditi (6). We thank Anne Martin for excellent technical assistance.

REFERENCES

1. Nayak,D.P. (1980). Ann. Rev. Microbiol. 34, 619.
2. Davis,A.R. and Nayak,D.P.(1979). Proc. Natl. Acad. U.S.A. 76, 3092.
3. Jennings,P.A., Finch,J.T., Winter,G. and Robertson,J.S. (1983). Cell 34, 619.
4. Inglis,S., McGeoch,D.J. and Mahy,B.W.J. (1977). Virology 78, 522.
5. Inglis,S. and Brown,C. (1981). Nucleic Acids Res. 9, 2727.
6. Roditi,I. (1983). Ph.D. Thesis, University of Cambridge.

TRANSPORT OF VIRAL GLYCOPROTEINS AND ITS MODULATION BY MONENSIN[1]

Firelli V. Alonso-Caplen, Yumiko Matsuoka
and Richard W. Compans

Department of Microbiology
University of Alabama in Birmingham
Birmingham, Alabama 35294

We have obtained evidence (1) that distinct pathways of membrane glycoprotein transport exist in polarized epithelial cell monolayers, differing in their sensitivity to monensin, a sodium ionophore. Monensin blocks the transport of secretory and membrane glycoproteins at the level of the Golgi complex, which appears dilated as an effect of the drug (2). The replication of vesicular stomatitis virus (VSV), which forms at basolateral membranes of Madin-Darby canine kidney (MDCK) cells, was markedly inhibited by monensin, with a block occurring in the transport of the membrane glycoprotein, G, to the cell surface. In contrast, influenza virus maturation, which occurs at the free apical surfaces of MDCK cells, was not affected significantly using similar concentrations of the ionophore. We have recently demonstrated that the influenza hemagglutinin (HA) protein continues to be transported to the surface of monensin-treated cells, is cleaved into HA_1 and HA_2, and is incorporated into budding influenza virus particles (3). On the other hand, VSV G protein appears to accumulate in membranes of the dilated Golgi apparatus of monensin-treated cells. In this report, we have investigated the effects of monensin on the kinetics of transport of viral membrane glycoproteins to the cell surface. We have also studied the effect of the ionophore on the directional transport of viral glycoproteins in MDCK cells doubly-infected with influenza virus and VSV, and on the replication of viruses that are assembled at intracellular membranes.

APPEARANCE OF NEWLY-SYNTHESIZED HEMAGGLUTININ AT THE APICAL SURFACE OF INFLUENZA VIRUS-INFECTED CELLS

We used a surface immune precipitation procedure (4) to examine the effect of monensin on the kinetics of transport of

[1]This work was supported by NIH grants AI 12680 and AI07150, and NSF grant PCM 80-06498.

Copyright © 1984 by Academic Press, Inc.
All rights of reproduction in any form reserved.
ISBN 0-12-183501-4

influenza virus HA protein to cell surfaces. The HA protein was detected on surfaces of untreated or monensin-treated cells infected with influenza virus (A/WSN strain) as early as 20 min after synthesis (Fig. 1). However, there was a dramatic difference in the rate of cleavage of HA since the HA_1 protein, a cleavage product, was present on the surface of monensin-treated cells only after 60 min, whereas the HA_1 protein in controls was detected after a 20-min chase period. In addition, the incorporation of ^{3}H-mannose into HA protein in monensin-treated cells was markedly higher than in controls, suggesting an inefficient trimming of mannose residues during processing of the HA protein in the presence of the ionophore. With another virus strain (A/USSR), HA had a faster electrophoretic mobility in monensin-treated cells (Fig. 2), but was detected at surfaces of control or monensin-treated cells after a 20-min chase. The incorporation of ^{3}H-mannose into HA protein in monensin-treated cells was also significantly higher when compared to untreated cells. These results indicate that the HA protein is transported to the apical surfaces of monensin-treated MDCK cells at a rate similar to that of controls.

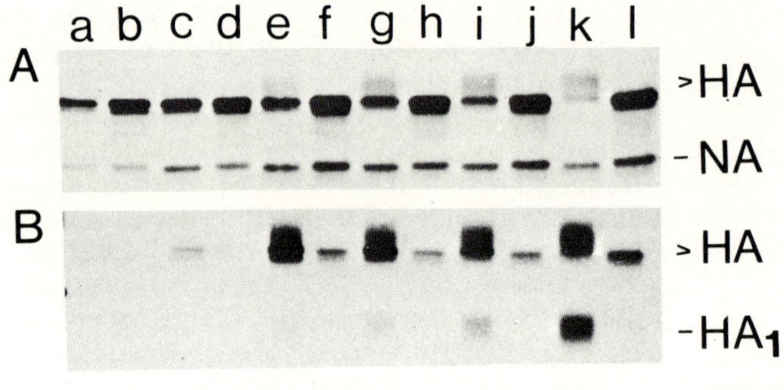

FIGURE 1. Cell surface appearance of influenza virus (A/WSN) HA protein. Monolayers of influenza virus-infected MDCK cells (MOI = 10) were untreated (lanes a, c, e, g, i, and k) or were treated with 10^{-5} M monensin (lanes b, d, f, h, j, and l) after adsorption. Cells were pulsed at 7 h postinfection for 15 min with ^{3}H-mannose (a and b), and chased with Eagle's medium containing 20 mM cold mannose for 10 min (c and d), 20 min (e and f), 30 min (g and h), 40 min (i and j), or 60 min (k and l). Cell surface glycoproteins were analyzed by exposure of intact monolayers to anti-HA monoclonal antibody, followed by lysis and immune precipitation. A. Lysates prior to immune precipitation. B. Immune precipitates.

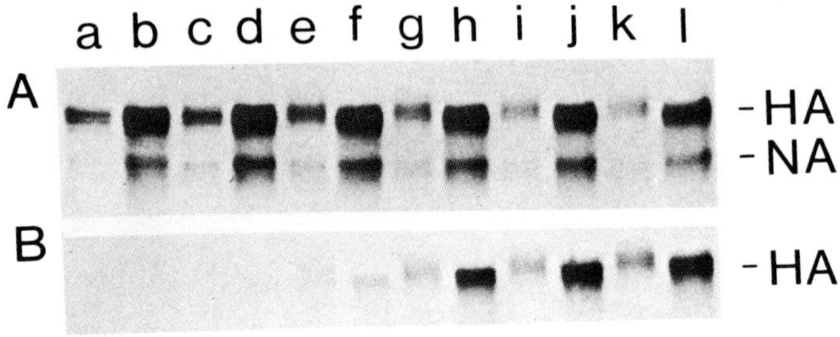

FIGURE 2. Cell surface appearance of influenza virus (A/USSR) HA protein. MDCK cell monolayers infected with influenza virus (MOI = 10) were maintained in the presence or absence of 10^{-5}M monensin. Labelling conditions for each lane correspond to those in Fig. 1.

TRANSPORT OF VIRAL MEMBRANE GLYCOPROTEINS IN DOUBLY-INFECTED CELLS

Previous reports indicated that the directional transport of viral glycoproteins is a characteristic not only of singly- but also of doubly-infected MDCK cells (5). We therefore wished to determine the effect of monensin on transport of HA and G proteins in MDCK cells simultaneously infected with influenza virus and VSV. No significant differences were observed in the levels of HA expression on surfaces of untreated or monensin-treated doubly-infected MDCK cells (Fig. 3A). The level of expression of neuraminidase (NA) protein on surfaces of control or monensin-treated cells was also found to be similar (Fig. 3B). In contrast, the amount of G protein present in the plasma membrane of controls was markedly greater than on surfaces of cells cultured in the presence of the ionophore (Fig. 3C). The level of fluorescence with antibody to G protein in monensin-treated doubly-infected cells was higher than in VSV-infected cells treated with the ionophore; however, similar fluorescence intensities were observed when cells singly infected with influenza virus were stained with anti-G antibody, indicating that the observed fluorescence represents a higher background level.

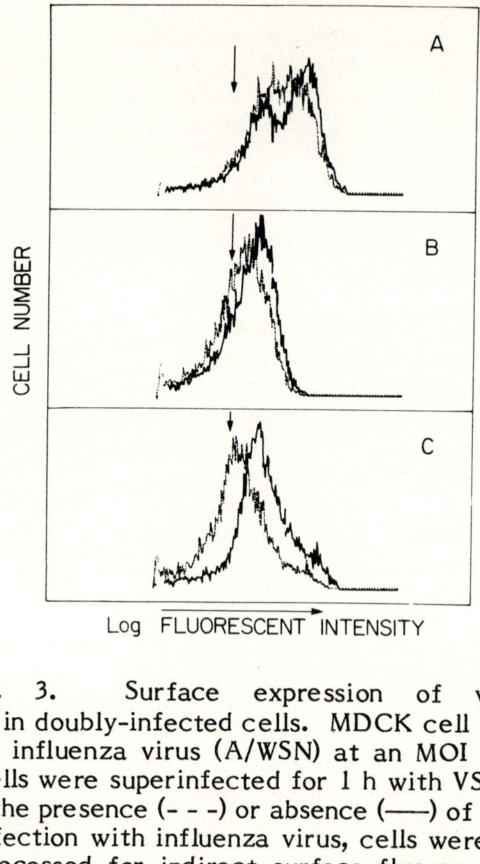

FIGURE 3. Surface expression of viral membrane glycoproteins in doubly-infected cells. MDCK cell monolayers were infected with influenza virus (A/WSN) at an MOI of 2. After 1 h adsorption, cells were superinfected for 1 h with VSV (MOI = 5), and incubated in the presence (- - -) or absence (——) of 10^{-5}M monensin. At 9 h postinfection with influenza virus, cells were suspended with EGTA and processed for indirect surface fluorescence (3) by first adding anti-HA, anti-NA, or anti-G monoclonal antibodies to aliquots in suspension, followed by a fluorescein-conjugated second antibody. Cell surface expression of influenza virus HA protein (A), and NA protein (B), and of VSV G protein (C) is shown. From Alonso-Caplen et al. (6).

We also examined the effect of monensin on infectious virus yields from doubly-infected MDCK monolayers (Table 1). Infectivity titers of influenza virus were not significantly affected even at a 10^{-5}M concentration of monensin. In contrast, yields of VSV showed greater than a 99% decrease at 10^{-6}M or 10^{-5}M monensin when compared to controls. These results indicate that the differential effect of monensin on influenza and VSV replication is similar in singly- and doubly-infected MDCK cells.

TABLE 1.
EFFECT OF MONENSIN ON PRODUCTION OF INFECTIOUS VIRUS PARTICLES IN DOUBLY-INFECTED MDCK CELLS[a]

Monensin concentration	Influenza virus titers (p.f.u./ml)	VSV titers (p.f.u./ml)
Control	3.8×10^7	2.0×10^8
10^{-6}M	2.5×10^7	2.3×10^6
10^{-5}M	1.5×10^7	9.0×10^5

[a]Confluent monolayers of MDCK cells were infected with influenza virus (MOI=10) and after 1 h, with VSV (MOI=10). Virus yields were measured at 24 h postinfection using MDCK cells for influenza virus plaque assays or BHK21 cells for VSV plaque assays.

EFFECT OF MONENSIN ON ENVELOPED VIRUSES THAT FORM AT INTRACELLULAR MEMBRANES

Avian infectious bronchitis virus (IBV) is a coronavirus which matures at the rough endoplasmic reticulum (7) and snowshoe hare virus (SSH) is a bunyavirus reported to assemble at membranes of the Golgi complex (8). We found that release of infectious IBV particles grown in African Green monkey kidney (Vero) cells was inhibited by 90% or more at 10^{-6} or 10^{-5}M monensin (Fig. 4A). Similarly, yields of cell-associated IBV virions were significantly reduced, indicating that monensin is not merely blocking exocytosis of infectious particles but the production of virus particles as well. As shown in Fig. 4B, viral protein synthesis was observed at 10^{-6}M monensin, where infectious virus yields were reduced by over 90% as compared to controls. This suggests that monensin inhibits the assembly and release of infectious IBV particles under conditions where no effect on viral protein synthesis is observed. At 10^{-5}M monensin, there was a significant decrease in viral protein synthesis, but this concentration may be toxic to the host cell. In a recent report (9), it was shown that monensin blocked the release of a murine coronavirus from infected cells. However, virus assembly at the membranes of the rough endoplasmic reticulum was not inhibited by monensin, and there was an apparent accumulation of cell-associated virions. It will be of interest to determine whether the differences in these results may be related to the differences in glycosylation processes of murine and avian coronaviruses (9, 12).

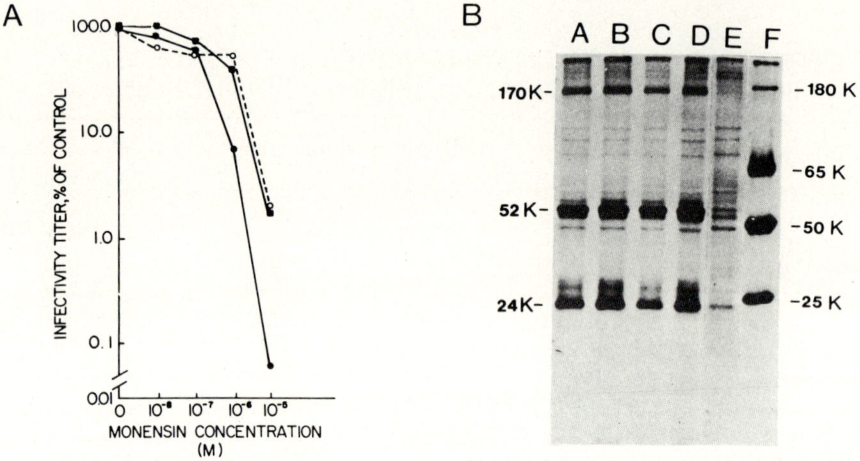

FIGURE 4. A. Effect of monensin on yields of IBV from Vero cells. Cells were infected with IBV (MOI = 0.5) or VSV (MOI = 2), monensin was added post-adsorption, and infectivity titers determined by plaque assays. Yields of both released and cell-associated IBV, obtained after freeze-thawing, were measured. Released IBV titers (●), cell-associated IBV titers (o), VSV infectivity titers (■). The control (100%) titer was 4.5 x 10^5 pfu/ml for released IBV, 1.1 x 10^5 pfu/ml for cell-associated IBV, and 2.7 x 10^9 pfu/ml for VSV. From Alonso-Caplen et al. (12).

B. Viral protein synthesis in monensin-treated Vero cells. At 19 h postinfection, cells were pulse-labeled for 1 h with ^{3}H-leucine. IBV-infected, untreated (A); 10^{-8} M (B), 10^{-7} M (C), 10^{-6} M (D), and 10^{-5} M monensin (E); lane F, VSV viral proteins used as molecular weight standards. From Alonso-Caplen et al. (12).

When yields of released and cell-associated SSH virions were examined (Fig. 5A), it was found that monensin did not block virus assembly or the release of SSH particles into culture supernatants. Polyacrylamide gel electrophoretic analysis of SSH-infected BHK21 cells treated with the ionophore (Fig. 5B) revealed the synthesis of all viral proteins. In contrast, other investigators (10, 11) have reported that monensin inhibits the replication of LaCrosse (LAC) virus, another bunyavirus. Infectious virus yields were greatly reduced but the synthesis of viral proteins was not affected (10); in addition, there was an accumulation in monensin-treated cells of an incompletely processed form of G1, one of the two envelope glycoproteins of LAC (11). The difference between these results may be related to the slower replication cycle of LAC in BHK21 cells as compared to SSH.

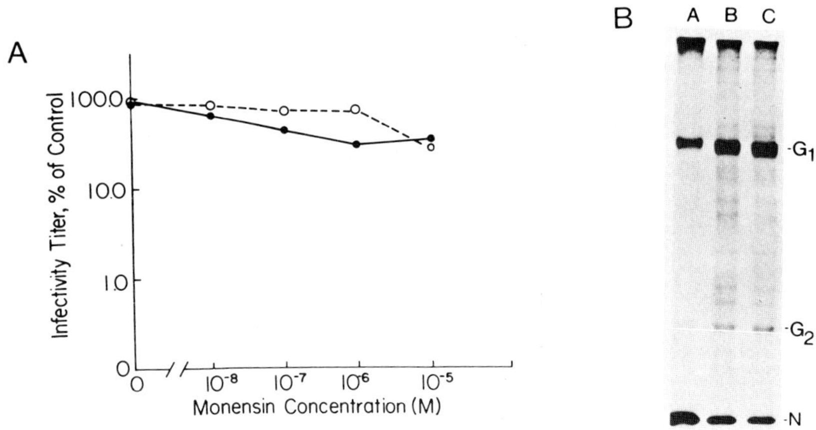

FIGURE 5. A. Yields of SSH from monensin-treated BHK21 cells. Monensin was added to cells after infection with SSH (MOI = 10). Yields of both released and cell-associated virions, obtained by exposure of intact cells to neutralizing antibody followed by one cycle of freeze-thawing, were measured at 24 h postinfection. Released SSH titers (o), intracellular SSH titers (•). The control (100%) titer was 7.6×10^6 pfu/ml for released SSH, and 6.0×10^4 pfu/ml for cell-associated SSH.

B. Viral protein synthesis in monensin-treated BHK21 cells. SSH-infected BHK21 cells were pulse-labeled with ^{3}H-leucine for 1 h at 16 h postinfection. Purified SSH virions (A); SSH-infected, untreated cells (B); SSH-infected cells treated with 10^{-5} M monensin (C).

CONCLUSIONS

Our results indicate that systems for the intracellular transport of viral membrane glycoproteins in MDCK cell monolayers differ in sensitivity to monensin. Influenza virus glycoproteins, HA and NA, are transported to the free apical surfaces via a monensin-resistant pathway, whereas the targeting of VSV G protein to the basolateral membrane domain involves a monensin-sensitive route. These distinct transport pathways are functional not only in singly-infected but also in doubly-infected MDCK monolayers. Our findings suggest that sorting of viral glycoproteins occurs at the stage when they exit from the Golgi apparatus enroute to the plasma membrane, since monensin is thought to block glycoprotein transport at the level of the Golgi complex. We have also observed a differential effect of monensin on enveloped viruses which assemble at intracellular membranes. The replication of IBV and its release from the membranes of the rough endoplasmic reticulum are inhibited, whereas monensin treatment of host cells has no effect on the formation of SSH virions. Possibly, the actual budding process of IBV may be more easily affected than that of SSH by changes in ionic gradients within the host cell that may occur as an effect of monensin treatment.

REFERENCES

1. Alonso, F. V., and Compans, R. W. (1981). J. Cell Biol. 89, 700.
2. Tartakoff, A. M. (1983). Cell 32, 1026.
3. Alonso-Caplen, F. V., and Compans, R. W. (1983). J. Cell Biol. 97, 659.
4. Herrler, G., Alonso-Caplen, F. V., and Compans, R. W. Manuscript in preparation.
5. Roth, M. G., and Compans, R. W. (1981). J. Virol. 40, 848.
6. Alonso-Caplen, F. V., Kearney, J. F., and Compans, R. W. Manuscript in preparation.
7. Robb, J. A., and Bond, C. W. (1979). In "Comprehensive Virology (H. Fraenkel-Conrat and R. R. Wagner, eds.), p. 193. Plenum Press, New York.
8. Obijeski, J. F., and Murphy, F. (1977). J. Gen. Virol. 37, 1.
9. Niemann, H., Boscheck, B., Evans, D. Rosing, M., Tamura, T., and Klenk, H.-D. (1982). EMBO J. 1, 1499.
10. Cash, P. (1982). J. Gen. Virol. 59, 193.
11. Madoff, D. H., and Lenard, J. (1982). Cell 28, 821.
12. Alonso-Caplen, F. V., Matsuoka, Y., Wilcox, G. E., and Compans, R. W. Submitted for publication.

INTERFERING WITH GLYCOPROTEIN PROCESSING IN INFLUENZA AND SINDBIS VIRUS-INFECTED CELLS

Pedro A. Romero, Roelf Datema,[1] and Ralph T. Schwarz

Institut für Virologie
Justus-Liebig-Universität Giessen
Giessen, FRG

Glycoproteins are glycosylated initially by an oligosaccharide, $Glc_3Man_9(GlcNAc)_2$, donated by a lipid-carrier (dolichol diphosphate). The oligosaccharide is assembled on this lipid-carrier by stepwise additions of sugar residues (the dolichol cycle). After the transfer of oligosaccharide to protein, it is processed by concerted action of glycosidases and glycosyl transferases to mature forms, and these processing steps generate diversity of oligosaccharide side-chains (1). The first step in the processing-sequence is the removal of the outermost glucose residue from the protein-linked oligosaccharide. Removal of the glucose residues and of some mannose residues gives rise to so-called high-mannose oligosaccharides. They can be endproducts. Further trimming, plus readdition of other sugar residues (GlcNAc, Gal, NeuAc, Fuc) gives rise to different kinds of so-called complex oligosaccharides.

The glucosidase removing the outermost glucose residue (glucosidase I) can be inhibited in cell-free assays by N-methyldeoxynojirimycin (2). In addition, this glucose analog inhibits the very same reaction in intact cells and does not affect the dolichol cycle (2). Therefore, proteins are fully glycosylated; however, the inhibition of glucose-trimming prevents further processing to the mature high-mannose and complex oligosaccharides. This has been observed in mammalian, avian and insect cells. The viral glycoproteins synthesized in these cells are therefore equipped with immature oligosaccharides. This may or may not inhibit the release of intact virus particles dependent on cell-type and virus (see below). We observed that mannose-trimming can

[1]Present address: Department of Antiviral Chemotherapy, Astra Läkemedel AB, Södertälje, Sweden.

Copyright © 1984 by Academic Press, Inc.
All rights of reproduction in any form reserved.
ISBN 0-12-183501-4

occur to some extent in absence of glucose-trimming because the composition of the protein-linked oligosaccharide is $Glc_3Man_x(GlcNAc)_2$, with x = 7, 8, 9, the tetradecasaccharide being a precursor of the trideca- and dodeca-saccharide (2).

N-methyldeoxynojirimycin, at concentrations blocking oligosaccharide processing, does not inhibit viral protein synthesis in the infected cells. Furthermore, it does not inhibit exponentially growing human embryo-cells and may, therefore, not be very toxic. Nevertheless, release of sindbis virus from infected chick-embryo or BHK cells is significantly inhibited. Fowl-plaque virus formation in chick-embryo cells is not inhibited by N-methyldeoxynojirimycin and fully-infectious virus particles containing glycoproteins with unprocessed oligosaccharides are formed. The hemagglutinin is cleaved normally into HA_1 and HA_2, although the cleavage rate may be decreased because uncleaved HA accumulates (3).

In sindbis virus-infected chick-embryo cells treated with N-methyldeoxynojirimycin the polyglycoprotein-precursor is proteolytically cleaved normally into the E_1 and E_2 proteins. In sindbis virus-infected BHK cells treated with the same drug, the cleavage of pE_2 to E_2 is prevented. So, this effect of N-methyldeoxynojirimycin (of inhibiting glucose-trimming) is cell-specific because also in sindbis-infected cells treated with bromoconduritol (an inhibitor of removal of the innermost glucose residue and of further oligosaccharide processing (3)) cleavage of pE_2 to E_2 is inhibited in BHK cells and not in chick-embryo cells.

To investigate whether release of sindbis virus is generally inhibited upon inhibition of formation of complex oligosaccharides, sindbis virus-infected insect cells were used. As suggested by other studies (4, 5), insect cells may be unable to make the complex-type of oligosaccharides because some glycosyl-transferases are lacking. Indeed, sindbis virus glycoprotein obtained from infected insect-cells (*Aedes albopictus*) have mainly, if not exclusively, high-mannose-type oligosaccharides. They differ however from the high-mannose oligosaccharides obtained from chick-embryo or BHK-grown sindbis glycoproteins because the "sindbis-oligosaccharides" from insect cells have a higher proportion of glucosylated high-mannose oligosaccharides. Thus, the sindbis virus-infected chick-embryo cells yield infectious virus with no complex-type oligosaccharides, whereas inhibition of oligosaccharide processing in avian or mammalian cells prevents virus release. Apparently, virus release in insect cells has different demands on the oligosaccharide structure. Indeed, N-methyldeoxynojirimycin does not prevent release of infectious sindbis virus from insect cells although the glucosidase I is inhibited. Furthermore, and maybe related to this, the mode of virus release seems to be different in insect and vertebrate cells (6).

ACKNOWLEDGEMENTS

Our work was supported by the Deutsche Forschungsgemeinschaft (Sonderforschungsbereich 47 and a Heisenberg Stipendium to RTS), the Stiftung Volkswagenwerk, the Justus-Liebig-University, the Alexander von Humboldt Stiftung (to PAR) and the Zentrum für International Migration und Entwicklung (to PAR). The technical assistance of Kirstin and Ulrike and the encouragement from Drs. R. Rott and C. Scholtissek are highly appreciated.

REFERENCES

1. Kornfeld, S. (1982). In "The Glycoconjugates" (M. I. Horowitz, ed.), Vol. III, p. 3. Academic Press, New York.
2. Romero, P. A., Datema, R., and Schwarz, R. T. (1983). Virology, in press.
3. Datema, R., Romero, P. A., Legler, G., and Schwarz, R. T. (1982). Proc. Natl. Acad. Sci. USA 79, 6787.
4. Butters, T. D., Hughes, R. C., and Vischer, P. (1981). Biochim. Prophys. Acta 640, 672.
5. Butters, T. D., and Hughes, R. C. (1981). Biochim. Biophys. Acta 640, 655.
6. Gliedman, J. B., Smith, J. F., and Brown, D. T. (1975). J. Virol. 16, 913.

Viral Proteins: Antigenic and Functional Analyses

ANTIGENIC AND STRUCTURAL STUDIES ON THE GLYCOPROTEINS OF LYMPHOCYTIC CHORIOMENINGITIS VIRUS[1]

Michael J. Buchmeier

Department of Immunology
Scripps Clinic and Research Foundation
La Jolla, California 92037

Lymphocytic choriomeningitis virus (LCMV) infection of the mouse presents biologists with a superb experimental model for host-virus interaction. By manipulating the age at infection and the genetic background of mice as well as the strain of infecting virus, one can reproducibly elicit diseases ranging from lifelong persistent infection to rapidly lethal acute choriomeningitis (1). Moreover, recent work (2) has demonstrated that LCMV may also be responsible for endocrine disease characterized by runting and hypoglycemia.

In addition to being an important experimental model, LCMV is the prototype virus of the arenavirus group (3). LCMV, and the serologically related African arenaviruses Lassa, Mozambique, and a newly described arenavirus of the Central African Republic (4) form a group informally known as the "Old World" arenaviruses. One of these agents, Lassa virus, is the cause of a severe and often fatal acute disease in man for which there is neither adequate means of prevention nor treatment.

We have directed our studies of the immunobiology and biochemistry of LCMV toward two goals: 1) development of molecular probes (Southern and Oldstone, this volume) and immunologic reagents (5) specific for LCMV to use in detailed studies of viral pathogenesis, and 2) a detailed antigenic study of LCMV and comparison with the pathogenic arenaviruses in order to develop rational strategies for immunization

[1]This work was supported by NIH grants AI-16102, NS-12428, and AI-09484 and by USAMRIID Contract C-3013. The findings in this report are not to be construed as an official Department of the Army position unless so designated by other authorized documents. The Author is an Established Investigator of the American Heart Association.

Copyright © 1984 by Academic Press, Inc.
All rights of reproduction in any form reserved.
ISBN 0-12-183501-4

against these agents. We have developed a library of monoclonal antibodies against the major structural proteins of LCMV; NP, GP-1 and GP-2 (5, 6) and this report describes the properties and some of the uses of these reagents.

MONOCLONAL ANTIBODIES TO LCMV GLYCOPROTEINS

We have raised a library of monoclonal antibodies to the LCMV structural proteins (5) including more than 20 directed against the glycoproteins. Reactions of these antibodies against viral polypeptides have been assessed by immunoprecipitation and Western immunoblotting (7, Fig. 1) and on the basis of these analyses we have identified 6 anti GP-1 and 6 anti GP-2 specific antibodies. The remaining antibodies reacted with virus specific cell surface antigens by immunofluorescence and most both immunoprecipitated GP-1 and GP-2, presumably as a complex (5), but did not react with denatured viral polypeptides in Western blotting.

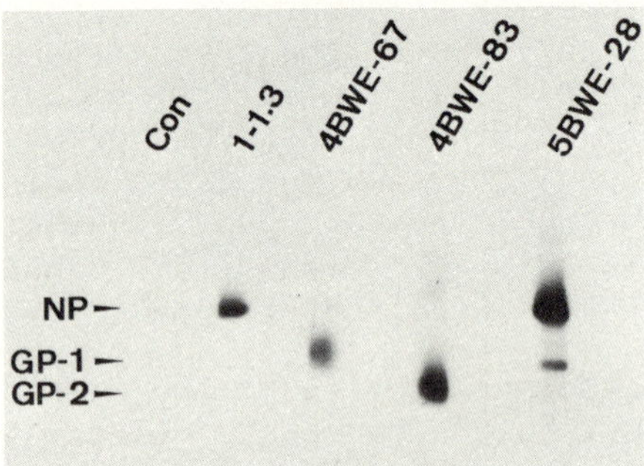

FIGURE 1. Western blotting of LCMV polypeptides by monoclonal antibodies. LCMV (10 µg/lane) was electrophoresed on a 10.5% gel and then separated proteins were transferred electrophoretically to a nitrocellulose sheet. Monoclonal ascites diluted 1/100 in 1% normal goat serum was reacted with the strips and bound immunoglobulin detected by incubation with ^{125}I labeled goat antimouse IgG. Control antibody was an unrelated monoclonal. Reactions are evident against NP, GP-1 and GP-2 as well as degradation products of NP in the last lane.

The virus specificity of this panel of antibodies was determined by immunofluorescence and ELISA assays against 6 commonly used laboratory strains of LCMV (CA1371, WE, UBC, Traub, Pasteur Inst. and 325a) as well as Lassa and Mozambique viruses kindly supplied by Dr. P. Jahrling (USAMRIID, Ft. Detrick). Three patterns of reactivity have emerged which are summarized in Table 1. One group of 9 antibodies, including 4 of confirmed anti GP-1 reactivity, bound to all of the 6 laboratory strains of LCMV but did not react with the African arenaviruses. These 9 antibodies defined framework antigens of LCMV which are highly conserved among strains. A second group of 8 antibodies reacted in a strain dependent fashion with specific viruses in the LCMV panel. That is, these reacted with one or more of the 6 LCMV strains, but none reacted with all strains. These antibodies defined type specific antigens among LCMV strains which have not previously been recognized. The third group consists of 6 anti-GP-2 antibodies. All of these 6 monoclonals reacted in immunofluorescence with all of the LCMV strains examined, but unlike the two groups described above, 5 of the 6 also reacted with either Lassa or Mozambique viruses or both substantiating our earlier finding of conservation of GP-2 among these viruses (8). Three of these 5 antibodies (4BWE-33.6, 4BWE-83.5 and 4BWE-57.6) also reacted strongly in ELISA assays against purified Pichinde virus suggesting for the first time that a portion of GP-2 may also be conserved between the New and Old World arenaviruses.

Three monoclonals among our panel show strong virus neutralizing activity in the absence of complement. In assays against the Armstrong (CA1371) and WE strains of LCMV (Table 2), one of the three antibodies (clone 2WE-6.2) neutralized both strains of virus (corresponding to previous results by ELISA and immunofluorescence assays) indicating a framework reactivity. Two other antibodies, 2-11.10 and 4BWE-40 selectively reacted with the CA1371 and WE strains respectively. These monoclonals define regions of the GP-1 molecule which are targets for virus neutralization. We are currently determining by competitive binding assay the number and relationships of functional antibody binding sites (epitopes) on the LCMV glycoproteins. It will be of interest to determine whether antibodies showing virus strain specificity recognize different regions on the viral glycoproteins or whether the differences in specificity reflect heterogeneity within epitopes.

TABLE 1. SUMMARY OF HYBRIDOMA CROSS REACTIVITY AMONG OLD WORLD ARENAVIRUSES

Hybridoma Antibody	Polypeptide Specificity	Reaction with LCMV Strain Arm CA1371	WE	UBC	TRAUB	PASTEUR	325a	African Arenaviruses Lassa	Mozambique
2WE-2	GP-1	+	+		+	+	+	-	-
2WE-6.2	GP-1	+	+	+	+	+	+	-	-
4BWE-67	GP-1	+	+	+	+	+	+	-	-
4AWE-18	GP-1	+	+	+	+	+	+	-	-
1AWE-7.5	GP	+	+		+	+	+	-	-
4AWE-103	GP	+	+	+	+	+	+	-	-
5AWE-2	GP	+	+		+	+	+	-	-
23A-6.15	GP	+	+		+	+	+	-	-
26-20.8	GP	+	+		+	+	+	-	-
2-11.10	GP-1	+	-	-	-	-	-	-	-
5BWE-24	GP	+	+	+	+	-	+	-	-
4BWE-36	GP	+	+	-	+	+	-	-	-
4BWE-60	GP	+	+	-	+	+	-	-	-
4AWE-40	GP-1	-	+	-	-	+	-	-	-
1-16.3	GP	+	-	-	-	+	-	-	-
2-8.1	GP	+	-	-	-	-	-	-	-
1AWE-1.2	GP	-	+	-	-	-	+	-	-
4BWE-28	GP-2	+	+	+	+	+	+	-	-
4BWE-83.6	GP-2	+	+	+	+	+	+	+	+
4BWE-33.6	GP-2	+	+	+	+	+	+	+	+
4BWE-57.6	GP-2	+	+	+	+	+	+	+	+
4BWE-11.4	GP-2	+	+	+	+	+	+	-	+
9-7.9	GP-2	+	+	+	+	+	+	-	+

TABLE 2
SUMMARY OF NEUTRALIZATION OF LCMV STRAINS
BY MONOCLONAL ANTIBODIES

Monoclonal Antibody	Specificity	PRD_{50} Titer vs. LCMV[a] Armstrong (CA1371-83.1)	WE
2WE-6.2	GP-1	525	1778
2-11.10	GP-1	> 10,000	< 20
4AWE-40	GP-1	< 20	562
9-7.9	GP-2	< 50	< 50

[a]Dilution of antibody giving 50% reduction of 100 pfu of virus.

Additional evidence of the presence of conserved framework and virus specific regions in the GP-1 glycoprotein has come from structural studies. GP-1 and GP-2 glycoproteins of each of 6 LCMV strains were analysed by tryptic peptide mapping. As shown in Figure 2 common peptides were observed in the GP-1 of each of the six strains (Fig. 2, lower left side of each panel). In addition, type specific tryptic peptides unique to each virus were found. Analysis of the GP2 glycoproteins of these viruses did not show similar variation.

It is clear from the results presented that LCMV is not a single virus. Rather, using monoclonal antibodies we have shown a degree of antigenic variation among commonly used laboratory strains of LCMV which was not previously appreciated. Similar antigenic variation has been observed among isolates of the African arenaviruses (4) and it remains to be seen whether this variation is an important factor in naturally occurring disease.

EXPRESSION OF LCMV GLYCOPROTEINS DURING INFECTION

Maintenance of LCM virus in nature depends on the ability of the virus to persistently infect the mouse (mus musculis). Natural infection is silent; infected mice behave and reproduce within normal limits (9). In the laboratory, persistent infection is initiated by neonatal inoculation of mice. Mice thus infected were initially thought to be immunologically tolerant to the virus, but we now know that LCMV carrier mice develop immune complex glomerulonephritis due to deposition of antigen-antibody complexes in the renal glomeruli (10). Subsequently, we found that LCMV carrier mice make circulating IgG antibody against the structural proteins of the virion (11).

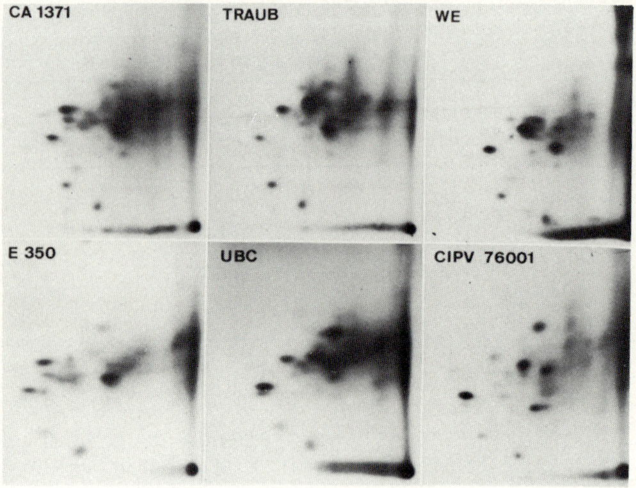

FIGURE 2. Tryptic peptide maps of the GP-1 glycoproteins of 6 strains of LCM virus. Purified virions were disrupted with 1% NP-40, labeled to high specific activity with ^{125}Iodine, then the polypeptides were separated by SDS-polyacrylamide gel electrophoresis (8). Isolated GP-1 bands were excised and digested with TPCK-trypsin. Maps were then prepared by two dimensional electrophoresis (right to left) followed by ascending chromatography. Common peptides shared among all of the virus isolates form an arc on the lower left hand side of each panel while virus specific peptides are predominantly clustered toward the upper right.

Persistence of LCMV in the presence of an active antibody response may be due in part to the ability of the virus to modulate the expression of glycoprotein and budding virions at the cell surface while maintaining nucleocapsid protein synthesis within the cell (12). In collaboration with Dr. Michael Oldstone we have studied the expression of NP and GP antigens in the brains of mice infected at birth using monoclonal antibodies to these polypeptides (13). Mice were infected at birth and sacrificed on days 5, 15, and 84 for study. Cryostat sections of brains were stained with monoclonal antibody to LCMV NP (clone 1-1.3) and with a mixture of antibodies to LCMV GP-1 (2WE-6.2; 2-11.10; 26-20.8) and GP-2 (9-7.9). Sections were scored for the presence of detectable viral antigen in neurons at these intervals, and the results are summarized in Table 3. With increasing time after persistent infection, the number of mice expressing viral GP diminished in spite of the continuous presence of viral NP within neurons. Thus, the antigenic target of cells in the CNS expressing LCMV glycoprotein progressively diminished during the course of persistent infection. Electron microscopic study by immuno-peroxidase staining revealed NP antigen intimately associated with polyribosomes in these neurons (14). It remains to be determined whether this association of NP with ribosomes results in functional deficits in these neurons similar to those reported in LCMV infected neuroblastoma cells *in vitro* (15). It is clear from these studies that persistent infection by LCMV is a complex process involving host response to virus and the adaptations that the virus has evolved to counter that response. Precise understanding of this complex interaction will require detailed analysis of the infection using the most specific molecular and immunologic probes available.

TABLE 3
DISSOCIATED EXPRESSION OF LCMV GP AND NP DURING PERSISTENT INFECTION *IN VIVO*

Neonatally Infected Mouse Strain	Expression of Viral Antigens in Neurons[a]					
	day 5[b]		day 15		day 84	
	GP	NP	GP	NP	GP	NP
C_3H/St	5/5[(c)]	5/5	9/15	15/15	1/8	8/8
BALB/W	5/5	5/5	nd	nd	1/20	20/20
SWR/J	5/5	5/5	nd	nd	3/25	25/25

[a]Four micron cryostat sections through the anterior third of the temporal lobe were fixed in ether:alcohol and stained with the appropriate monoclonal antibodies. More than 100 individual neurons in each section were studied.

[b]Mice were infected intracerebrally with 60 pfu of LCMV strain CA 1371 at birth then sacrificed for study on the days indicated.

[c]Number of mice studied that contained one or more cells expressing a specific polypeptide/total number of mice examined.

ACKNOWLEDGMENTS

This is Publication No. 3201-IMM from the Department of Immunology, Scripps Clinic and Research Foundation, La Jolla, California 92037. I thank Hanna Lewicki and Paul Blount for excellent technical help, Lisa Flores and Gay Wilkins for manuscript preparation and Michael B. A. Oldstone for stimulating discussions.

REFERENCES

1. Buchmeier, M. J., Welsh, R. M., Dutko, F. J., and Oldstone, M. B. A. (1980). Adv. Immunol. 30, 275.
2. Oldstone, M. B. A., Sinha, Y., Blount, P., Tishon, A., Rodriguez, M., von Wedel, R., and Lampert, P. (1982). Science 218, 1125.
3. Rowe, W., Pugh, W., Webb, P. and Peters, C. (1970). J. Virol. 5, 289.
4. Gonzalez, J., McCormick, J., Saluzzo, J., Herve, J., Georges, A. and Johnson, K. (1983). Intervirology 19, 105.
5. Buchmeier, M. J., Lewicki, H. A., Tomori, O., and Oldstone, M. B. A. (1981). Virology, 113, 73.
6. Buchmeier, M. and Oldstone, M. (1979). Virology 99, 111.
7. Renert, J., Reiser, J. and Stark, G. (1979). Proc. Natl. Acad. Sci. USA 76, 3116.
8. Buchmeier, M., Lewicki, H., Tomori, O. and Johnson, K. (1980). Nature 288, 486.
9. Lehmann-Grube, F. (1971). In "Lymphocytic Choriomeningitis Virus. Virology Monographs. Vol. 10" Springer-Verlag, Vienna.
10. Oldstone, M. B. A., and Dixon, F. J. (1969). J. Exp. Med., 129, 483.
11. Buchmeier, M. J., and Oldstone, M. B. A. (1978). J. Immunol., 120, 1297.
12. Welsh, R. M., and Oldstone, M. B. A. (1977). J. Exp. Med., 145, 1449.
13. Oldstone, M. B. A., and Buchmeier, M. J. (1982). Nature, (in press).
14. Rodriguez, M., Buchmeier, M., Oldstone, M. and Lampert, P. (1983). Am. J. Pathol. 110, 95.
15. Oldstone, M. B. A., Holmstoen, J., and Welsh, R. M. (1977). J. Cell. Physiol., 91, 459.

COMPARATIVE ANALYSIS OF LASSA AND LASSA-LIKE ARENAVIRUS ISOLATES FROM AFRICA.

J.P.GONZALEZ[1,3], M.J.BUCHMEIER[2],
J.B.McCORMICK[1], S.W.MITCHELL[1],
L.H.ELLIOTT[1], M.P.KILEY[1]

[1]Special Pathogens Branch, Division of Viral Disease, Centers for Disease Control, Atlanta, GA 30333
[2]Department of Immunopathology, Scripps Clinic and Research Foundation, La Jolla, Cal 92037

Lassa Fever is a hemorrhagic disease which causes significant morbidity and mortality in Western Africa. The causative agent, Lassa Virus, was first isolated in 1969 from humans (1), in 1972 from a rodent Mastomys (2) in Sierra Leone and subsequently in other african countries (3). The virus isolated by Buckley and colleagues was identified as an arenavirus by electron microscopy (4). Serosurveys have indicated a prevalence of anti-Lassa antibody in various human populations in West and Central Africa (3). In addition anti-Lassa antibodies were also detected in wild rodents in the Central African Republic (CAR) (5). Consequently, virus isolation from these hosts has been attempted at several locations in Central and Southern Africa. To date, Lassa-like viruses have been isolated in three different geographical areas: In Mozambique (6) and Zimbabwe (7) from Mastomys natalensis and in CAR (5) from Praomys sp. and Mastomys sp..

We have established a program to investigate and differentiate African arenaviruses. We are currently investigating their biology, chemistry, ecology, immunology and pathology. We have previously reported that infection of the Rhesus monkey with Mozambique virus protects the monkey from subsequent infection with virulent Lassa Virus (8). We previously examined the chemical composition of LAS and MOZ virus to determine differences which may affect pathogenicity (9). We also demonstrated the immune precipitation of Mozambique virion proteins by monoclones produced against lymphocitic choriomeningitis virus (LCM). We have now expanded our comparison to include an arenavirus

[3]Guest researcher from the Office de la Recherche Scientifique et Technique Outre Mer (Paris, France)

SEGMENTED NEGATIVE STRAND VIRUSES

Copyright © 1984 by Academic Press, Inc.
All rights of reproduction in any form reserved.
ISBN 0-12-183501-4

recently isolated in the CAR.

Here we present some of the physicochemical and antigenic characteristics of the three viruses including a comparison of virus protein and RNA. We also demonstrate their differentiation by use of monoclonal antibodies produced against LAS and MOZ viruses (5).

VIRUS STRAINS

The Sierra Leone strain of Lassa virus (LAS) has been described previously (9). Because of the convention of naming virus strains after local geographic landmarks we have provisionally designated the virus from Mozambique (6) as Mopeia (MOP) virus. The CAR virus has previously been designated as Mobala (MOB) virus, from the local name of the rodent host (10). Each virus strain was plaque purified 3 times in Vero E6 cells before use.

VIRION PROTEINS

Polyacrylamide gel analysis of virion proteins of the three viruses is presented in Fig.1. The major protein species present in all three viruses is the nucleoprotein (N) (Fig.1a). For each virus the two virion glycoproteins G1 and G2, appear as broad bands beneath the N protein band. The amount of G2 protein always exceeded that of G1 protein and the latter was often difficult to detect. A large protein is often seen in preparations of purified virus but its origin and function are unknown.

From the electrophoregram in Fig.1a the N protein of LAS virus appears to migrate slightly faster than the N protein of MOP, and possibly MOB viruses. This finding was quite reproducible and was confirmed by the experiment depicted in Fig.1b where ^{35}S-methionine labelled LAS and MOP virion protein were electrophoresed either separately or together.

When compared to the series of molecular weight standards the LAS virion proteins have molecular weights of 180,000 (L), 61,000 (N), 45,000 (G1) and 38,000 (G2). The corresponding values for MOP virus are 180,000 (L), 61,000 (N), 48,000 (G1) and 35,000 (G2) while those for MOB are 180,000 (L), 60,000 (N), 48,000 (G1) and 37,000 (G2). The band breadth does not allow accurate comparison of size differences among corresponding glycoproteins of the different viruses.

The molecular weight values reported here are somewhat smaller than those previously reported (9). This is due to a difference in the gel employed. The molecular weight for LAS which we report here are in good agreement with results recently published by Clegg and Lloyd who also employed the Laemmli gel system (11).

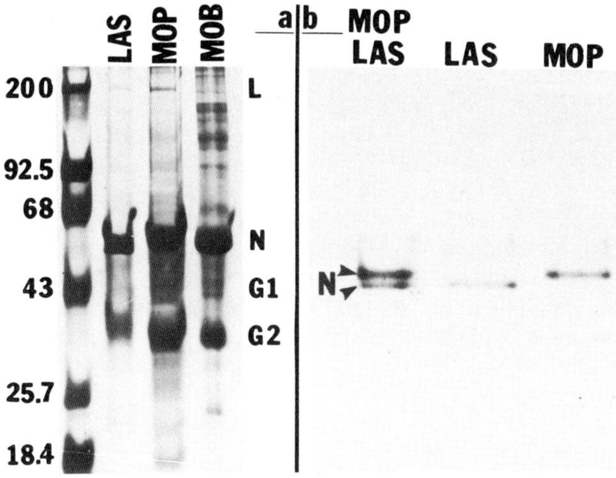

Figure 1. SDS-PAGE analysis of LAS, MOP and MOB virion proteins. Supernatant fluids from virus infected E6 cells (m.o.i.=.001) were collected at 3 days postinfection and virus was precipitated by polyethylene glycol and purified in KT-Glycerol and then sucrose gradients as previously described (9). SDS-PAGE in slab gels was according to the method of Laemmli (12). In 1a the gel was stained with Coomassie Blue while 1b is an autoradiogram where ^{35}S methionine labelled virus was analysed. Molecular weight markers were prestained standards (BRL) and represent (in decreasing order of MW) Myosin, Phosphorylase B, BSA, Ovalbumine and L Chymotripsinogen.

DIFFERENTIATION OF LAS, MOP AND MOB BY MONOCLONAL ANTIBODIES.

The reactivities of a selected population of monoclonal antibodies produced against either LAS (52 series) or MOP (53 series) viruses is shown in Table 1. These monoclones have shown differencies among the three viruses in both theN and the G proteins. For example, both an anti N monoclone (53-29-1) and an anti GP monoclone (53-16-11) are able to distinguish between MOP and MOB viruses. The monoclonal studies also indicate that all three viruses share many common antigenic sites in the N protein (ie. 52-93-4) as well as the glycoproteins (ie. 52-85-6). The specific reactivity of the monoclones was determined by immunoprecipitation of radiolabelled virus proteins from either purified virions or virus infected cells. An example of an SDS-PAGE slab gel analysis of a radioimmuno-precipitation experiment using infected cells is shown in the insert with Table 1. The immunoprecipitation was done as described previously (9), except that cytosols were pelleted at 100,000xg and made .1% SDS before reacting with antibody (In the insert: L=LAS, M=MOP, A=MOB). In this case the monoclone, 53-177-10-Q, reacts with the GP2 proteins of MOZ and ACAR but not with the LAS GP2.

TABLE 1
MONOCLONAL ANTIBODIES USED IN THE DIFFERENTIATION OF LASSA-LIKE ARENAVIRUSES BY INDIRECT IMMUNOFLUORECENCE

Antibody	Virus antigen[1] LAS	Virus antigen[1] MOP	Virus antigen[1] MOB	Protein[2] Specificity
52-93-4[3]	+	+	+	N
52-85-6	+	+	+	G2, GPC
52-158-3	+	-	-	N
52-135-17	+	-	-	G2, GPC
53-177-10-Q	-	+	+	G2, GPC
53-16-11(3)	-	+	-	G
53-29-1	-	+	-	N

[1]Antigens were acetone fixed, gamma irradiated infected cells fixed onto slides. [2]Determined by radio-immunoprecipitation. [3]52=From mice infected with LAS antigen; 53=From mice infected with MOZ antigen.

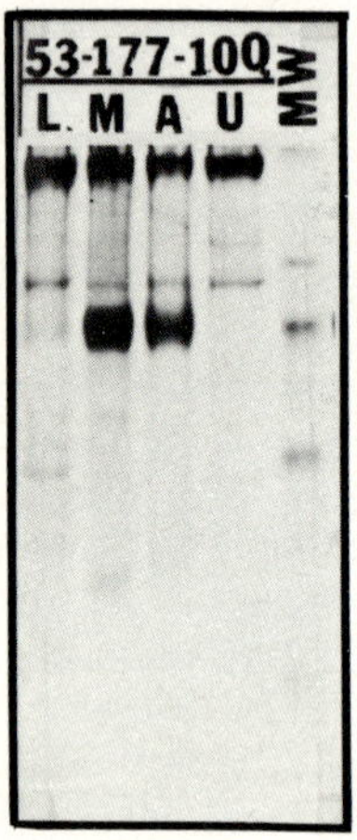

In addition to its reaction with GP2 the monoclone also reacts with the precursor glycoprotein GPC wich we find to be the major glycoprotein species present in cells infected with all three viruses. This precursor protein has previously been demonstrated in LCM (13) and Lassa (11) virus infected cells.

T1 OLIGONUCLEOTIDE MAPPING

Because the three virus strains can be differentiated by immunological (5) means, we measured the genetic relatedness of the viruses by the technique of T1 oligonucleotide "finger printing" (Fig.2). The methods used were identical to those used in our previous studies with Ebola virus (14). The ribosomal RNA present in purified virions did not appear to influence the Tl patterns.

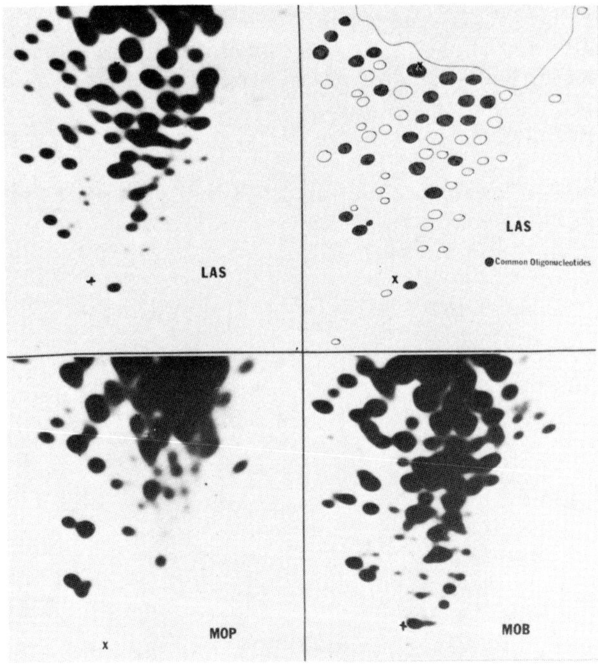

Figure 2. Tl oligonucleotide maps of virion RNA. Virus was grown and purified as in Fig.1. RNA was extracted from purified virions, labelelled in vitro with ^{32}P.

Table 2- Tl OLIGONUCLEOTIDE ANALYSIS OF VIRION RNA

Virus	*Lassa"Josiah"	MOP	MOB
LAS	**(60/60)100%	(27/43) 63%	(40/60) 67%
MOP	(25/60) 47%	(43/43)100%	(25/60) 42%
MOB	(45/60) 75%	(30/43) 70%	(60/60)100%

* Reference strain
** (xx/xx) = common oligonucleotides/total counted

The degree of homology among the viruses as determined by analysis of oligonucleotides spots, is presented in Table 2. By this method, regardless of which virus is used as the baseline, the ACAR virus has more sequences in common with

LAS virus than with MOP virus. This is contrary to immunologic data wich indicates a much stronger relationship between the MOP and MOB viruses than between LAS and either MOP or MOB virus.
It is possible that the sequences present in the T1 maps represent regions of the genome that are either not translated or whose protein products are not involved in the immunological functions measured by the other methods.

PEPTIDE MAPPING

For these experiments unlabelled proteins from purified virions were separarted by SDS-PAGE.

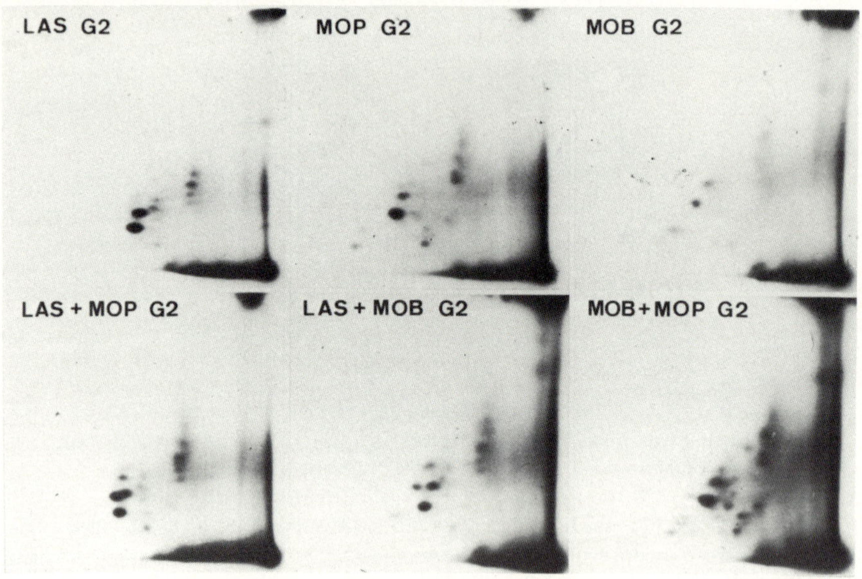

Figure 3. Peptide maps of virion G2 glycoproteins. Unlabelled purified virions were analysed by SDS-PAGE in slab gels. Proteins were visualized by Coomassie Blue staining, excised, labelled with ^{125}I and peptide maps prepared as previously described (15).

Virus preparations did not contain sufficient amounts of GP1 proteins to prepare reproducible peptides maps. Fig.3 presents peptide maps (individual proteins and mixtures) of the GP2 protein of the three viruses. The data demonstrate a greater degree of identity between MOP and MOB than between LAS and either of the other viruses, although all three viruses share some identity. These results are in agreement with monoclonal antibody data. Similar degrees of identity were also observed when virion N protein peptide maps were compared (data not shown).

DISCUSSION

Our data supports other recently published data (3,5,6,7,10)), strongly suggesting that there is a Lassa complex of arenaviruses on the African continent analogous to the Tacaribe complex in the New World.

All three viruses contain major polypeptides N,GP1 and GP2, and sometimes a large protein is seen. These are comparable to those reported for LCM virus (16) and reported for LAS virus (11). In our studies the MOP N protein is sligtly larger than the N protein of LAS or MOB and this is useful as a marker for the presence of MOP nucleocapsid. In addition to the three virion proteins infected cells also contain a precursor to the virion glycoproteins, GPC, wich is often the most abundant polypeptide in infected cells. The GPC protein is also found in cells infected with LCM (13) or Lassa (11) viruses.

Comparison of the viruses using monoclonal antibodies and peptide mapping of isolated proteins suggests that the MOP and MOB virus are more closely related to each other than either virus is to LAS. Experiments testing virulence of the viruses for suckling mice seem to confirm the relatedness of MOP and MOB (Gonzalez et al.,unpublished).

It has yet to be determined whether the degree of relatedness determined by the above techniques is the result of geographic distance, host of origin, or additional factors. The fact that the MOP and MOB viruses were both isolated from rodents where the LAS virus is a human isolate must be considered in any analysis.

By T1 oligonucleotide mapping analysis LAS virus is more closely related to MOB virus than MOB is to MOP. If this observation is correct then geographic distance may be the most important factor in determining genetic relatedness. This technique may involve only about 10% of the genome and may be measuring differences in base sequences which do not affect antigenicity or virulence.

We have determined some biological and physicochemical properties of the three viruses and now have means to distinguish among the viruses. Our studies have not pinpointed the factor or factors responsible for the diofference in virulence between LAS virus and the MOP and MOB isolates. We hope that our studies on the pathology of arenavirus infection (22,23) combined with additional physicochemical characterization of the viruses will allow us to better understand the virulence factors associated with LAS virus.

ACKNOWLEDGMENTS

We thank Dr R.Swanepoel for helpful discussions on nomenclature and all Branch members for their excellent cooperation. We also thank Dr N.J.Cox and R.Black for their help with the T1 mapping.

REFERENCES

1. Buckley,S.M.,CASALS,J.(1970).Amer.J.trop.Med.Hyg.19,680
2. Monath,T.P., Newhouse,V.F., Kemp,G.E., Setzer,H.W. and Cacciapuoti,A.(1974).Science 185,263
3. Monath,T.P.(1975)Bull.World Health Organ.52,577.
4. Speir,R.W., Wood,O., Liebhaber,H., Buckley,S.M.(1970). Amer.J.trop.Med.Hyg.19,692.
5. Gonzalez,J.P.,McCormick,J.B., Herve,J.P., Johnson,K.M. and Georges,A.J. (1983).Intervirol.19,105.
6. Wulff,H., McIntosh,B.M., Hammer,D.B., Johnson,K.M.(1977) Bull.World Health Organ.55,441.
7. Johnson,K.M., Taylor,P., Elliott,L.H. and Tomori,O. (1981).Amer.J.trop.Med.Hyg.30,1291.
8. Kiley,M.P., Lange,J.V. and Johnson,K.M.(1979)Lancet2,738
9. Kiley,M.P., Tomori,O., Regnery,R.L. and Johnson,K.M. (1981). In"the replication of Negative Stand Viruses" (D.H.L. Bishop and R.W. Compans eds.) p7.Elsevier/North Holland,New York.
10. Gonzalez,J.P.,McCormick,J.B.,Saluzzo,J.F. and Georges ,A.J.(1982)Cah.ORSTOMser.Ent.med.Parasitol,20
11. Clegg,J.C.S., and Lloyd,G. (1983).J.Gen.Virol.64,1127
12. Laemmli,U.K.(1970)Nature 227,680.
13. Buchmeier,MJ. and Oldstone,M.B.A.(1979)Virol.99,111.
14. Cox,N.J.,McCormick,J.B.,Johnson,K.M. and Kiley,M.P. (1983)J.Infect.Dis.147,272.
15. Buchmeier,M.J.,DeFries,R.U.,McCormick,J.B. and Kiley, M.P.(1983)J.Infect,Dis,147,276.
16. Buchmeier,M.J., Elder,J.H. and Oldstone,M.B.A.(1978) .Virology 89,133.
17. Walker,d.h.,McCormick,J.B.,Johnson,K.M.,Webb,P.A.,Kombo-Kono,G.,Elliot,L.H. and Gardner,J.J.(1982)Amer.J.Pathol. 107,349
18. Walker,D.H.,Johnson,K.M.,Lange,J.V.,Gardner,J.J.,Kiley,M.P. and McCormick,J.B.(1982)J.infect.Dis.146,360.

NEUTRALIZATION OF ARENAVIRUSES: REACTION OF TACARIBE VIRUS AND VARIANTS WITH MONOCLONAL ANTIBODIES[1]

Lisa M. Allison[2], Mark Salter[2], Michael J. Buchmeier[3]
Hanna Lewicki[3] and Colin R. Howard[2]

[2]Department of Medical Microbiology
London School of Hygiene and Tropical Medicine
London WC1E 7HT

and

[3]Department of Immunology
Scripps Clinic and Research Foundation
La Jolla, California 92037

The Tacaribe complex of the family Arenaviridae presently contains eight viruses, all of which serologically cross-react to varying degrees with Tacaribe virus, the type member of the group. The extent of relatedness between individual members has been difficult to assess using convalescent or hyperimmune antisera (1,2). Of interest is the finding of a close relationship between Tacaribe and Junin viruses; both guinea pigs and marmosets inoculated with Tacaribe show no signs of disease and are protected against challenge with the normally lethal Junin virus (3,4). It has been suggested that Tacaribe virus primes a specific immune response to Junin as a result of the antigenic similarity between the two viruses (5) and there is some evidence to suggest that Junin virus is weakly neutralized by Tacaribe immune serum (3,6).

The development of monoclonal antibodies against LCM and Pichinde viruses has been pursued in order to obtain defined reagents suitable for the dissection of the interrelationships between arenaviruses (7,8). In this report, a panel of monoclonal antibodies to Tacaribe virus have been characterised with respect to their capacity to neutralize virus infectivity and for reactivity against other arenaviruses within the Tacaribe complex.

[1] This work supported by the Medical Research Council[2] and NIH grants AI 16102 and NS 12428[3], and is publication number from the Department of Immunology, Scripps Clinic and Research Foundation. MJB is an Established Investigator of the American Heart Association.

Copyright © 1984 by Academic Press, Inc.
All rights of reproduction in any form reserved.
ISBN 0-12-183501-4

RESULTS

Serological properties of monoclonal antibodies to Tacaribe virus. A total of 21 cell lines producing antibodies to Tacaribe virus were obtained by fusion of immune spleen cells with P3 x Ag 8.6531 plasmacytoma cells (7). Of these 18 were selected for further study. The initial screening of cultures by immunoflourescence suggested the antibody-producing hybrids fell into two groups, on the basis of staining patterns and reactions with unfixed infected cells. The first was characterised by the absence of surface fluorescence. These antibodies were further differentiated on the basis of reaction with fixed substrates infected with heterologous arenaviruses. (Table 1). Among the 5 antibodies which additionally stained Junin-infected cells a cross-reaction against Amapari virus was also found with the same 5 antibodies. Only 2 were positive against Machupo virus, despite the close relationship between this virus and Junin. More limited cross-reactivity was seen against Pichinde virus, regarded as being more distantly related. Representative antibodies from this group reacted with the internal N polypeptide as assessed by analysis of immune complexes obtained by reaction of antibody with radiolabelled cell extracts. None of the Tacaribe antibodies reacted with LCM or other Old World arenaviruses.

Table 1 Cross-reactivity by immunofluorescence of monoclonal antibodies[1] directed against the Tacaribe virus nucleocapsid

	JUNIN	MACHUPO	PICHINDE	AMAPARI
2.16.2	+	-	+	+
2.48.3	+	+	+	+
2.52.2	+	-	-	+
2.74.3	+	-	-	+
2.100.3	+	+	-	+

[1]A further 8 antibodies did not cross-react with the heterologous substrates. No reactions were recorded against either LCM or Lassa viruses

The second group of 5 antibodies reacted specifically with the plasma membrane of unfixed cells infected with Tacaribe virus. No reactions were recorded against heterologous arenaviruses. Competitive radioimmunoassays revealed that one antibody (2.25.4) failed to compete for sites on intact virus with any of the remaining antibodies in this group (2.2.1, 2.12.5, 2.14.D4, and 2.82.2), thereby indicating that these reagents distinguished at least two viral epitopes expressed on the surface of infected cells.

Neutralization of Tacaribe virus infectivity with monoclonal antibodies. Aliquots of supernatant fluid from antibody-producing hybridoma cell cultures were screened for *in vitro* neutralization by measurement of virus titre reduction in the presence of a constant amount of antibody. The extent of virus neutralization was then recorded as the $\log_{10}$ reduction in virus titre observed after plaque visualisation 6 days later. All 5 antibodies in the group previously found to react with the surface of Tacaribe-infected cells reduced viral infectivity. No reactivity was found against the XJ-Cl_3 strain of Junin virus. These antibodies were therefore tentatively assigned as being reactive for the 44,000 mol. wt. envelope glycoprotein (G) of the virus, although immunoprecipitation studies followed by SDS-PAGE frequently resulted in the appearance of both G and N polypeptides.

Antibodies against Tacaribe virus found positive for homologous neutralization were examined further in order to quantitate the rate of reaction between antibody and homologous virus. Neutralization kinetics were measured by incubating dilutions of antibody-containing ascites previously inactivated for 30 minutes at 56°C with approximately 5×10^5 p.f.u. of Tacaribe virus in a 30°C water-bath. Aliquots of each mixture were withdrawn at various times after mixing and virus-antibody reactions stopped by immediate dilution in cold medium. Residual virus infectivity was quantitated on Vero cells and the results expressed as $\log_{10}$ of the fraction of infectivity remaining compared to time 0 (V/Vo). The antibodies differed according to the degree of neutralization achieved after 20 minutes of incubation. At high dilution antibody 2.25.4 effectively neutralised all infectious virus within the limit of sensitivity of the assay system (Figure 1). Extrapolation of the linear portion of the graphs obtained at three different temperatures all indicated that an initial lag was present in the reaction indicative of two molecules of antibody being required for neutralization.

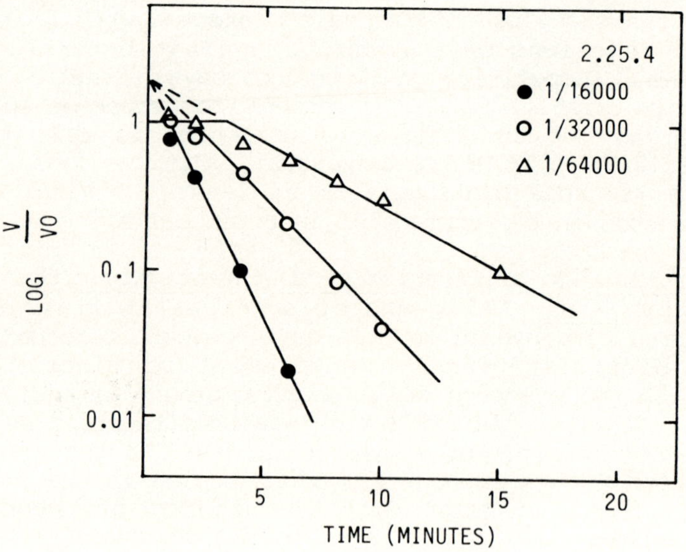

FIGURE 1 Neutralization of Tacaribe virus with antibody 2.25.4 at 30°C. Dilutions of antibody examined are indicated in this and subsequent figures.

The remaining four neutralizing antibodies gave significant non-neutralized fractions after an initial linear response (Figure 2). Residual virus infectivity was decreased by addition of either antibody 2.25.4 or anti-mouse IgG to the virus-antibody mixtures. A similar reduction was also observed by addition of fresh guinea-pig complement. Given the monoclonal nature of the antibodies being examined it is unlikely that the persistent fraction is due to steric hindrance from non-neutralizing antibodies.Although the possibility of a heterogenous population of virus particles reactive and non-reactive for antibody cannot be excluded, all virus stocks were repeatedly recloned during the course of these experiments.

Antibody 2.82.2 was examined in combination with antibody 2.25.4 . The reduction in titre produced by the two antibodies was found to be equivalent to the sum of the reduction in infectivity produced by each antibody tested singly (figure 3). The persistent non-neutralised fraction

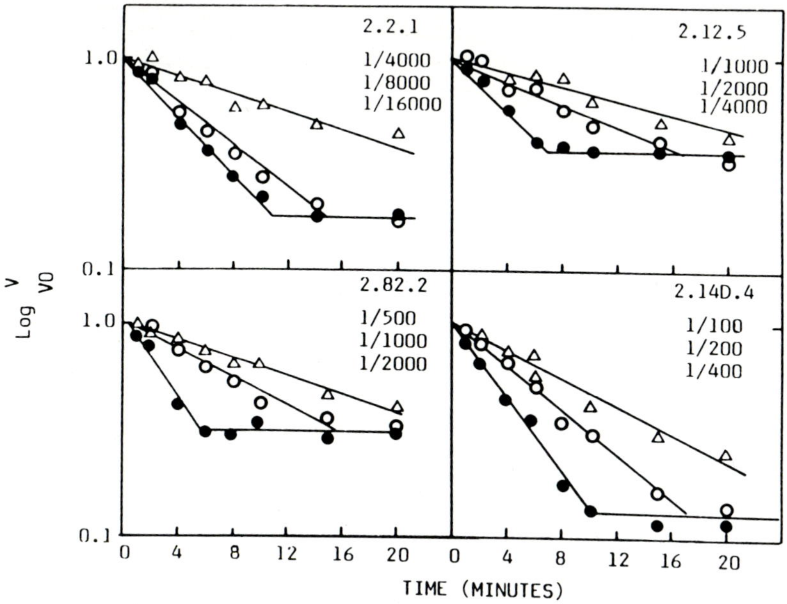

FIGURE 2 Neutralization of Tacaribe Virus with antibodies recognising a second functionally-distinct epitope on the viral glycoprotein. Each antibody was examined at three dilutions.

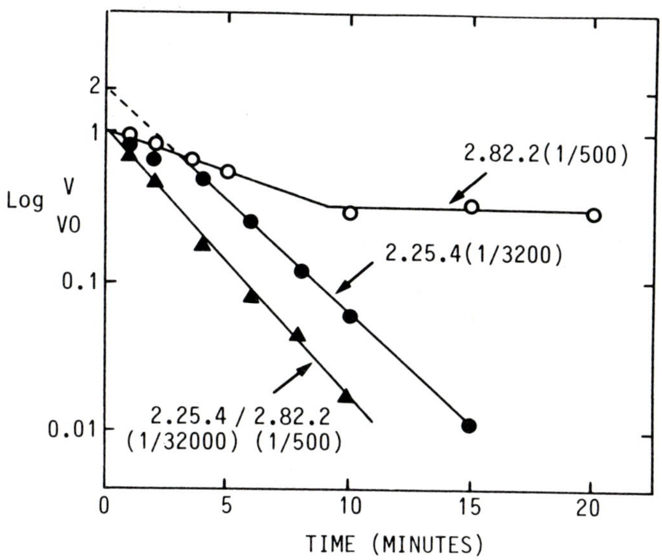

FIGURE 3 Combined kinetics of neutralization with antibodies 2.25.4 and 2.82.2.

obtained with antibody 2.82.2 alone is thus unlikely to be due to virus aggregation and provides further evidence of two distinct epitopes playing a role in neutralization

Table 2 Comparative Neutralization of Tacaribe virus variants.

VARIANT	log_{10} reduction in virus titre[1]	
	Immune ascites	Antibody 2.25.4
1	1.47	0.01
2	1.69	0.37
3	1.64	0.35
4	1.46	0.13
5	1.21	0.00
6	1.36	0.04
7	1.50	0.02
Control	2.70	>5.32

[1] Initial titre of virus stocks in range 4.3 to 6.6 log_{10}

Selection of Tacaribe variants resistant to antibody 2.25.4. In order to analyse further the nature of reactions with antibody, Tacaribe virus was mixed with a 1:500 final dilution of antibody 2.25.4 for 15 min. and the mixture plated on Vero cells at high dilution under an agar overlay containing the same antibody. Well-separated plaques were then recloned three times and stocks grown in BHK-21 cells. Variants were selected on the basis of retention of reactivity with a mouse hyperimmune antiserum but continuing resistance to 2.25.4. Results from one experiment are shown in Table 2. Five of these variants were selected for further study (Nos. 1, 3, 5, 6 and 7). using heterologous antibodies. All variants reacted as expected with antibody 2.2.1 to produce the large persistent fraction as seen with the original virus (cf. figure 2). However the variants were clearly differentiated into two groups by the complete absence of neutralization of variants 3, 5 and 7 by antibody

2.14.D4 (Table 3) whereas a large non-neutralized fraction was still present when variants 1 and 6 were similarly examined.

Table 3 Comparative Neutralisation of Tacaribe Virus and Variants by Monoclonal Antibody 2.14.D4.

VARIANT	DILUENT*	2.14.D4*
1	3.81	2.57 (1.34)
3	2.72	2.73 (0.00)
5	3.53	3.50 (0.03)
6	2.53	<0.83 (>1.72)
7	3.66	3.69 (0.00)
Control	3.50	1.73 (1.72)

*Log_{10} pfu/ml

DISCUSSION

Antibody 2.25.4 Tacaribe virus recognizes a distinct epitope to the remaining four glycoprotein antibodies as shown by (a) competition radioimmunoassay, and (b) the reaction of virus with antibody 2.25.4 is unaffected by prior addition of antibody 2.82.2. The large persistent fraction produced by the remaining antibodies may indicate incomplete neutralization resulting from an equilibrium between free and bound antibody molecules at a point prior to complete neutralization (9). Accurate measurements of antibody affinity are needed to correlate the extent of binding with the size of the non-neutralized fraction. Variants resistant to antibody 2.25.4 were readily selected at a frequency greater that 10^{-5}. Reaction of these variants with the heterologous antibodies manifesting incomplete neutralization suggests that point mutations occurring in or around the epitope recognized by antibody 2.25.4 may alter the configuration of non-overlapping distinct epitopes recognized by the second subset of antibodies. Some variants lost their reactivity with antibody 2.14.D4: this may be explained on the basis that under the selection pressure of the monoclonal antibody such mutations may not necessarily result from alterations of any one particular amino particular amino acid and that the nature of such changes may have variable effects on the ability of the glycoprotein to bind other antibodies. Alternatively, antibody 2.14.D4 may recognize an epitope which overlaps partially with both sites defined by the remaining antibodies, but owing to a high affinity is not clearly differentiated by competitive radioimmunoassay from low affinity antibodies recognizing a site distinct from the epitope reactive for antibody 2.25.4. Further analysis of these monoclonal antibodies will aid in clarifying the role of antibody and virus persistence in vivo.

Acknowledgements.

We are indebted to Dr. K. Johnson for supply of Machupo virus antigens and Mr. G. Lloyd and Miss S. Al-Mufti for immunoflourescent screening of antibodies to Lassa and Amapari viruses respectively. We should also like to express our thanks to Miss S Rumens for the typing of this manuscript.

References

1. Casals, J., Buckley, S.M. and Cedeno, R. (1975). Bull. Wld. Hlth. Organ. 52, 421
2. Wulff, H., Large, J.V., and Webb, P.A. (1978). Intervirology 9. 344
3. Weissenbacher, M.C., Coto, C.E., and Calello, M.A. (1975/1976). Intervirology 6, 42
4. Weissenbacher, M.C., Calello, M.A., Quintans, C.J., Panisse, H., Woyskowsky, N.M., and Zannoli, V.H. (1983). Intervirology 19, 1
5. Coto, C.E., Damonte, E.B., Calello, M.A., and Weissenbacher, M.C. (1980). J. inf. Dis. 141, 389
6. Henderson, J.R., and Downs, W.G. (1965). Amer. J. Trop. Med. Hyg. 14, 796
7. Buchmeier, M.J., Lewicki, H.A., Tomori, O., and Oldstone, M.B.A. (1981). Virology 113, 73
8. Buchmeier, M.J., Lewicki, H.A., Tomori, O., and Johnson, K., (1980). Nature, Lond., 288, 486
9. Volk, W.A., Snyder, R.M., Benjamin, D.C., and Wagner, R.J. (1982). J. of Virology, 42, 220

ANTIGENIC SITES ON THE G1 GLYCOPROTEIN OF LA CROSSE VIRUS THAT ARE INVOLVED IN NEUTRALIZATION[1]

Laura Kingsford and Lori D. Ishizawa

Department of Microbiology
California State University
Long Beach, California 90840

La Crosse (LAC) virus, a medically important Bunyavirus (1), has been chosen as a model to study the molecular mechanism of neutralization. This virus has four structural proteins; L (180K), G1 (120K), G2 (34K), and N (23K) (2). Of these four proteins, only the G1 glycoprotein is involved in attachment to cells and induces and reacts with antibody that results in neutralization of the virus (3).

When LAC virus is incubated with trypsin or chymotrypsin, a 20K to 53K Da region at the presumed amino terminus of the G1 glycoprotein is proteolytically removed. However, a significant amount of viral infectivity remains after the protease treatment. Polyclonal antibody binds to the part of G1 remaining virion associated, but only 1 to 1.5 logs of virus are neutralized in 60 min compared with over 4 logs of control virus in 20 min (3,4). This implies there are multiple antibody binding sites on G1 and that the ones critical for neutralization are either removed or altered by the proteases. Another interpretation previously proposed is that antibody molecules must bind to at least two different antigenic sites on the G1 glycoprotein (4). When one of these sites is removed by proteases, antibody binding to the remaining site on the G1 polypeptide does not completely block virus-cell interaction. We now present data to indicate there are sites critical for neutralization and that when antibody is bound to two of these sites, enhancement of neutralization does occur.

TOPOLOGICAL MAPPING OF ANTIGENIC SITES ON G1

To investigate the location and nature of the antibody binding sites on the G1 glycoprotein, we prepared a panel of

[1]This work was supported by Public Health Service Grant AI17931 from the National Institutes of Allergy and Infectious Diseases.

Copyright © 1984 by Academic Press, Inc.
All rights of reproduction in any form reserved.
ISBN 0-12-183501-4

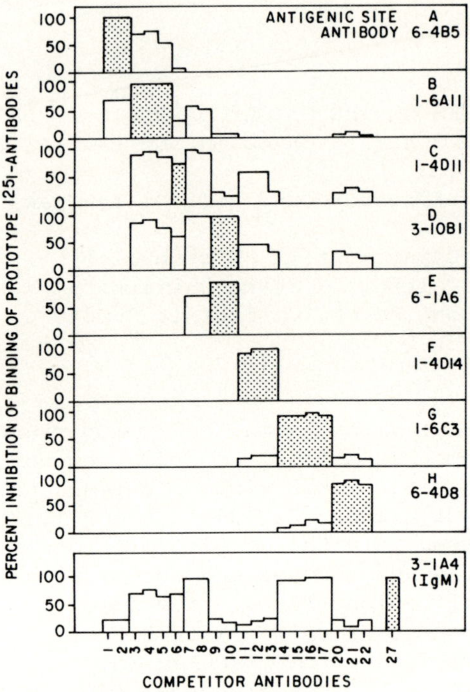

FIGURE 1. Antigenic site specificities of monoclonal antibodies to the Gl glycoprotein. Twenty IgG-class and one IgM-class antibodies were iodinated and used in a reciprocal competitive radioimmunoassay. Shown is the inhibition of binding which occurred when a prototype ^{125}I-antibody from each antigenic site was mixed with each of the monoclonal antibodies. The competitor antibody clone numbers are those arbitrarily assigned to the monoclonal cell lines (Table 1). The shaded area in each bar graph indicates the antibodies which are specific for the designated site. Source: Kingsford et al. (4)

monoclonal antibodies. When these antibodies were purified, iodinated, and used in a two-way competition binding assay, 20 IgG-class antibodies could be assigned to eight antigenic sites on Gl (5). A high degree of partial competition occurred among antibodies to five of these sites (Fig. 1). This was observed both ways in reciprocal assays and all antibodies used had high avidities. This indicated that these five antigenic sites were distinct but closely adjacent. The pattern of partial competition also allowed us to tentatively order these antigenic sites linearly A through E. Likewise, because there was some competition (30% or less) between antibodies to the remaining sites, they were tentatively ordered F through H. This partial competition has been observed by others and interpreted to indicate the sites are in close proximity (6). However, it could also be due to allosteric inhibition of binding. If so, the sites may be located on various parts of the glycoprotein and not in a linear sequence.

To determine if we could detect overlapping epitopes within these reactive sites, each monoclonal antibody was assayed in an enzyme-linked immunosorbent assay (ELISA) incorporating each of the subtypes of the California encephalitis complex of viruses (7). Five of the previously defined antigenic sites (A, D, F, G, and H) contained within them overlapping epitopes. In all, 15 epitopes were identified using the serologically related viruses. Shown in Table 1 are the reactions of the monoclonal antibodies with LAC, showshoe hare, California encephalitis, and Tahyna viruses. These four viruses alone could be used to differentiate all 15 epitopes. The only epitope we detected that was common to all California viruses was g_1 while epitope c_1 was unique to LAC virus.

PHYSICAL MAPPING OF THE EPITOPES ON G1

As previously reported, trypsin cleaves 25K or 53K Da of protein from G1 leaving two polypeptides of 95K and 67K, respectively (3). As the 67K polypeptide is part of the 95K one, an ELISA using trypsinized virus as antigen essentially measures the antibody binding sites on the 95K polypeptide. By difference, those antibodies which do not bind to trypsinized virus are specific for sites on the 25K Da part of G1. When assayed, antibodies to epitopes h_1 and h_2 were the only ones that bound to the trypsinized virus. In an experiment of similar design, virus pretreated with chymotrypsin was used as antigen in an ELISA. Like trypsin, chymotrypsin cleaves the G1 glycoprotein *in situ* leaving two size polypeptides. The resulting polypeptides, however, are 70K and 100K Da. In the ELISA, antibodies to epitopes h_1, h_2, g_2, g_3, and g_4 bound to chymotrypsin-treated virus. This indicated that most of antigenic site G was located between the trypsin 95K and the chymotrypsin 100K cleavage sites. We cannot totally rule out the possibility that some of the epitopes in antigenic sites A through F as well as epitope g_1 are also on the trypsin or chymotrypsin resistant polypeptides. Conformational changes due to cleavage could alter these epitopes so their respective antibodies could not bind. These data, however, along with the pattern of partial competition in the radioimmunoassay, allowed us to propose a tentative order for antigenic sites A through H, with H being closest to the viral membrane and A closest to the presumed amino terminal end of the glycoprotein.

TABLE 1
OVERLAPPING EPITOPES ON THE LAC G1 GLYCOPROTEIN

Antigenic site[a]	Epitope[b]	Clone number	ELISA reaction on:[b] LAC	SSH	TAH	CE
A	a_1	1	+	+	+	-
	a_2	2	+	+	-	-
B	b_1	3-5	+	-	-	+
C	c_1	6	+	-	-	-
D	d_1	7	+	-	-	+
	d_2	8	+	+	+	+
E	e_1	9,10	+	+	-	-
F	f_1	13,23	+	-	+	-
	f_2	11,12	+	-	-	-
G	g_1	17,19	+	+	+	+
	g_2	14,15	+	+	+	-
	g_3	16	+	-	-	-
	g_4	18	+	-	+	-
H	h_1	21,22	+	+	-	-
	h_2	20	+	+	+	-

[a]Antigenic site designations are based on a two-way competition binding assay (Fig. 1).

[b]Epitope designations are based on the assay of LAC anti-G1 monoclonal antibodies in an ELISA using LAC, showshoe hare (SSH), Tahyna (TAH), and California encephalitis (CE) viruses.

HIGHER ORDER STRUCTURE OF G1

When an IgM antibody was used in the competition binding assay, there was reciprocal competition between it and antibodies to antigenic sites B, C, D, and G (Fig. 1) (5). Thus, the epitope for the IgM antibody could not be determined, but the data indicate these four competing sites have to be in close proximity. If the proposed sequence of antigenic sites A to H is correct, the only way the IgM antibody could block antibodies binding to these four sites is if the linear amino acid sequence comprising site G is antiparallel to that of sites B,

C, and D. Thus, this IgM clone proved of value in giving some idea of a possible higher order relationship of antigenically reactive sites. Of interest, enhancement of binding also occurred between antibodies to these same four antigenic sites (5). Antibodies specific for site G showed reciprocal synergistic binding with antibodies to sites B, C, and D.

MODEL OF LOCATION OF EPITOPES ON THE G1 GLYCOPROTEIN

In an attempt to synthesize the proceding data, we propose the model shown in Fig. 2. The linear sequence of antigenic sites A through H is based on the competition binding data and the high degree of competition exhibited by antibodies to sites A through E (and F through G). The orientation of sites A through H is based on the assay of monoclonal antibodies on LAC virus pretreated with trypsin or chymotrypsin. The order of overlapping epitopes within each antigenic site is arbitrary at present. Only in antigenic site G is there some evidence based on assay of antibody on chymotrypsin-treated virus to indicate that epitope g_1 is closer to the amino terminal end of G1 than the other three "g" epitopes. The proposed foldback of the glycoprotein molecule is based on the competition binding with the IgM antibody which competed with antibodies to sites B, C, D, and G. This model is not intended to represent the tertiary structure of G1, but to provide a working model to explain our data and to be used in the study of the molecular mechanism of virus neutralization.

BIOLOGICAL ACTIVITIES OF MONOCLONAL ANTIBODIES

To relate the defined regions of G1 with biological functions, the respective antibodies were assayed for neutralizing and hemagglutination inhibition activities beginning at a protein concentration of 400 µg/ml. Antibodies to sites B, C, and

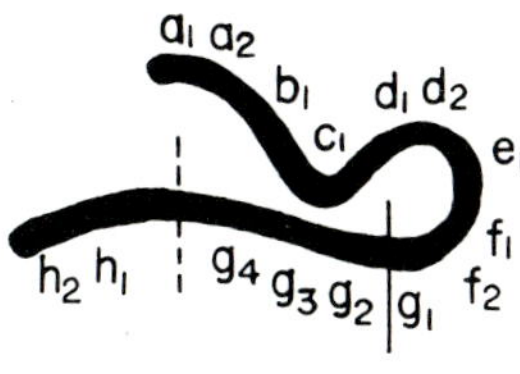

FIGURE 2. Topographic model for the location of epitopes on the 25K Da trypsin-sensitive region of the G1 glycoprotein. The proposed *in situ* cleavage sites for trypsin (--) and chymotrypsin (—) are shown. Epitopes a_1 and a_2 are located closest to the presumed amino terminus while epitopes h_1 and h_2 map on the 95K trypsin-resistant part of G1.

H had low or negative titers in both assays. If Gl does fold back as proposed, then these three sites may be in close proximity and located in an area not involved in attachment to cells or neutralization. Antibodies to the other epitopes varied in their capacities to neutralize or inhibit hemagglutination. Titers in both assays were highest for those antibodies binding to sites A, F, and G. In addition, one-third of the monoclonal antibodies used in this study were specific for these three sites. It may be that the tertiary folding of the glycoprotein places them near each other in the globular molecule and they comprise an immunodominant area as has been described for other proteins (8). These data also imply that a primary site (or sites) for attachment to erythrocytes and possibly other cell receptors is located near these three antigenic sites. From the competition binding data, we cannot determine if the sequence of sites A through E is as proposed or in the reverse order. If reversed, site E would be on the presumed amino terminal end and site A would be adjacent to F. This would place sites A, F, and G in a linear sequence, perhaps more compatible with the neutralization data and the idea they may comprise an immunodominant area.

NEUTRALIZATION KINETICS AND ENHANCEMENT OF NEUTRALIZATION

In order to study the neutralization of LAC virus in more detail, its reaction with each monoclonal antibody was examined in a kinetic assay. We found that all of the neutralization curves were biphasic. In addition, the maximum amount of neutralization in antibody excess ranged from 1.5 to 3 logs for the various monoclonal antibodies. This is compared with polyclonal anti-Gl antibody which neutralizes over 4 logs of virus.

Although the addition of homologous monoclonal antibody would not reduce the large persistent fractions observed, the additional polyclonal anti-Gl antibody would. This indicated that antibodies binding to multiple sites on Gl were collectively needed to give a greater degree of neutralization. To examine this, purified prototype monoclonal antibodies to each antigenic site were used in all combinations of two in a kinetic assay. Significantly, certain antibody pairs did exhibit an increased neutralizing capacity as well as an increased rate. Enhanced neutralization did occur for some of the antibodies which were synergistic in binding. Antibody to site G when mixed with antibody to sites B or D neutralized over 4 logs of virus in about 5 min (Fig. 3). One exception to this was antibody to site C, a non-neutralizing antibody which showed reciprocal synergistic binding with neutralizing antibodies to sites A and G, but no enhanced neutralization.

We found that for enhancement of neutralizaion, much less total antibody was needed to achieve the same level of

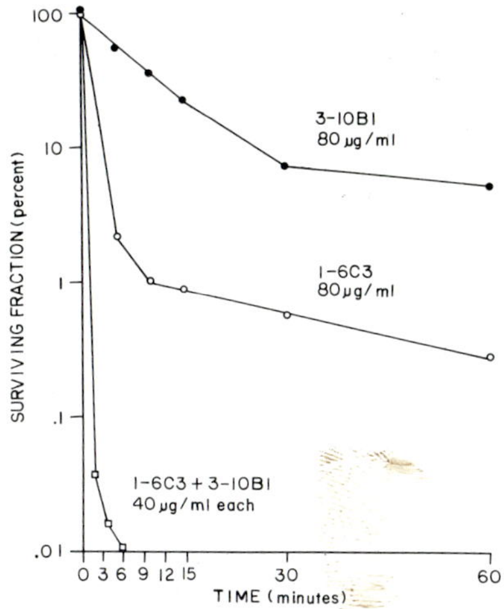

FIGURE 3. Enhanced neutralization of LAC virus exhibited by a pair of monoclonal antibodies (3-10B1 to site D and 1-6C3 to site G). Antibody was mixed with an equal volume of virus at a concentration of 6 x 10^6 PFU/ml. The reaction was carried out at 37°. Aliquots of 0.1 ml were taken at each time point, diluted 1:100 to stop neutralization and assayed for residual infectivity by plaque assay.

neutralization exhibited by either antibody alone. This would support a mechanism proposed by Lubeck and Gerhard (9) that binding of two synergistic antibodies in a reciprocal manner results in a conformational change such that it leads to an increased affinity between the antibody and glycoprotein molecules. This shift to a higher association constant would result in a greater proportion of bound virus at equilibrium and thus, a greater degree of neutralization.

However, enhancement of neutralization must involve more than the additional binding of antibody molecules due to a higher association constant because we found that certain pairs of non-synergistic antibodies also exhibited enhanced neutralization. Antibody to epitope g_2 when used in combination with antibody to sites A, E, or F resulted in a significant increase in the rate and degree of neutralization such that 4 logs of virus would be neutralized within 15 min. It may be that enhanced binding is also involved in these antibody pairs, but that we just did not detect it in our competition binding assay. Another possibility is that it is the conformational change induced by two antibodies binding to G1 that is needed to effectively neutralize LAC virus and this allosteric change may or may not lead to enhanced binding of antibody. A conformational change needed for neutralization of virus has been proposed for polio virus (10). For LAC virus, antibodies to site G seem to be the critical ones. These antibodies used in combination with any other neutralizing antibody resulted in a rate and degree of neutralization equal to or greater than that exhibited by polyclonal anti-Gl or anti-LAC antibody. The only other pair of antibodies exhibiting enhanced neutralization bound to sites F and B.

CONCLUSIONS

These data clearly support the need for antibody to bind to at least two different but specific antigenic sites for effective neutralization. These data also indicate that some concepts about antibody neutralization based on conventional antisera need to be revised. For example, the need for more than one antibody molecule to inactivate virus means that neutralization is not a first order reaction. Monoclonal antibodies now provide a way to isolate and identify specific reactions and interactions between virus and antibody as well as between virus-antibody complexes and cells. The studies reported here are a beginning in our understanding of a molecular mechanism for virus neutralization.

REFERENCES

1. Thompson, W. H., Kalfayan, B., and Anslow, R. O. (1965). Amer. J. Epidemiol. 81, 245.
2. Obijeski, J. F., Bishop, D. H. L., Murphy, F. A., and Palmer, E. L. (1976). J. Virol. 19, 985.
3. Kingsford, L., and Hill, D. W. (1983). J. Gen. Virol. 64 (in press).
4. Kingsford, L., and Hill, D. W. (1981). In "The Replication of Negative Strand Viruses" (D. H. L. Bishop and R. W. Compands, eds.), p. 111. Elsevier North Holland, New York.
5. Kingsford, L., Ishizawa, L. D., and Hill, D. W. (1983). Virology 130 (in press).
6. Stone, M. R., and Nowinski, R. C. (1980). Virology 100, 370.
7. Calisher, C. H. (1983). In "California Serogroup Viruses" (C. H. Calisher and W. H. Thompson, eds.). p. 1. Alan R. Liss, New York.
8. Maron, E., Shiozawa, C., Arnon, R., and Sela, M. (1971). Biochem. 10, 763.
9. Lubeck, M. and Gerhard, W. (1982). Virology 118, 1.
10. Emini, E. A., Shaw-Yi, K., Lewis, A. J., Crainic, R. and Wimmer, E. (1983). J. Virol. 46, 466.

INFLUENZA VIRUS NEURAMINIDASE: STRUCTURE AND VARIATION

W.G. Laver, P.M. Colman*, C.W. Ward*, J.M. Varghese*, G.M. Air†, and R.G. Webster§

John Curtin School of Medical Research,
PO Box 334, Canberra City ACT 2601
Australia

Influenza virus neuraminidase exists as a mushroom-shaped spike on the surface of the virion. It has a box-like head made out of four coplanar and roughly spherical subunits and a centrally attached stalk containing a hydrophobic region by which it is embedded in the viral membrane.[1,2,3]

The neuraminidase molecule is composed of a single polypeptide chain coded by gene segment 6 and is oriented in the virus membrane in the opposite way to the hemagglutinin (Figure 1). No post-translational cleavage of the NA polypeptide occurs, no signal peptide is split off and even the initiating methionine is retained.[4] No processing at the C-terminus takes place - the C-terminal sequence, -MET-PRO-ILE predicted from the gene sequence is found in intact NA molecules isolated from virus, and in the pronase-released NA heads.[4] A sequence of six polar amino acids at the N-terminus of the NA polypeptide, which are totally conserved in the 9 different NA subtypes, is followed by a sequence of hydrophobic amino acids which probably represent the transmembrane region of the NA stalk.[4] This sequence is not conserved at all between subtypes (apart from conservation of hydrophobicity). Pronase cleaves the polypeptide in the positions shown (Figure 1), removing the stalk and releasing the enzymically and antigenically active head of the NA which, in some cases, can be crystallized (Figure 2).

---

* CSIRO Division of Protein Chemistry, 343 Royal Parade, Melbourne Vic 3052, Australia

† Laboratory for Special Cancer Research, University of Alabama in Birmingham, 520 11th Street South, Birmingham Alabama 35294, USA

§ Division of Virology, St Jude Children's Research Hospital, PO Box 318, Memphis Tennessee 38101, USA

Copyright © 1984 by Academic Press, Inc.
All rights of reproduction in any form reserved.
ISBN 0-12-183501-4

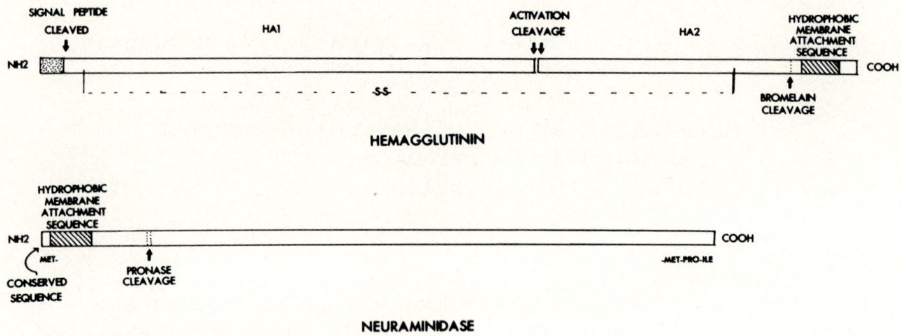

Figure 1. Diagramatic representation contrasting certain features of the hemagglutinin and neuraminidase polypeptides.

ANTIGENIC VARIATION IN THE NEURAMINIDASE

Antigenic shift and antigenic drift occur in the NA; among the human influenza viruses antigenic shift occurred in 1957, with the emergence of the H2N2 subtype and the replacement of N1 by N2.

Gene sequences (and the deduced amino acid sequences) of the NA from five N2 strains[6,7,8,9] and two N1 strains[10,11] have been obtained. The overall homology between the amino acid sequences of the N1 and N2 neuraminidase was estimated to be 39%.

Antigenic drift in N2 neuraminidase has been investigated most extensively. Neuraminidase-inhibition tests using polyclonal hyperimmune rabbit antisera showed that the neuraminidases of the 1957 and 1975 H2N2 and H3N2 strains were antigenically very different.

These antigenic differences are associated with a number of sequence changes in the NA polypeptide (Figure 3) but which of these are located in antigenic sites and which have no effect on antigenicity is not precisely known.

SEQUENCE CHANGES IN THE NEURAMINIDASE OF 'MONOCLONAL VARIANTS'

Antigenic variants of neuraminidase have been selected with a frequency of approximately 10^{-5} after a single passage

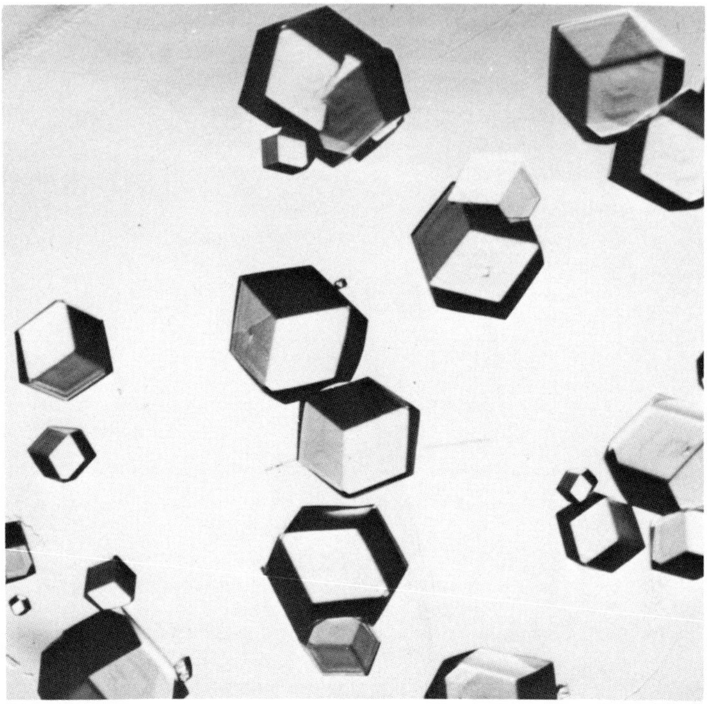

Figure 2. Crystals of pronase released neuraminidase heads of NA subtype N9 (Intact N9 neuraminidase molecules possess hemagglutinin activity).

Figure 3.(next pages) Amino acid sequence changes associated with antigenic drift in the N2 neuraminidase. The data sources are A/RI/5⁻/57 (Elleman et al, 1982), A/Tokyo/3/67 (Ward et al, 1982), NT/60/68 (Bentley & Brownlee, 1982), A/Udorn/72 (Markoff & Lai, 1982) and A/Victoria/75 (van Rompuy et al, 1982). The amino acid substitutions that have accumulated since 1957 are shaded.

RI/5$^-$/57	Met	Asn	Pro	Asn	Gln	Lys	Thr	Ile	Thr	Ile	Gly	Ser	Val	Ser	Leu	Thr	Ile	Ala	Thr	Val
Tokyo/3/67	Met	Asn	Pro	Asn	Gln	Lys	**Ile**	Ile	Thr	Ile	Gly	Ser	Val	Ser	Leu	Thr	Ile	Ala	Thr	Val
NT/60/68	Met	Asn	Pro	Asn	Gln	Lys	**Ile**	Ile	Thr	Ile	Gly	Ser	Val	Ser	Leu	Thr	Ile	Ala	Thr	Val
Udorn/307/72	Met	Asn	Pro	Asn	Gln	Lys	**Ile**	Ile	Thr	Ile	Gly	Ser	Val	Ser	Leu	Thr	Ile	Ala	Thr	**Ile**
Vic/3/75	Met	Asn	Pro	Asn	Gln	Lys	**Ile**	Ile	Thr	Ile	Gly	Ser	Val	Ser	Leu	Thr	Ile	Ala	Thr	**Ile**
										10										20

RI/5$^-$/57	Cys	Phe	Leu	Met	Gln	Ile	Ala	Ile	Leu	Ala	Thr	Thr	Val	Thr	Leu	His	Phe	Lys	Gln	His
Tokyo/3/67	Cys	Phe	Leu	Met	Gln	Ile	Ala	**Met**	Leu	**Val**	Thr	Thr	Val	Thr	Leu	His	Phe	Lys	Gln	His
NT/60/68	Cys	Phe	Leu	Met	Gln	**Thr**	Ala	Ile	Leu	**Val**	Thr	Thr	Val	Thr	Leu	His	Phe	Lys	Gln	**Tyr**
Udorn/307/72	Cys	Phe	Leu	Met	Gln	Ile	Ala	Ile	**Gln**	**Val**	Thr	Thr	Val	Thr	Leu	His	Phe	Lys	Gln	**Tyr**
Vic/3/75	Cys	Phe	Leu	Met	Gln	Ile	Ala	Ile	Leu	**Val**	Thr	Thr	Val	Thr	Leu	His	Phe	Lys	Gln	**Tyr**
										30										40

RI/5$^-$/57	Glu	Cys	Asp	Ser	Pro	Ala	Ser	Asn	Gln	Val	Met	Pro	Cys	Glu	Pro	Ile	Ile	Ile	Glu	Arg
Tokyo/3/67	**Asp**	Cys	Asp	Ser																
NT/60/68	Glu	Cys	Asp	Ser	Pro	Ala	Ser	Asn	Gln	Val	Met	Pro	Cys	Glu	Pro	Ile	Ile	Ile	Glu	Arg
Udorn/307/72	Glu	Cys	Asp	Ser	Pro	Ala	**Asn**	Asn	Gln	Val	Met	Pro	Cys	Glu	Pro	Ile	Ile	Ile	Glu	Arg
Vic/3/75	Glu	Cys	Asp	Ser	Pro	Ala	**Asn**	Asn	Gln	Val	Met	Pro	Cys	Glu	Pro	Ile	**Ser**	Ile	Glu	Arg
										50										60

RI/5$^-$/57	Asn	Ile	Thr	Glu	Ile	Val	Tyr	Leu	Asn	Asn	Thr	Thr	Ile	Glu	Lys	Glu	Ile	Cys	Pro	Glu
Tokyo/3/67														Glu	Lys	Glu	Ile	Cys	Pro	**Lys**
NT/60/68	Asn	Ile	Thr	Glu	Ile	Val	Tyr	Leu	Asn	Asn	Thr	Thr	Ile	Glu	Lys	Glu	Ile	Cys	Pro	**Lys**
Udorn/307/72	Asn	Ile	Thr	Glu	Ile	Val	Tyr	Leu	**Thr**	Asn	Thr	Thr	Ile	Glu	Lys	Gly	Ile	Cys	Pro	**Lys**
Vic/3/75	Asn	Ile	Thr	Glu	Ile	Val	Tyr	Leu	**Thr**	Asn	Thr	Thr	Ile	Glu	Lys	Gly	Ile	Cys	Pro	**Lys**
										70										80

RI/5$^-$/57	Val	Val	Glu	Tyr	Arg	Asn	Trp	Ser	Lys	Pro	Gln	Cys	Gln	Ile	Thr	Gly	Phe	Ala	Pro	Phe
Tokyo/67	Val	Val	Glu	Tyr	Arg	Asn	Trp	Ser	Lys	Pro	Gln	Cys	Gln	Ile	Thr	Gly	Phe	Ala	Pro	Phe
NT/60/68	Val	Val	Glu	Tyr	Arg	Asn	Trp	Ser	Lys	Pro	Gln	Cys	Gln	Ile	Thr	Gly	Phe	Ala	Pro	Phe
Udorn/72	**Leu**	Val	Glu	Tyr	Arg	Asn	Trp	Ser	Lys	Pro	Gln	Cys	**Lys**	Ile	Thr	Gly	Phe	Ala	Pro	Phe
Vic/3/75	**Leu**	Val	Glu	Tyr	Arg	Asn	Trp	Ser	Lys	Pro	Gln	Cys	**Lys**	Ile	Thr	Gly	Phe	Ala	Pro	Phe
										90										100

RI/5$^-$/57	Ser	Lys	Asp	Asn	Ser	Ile	Arg	Leu	Ser	Ala	Gly	Gly	Asp	Ile	Trp	Val	Thr	Arg	Glu	Pro
Tokyo/3/67	Ser	Lys	Asp	Asn	Ser	Ile	Arg	Leu	Ser	Ala	Gly	Gly	Asp	Ile	Trp	Val	Thr	Arg	Glu	Pro
NT/60/68	Ser	Lys	Asp	Asn	Ser	Ile	Arg	Leu	Ser	Ala	Gly	Gly	Asp	Ile	Trp	Val	Thr	Arg	Glu	Pro
Udorn/72	Ser	Lys	Asp	Asn	Ser	Ile	Arg	Leu	Ser	Ala	Gly	Gly	Asp	Ile	Trp	Val	Thr	Arg	Glu	Pro
Vic/3/75	Ser	Lys	Asp	Asn	Ser	Ile	Arg	Leu	Ser	Ala	Gly	Gly	Asp	Ile	Trp	Val	Thr	Arg	Glu	Pro
										110										120

RI/5$^-$/57	Tyr	Val	Ser	Cys	Asp	Pro	Gly	Lys	Cys	Tyr	Gln	Phe	Ala	Leu	Gly	Gln	Gly	Thr	Thr	Leu
Tokyo/67	Tyr	Val	Ser	Cys	Asp	Pro	**Val**	Lys	Cys	Tyr	Gln	Phe	Ala	Leu	Gly	Gln	Gly	Thr	Thr	Leu
NT/60/68	Tyr	Val	Ser	Cys	Asp	**His**	Gly	Lys	Cys	Tyr	Gln	Phe	Ala	Leu	Gly	Gln	Gly	Thr	Thr	Leu
Udorn/72	Tyr	Val	Ser	Cys	Asp	Pro	Gly	Lys	Cys	Tyr	Gln	Phe	Ala	Leu	Gly	Gln	Gly	Thr	Thr	Leu
Vic/75	Tyr	Val	Ser	Cys	Asp	Pro	**Arg**	Lys	Cys	Tyr	Gln	Phe	Ala	Leu	Gly	Gln	Gly	Thr	Thr	Leu
										130										140

RI/5$^-$/57	Asp	Asn	Lys	His	Ser	Asn	Gly	Thr	Ile	His	Asp	Arg	Ile	Pro	His	Arg	Thr	Leu	Leu	Met
Tokyo/67	Asp	Asn	Lys	His	Ser	Asn	**Asp**	Thr	**Val**	His	Asp	Arg	Ile	Pro	His	Arg	Thr	Leu	Leu	Met
NT/60/68	Asp	Asn	Lys	His	Ser	Asn	**Asp**	Thr	Ile	His	Asp	Arg	Ile	Pro	His	Arg	Thr	Leu	Leu	Met
Udorn/72	Asp	Asn	Lys	His	Ser	Asn	**Asp**	Thr	Ile	His	Asp	Arg	**Thr**	Pro	His	Arg	Thr	Leu	Leu	Met
Vic/75	**Glu**	Asn	Lys	His	Ser	Asn	**Asp**	Thr	Ile	His	Asp	Arg	**Thr**	Pro	His	Arg	Thr	Leu	Leu	Met
										150										160

RI/5$^-$/57	Asn	Glu	Leu	Gly	Val	Pro	Phe	His	Leu	Gly	Thr	Lys	Gln	Val	Cys	Val	Ala	Trp	Ser	Ser
Tokyo/67	Asn	Glu	Leu	Gly	Val	Pro	Phe	His	Leu	Gly	Thr	**Arg**	Gln	Val	Cys	**Ile**	Ala	Trp	Ser	Ser
NT/60/68	Asn	Glu	Leu	Gly	Val	Pro	Phe	His	Leu	Gly	Thr	**Arg**	Gln	Val	Cys	**Ile**	Ala	Trp	Ser	Ser
Udorn/72	Asn	Glu	Leu	Gly	Val	Pro	Phe	His	Leu	Gly	Thr	**Arg**	Gln	Val	Cys	**Ile**	**Gly**	Trp	Ser	Ser
Vic/75	Asn	Glu	Leu	Gly	Val	Pro	Phe	His	Leu	Gly	Thr	**Arg**	Gln	Val	Cys	**Ile**	Ala	Trp	Ser	Ser
										170										180

RI/5$^-$/57	Ser	Ser	Cys	His	Asp	Gly	Lys	Ala	Trp	Leu	His	Val	Cys	Val	Thr	Gly	Asp	Asp	Arg	Asn
Tokyo/67	Ser	Ser	Cys	His	Asp	Gly	Lys	Ala	Trp	Leu	His	Val	Cys	**Ile**	Thr	Gly	Asp	Asp	**Asp**	Asn
NT/60/68	Ser	Ser	Cys	His	Asp	Gly	Lys	Ala	Trp	Leu	His	Val	Cys	**Ile**	Thr	Gly	Asp	Asp	**Lys**	Asn
Udorn/72	Ser	Ser	Cys	His	Asp	Gly	Lys	Ala	Trp	Leu	His	Val	Cys	Val	Thr	Gly	**Tyr**	Asp	**Lys**	Asn
Vic/3/75	Ser	Ser	Cys	His	Asp	Gly	Lys	Ala	Trp	Leu	His	Val	Cys	Val	Thr	Gly	**Tyr**	Asp	**Lys**	Asn
										190										200

RI/5$^-$/57	Ala	Thr	Ala	Ser	Phe	Ile	Tyr	Asp	Gly	Arg	Leu	Val	Asp	Ser	Ile	Gly	Ser	Trp	Ser	Gln
Tokyo/67	Ala	Thr	Ala	Ser	Phe	Ile	Tyr	Asp	Gly	Arg	Leu	Val	Asp	Ser	Ile	Gly	Ser	Trp	Ser	Gln
NT/60/68	Ala	Thr	Ala	Ser	Phe	Ile	Tyr	Asp	Gly	Arg	Leu	Val	Asp	Ser	Ile	Gly	Ser	Trp	Ser	Gln
Udorn/72	Ala	Thr	Ala	Ser	Phe	Ile	Tyr	Asp	Gly	Arg	Leu	Val	Asp	Ser	Ile	Gly	Ser	Trp	Ser	Gln
Vic/3/75	Ala	Thr	Ala	Ser	Phe	Ile	Tyr	Asp	Gly	Arg	Leu	Val	Asp	Ser	Ile	Gly	Ser	Trp	Ser	Gln
										210										220

RI/5$^-$/57	Asn	Ile	Leu	Arg	Thr	Gln	Glu	Ser	Glu	Cys	Val	Cys	Ile	Asn	Gly	Thr	Cys	Thr	Val	Val
Tokyo/67	Asn	Ile	Leu	Arg	Thr	Gln	Glu	Ser	Glu	Cys	Val	Cys	Ile	Asn	Gly	Thr	Cys	Thr	Val	Val
NT/60/68	Asn	Ile	Leu	Arg	Thr	Gln	Glu	Ser	Glu	Cys	Val	Cys	Ile	Asn	Gly	Thr	Cys	Thr	Val	Val
Udorn/72	Asn	Ile	Leu	Arg	Thr	Gln	Glu	Ser	Glu	Cys	Val	Cys	Ile	Asn	Gly	Thr	Cys	Thr	Val	Val
Vic/3/75	Asn	Ile	Leu	Arg	Thr	Gln	Glu	Ser	Glu	Cys	Val	Cys	Ile	Asn	Gly	Thr	Cys	Thr	Val	Val
										230										240

RI/5^{-}/57	Met	Thr	Asp	Gly	Ser	Ala	Ser	Gly	Arg	Ala	Asp	Thr	Arg	Ile	Leu	Phe	Ile	Lys	Glu	Gly
Tokyo/67	Met	Thr	Asp	Gly	Ser	Ala	Ser	Gly	Arg	Ala	Asp	Thr	Arg	Ile	Leu	Phe	Ile	**Glu**	Glu	Gly
NT/60/68	Met	Thr	Asp	Gly	Ser	Ala	Ser	Gly	Arg	Ala	Asp	Thr	Arg	Ile	Leu	Phe	Ile	**Glu**	Glu	Gly
Udorn/72	Met	Thr	Asp	Gly	Ser	Ala	Ser	Gly	Arg	Ala	Asp	Thr	**Lys**	Ile	Leu	Phe	Ile	**Glu**	Glu	Gly
Vic/3/75	Met	Thr	Asp	Gly	Ser	Ala	Ser	Gly	Arg	Ala	Asp	Thr	**Lys**	Ile	Leu	Phe	Ile	**Glu**	Glu	Gly
										250										260

RI/5^{-}/57	Lys	Ile	Val	His	Ile	Ser	Pro	Leu	Ser	Gly	Ser	Ala	Gln	His	Ile	Glu	Glu	Cys	Ser	Cys
Tokyo/67	Lys	Ile	Val	His	Ile	Ser	Pro	Leu	**Ala**	Gly	Ser	Ala	Gln	His	**Val**	Glu	Glu	Cys	Ser	Cys
NT/60/68	Lys	Ile	Val	His	Ile	Ser	Pro	Leu	Ser	Gly	Ser	Ala	Gln	His	**Val**	Glu	Glu	Cys	Ser	Cys
Udorn/72	Lys	Ile	Val	His	Ile	Ser	Pro	Leu	Ser	Gly	Ser	Ala	Gln	His	**Val**	Glu	Glu	Cys	Ser	Cys
Vic/3/75	Lys	Ile	Val	His	Ile	Ser	Pro	Leu	Ser	Gly	Ser	Ala	Gln	His	**Val**	Glu	Glu	Cys	Ser	Cys
										270										280

RI/5^{-}/57	Tyr	Pro	Arg	Tyr	Pro	Asp	Val	Arg	Cys	Ile	Cys	Arg	Asp	Asn	Trp	Lys	Gly	Ser	Asn	Arg
Tokyo/67	Tyr	Pro	Arg	Tyr	Pro	**Gly**	Val	Arg	Cys	Ile	Cys	Arg	Asp	Asn	Trp	Lys	Gly	Ser	Asn	Arg
NT/60/68	Tyr	Pro	Arg	Tyr	Pro	**Gly**	Val	Arg	Cys	Ile	Cys	Arg	Asp	Asn	Trp	Lys	Gly	Ser	Asn	Arg
Udorn/72	Tyr	Pro	Arg	Tyr	Pro	**Gly**	Val	Arg	Cys	Ile	Cys	Arg	Asp	Asn	Trp	Lys	Gly	Ser	Asn	Arg
Vic/3/75	Tyr	Pro	Arg	Tyr	Pro	**Gly**	Val	Arg	Cys	Ile	Cys	Arg	Asp	Asn	Trp	Lys	Gly	Ser	Asn	Arg
										290										300

RI/5^{-}/57	Pro	Val	Ile	Asp	Ile	Asn	Met	Glu	Asp	Tyr	Ser	Ile	Asp	Ser	Ser	Tyr	Val	Cys	Ser	Gly
Tokyo/67	Pro	Val	**Val**	Asp	Ile	Asn	Met	Glu	Asp	Tyr	Ser	Ile	Asp	Ser	Ser	Tyr	Val	Cys	Ser	Gly
NT/60/68	Pro	Val	**Val**	Asp	Ile	Asn	Met	Glu	Asp	Tyr	Ser	Ile	Asp	Ser	Ser	Tyr	Val	Cys	Ser	Gly
Udorn/72	Pro	Val	**Val**	Asp	Ile	Asn	**Val**	**Lys**	Asp	Tyr	Ser	Ile	Asp	Ser	Ser	Tyr	Val	Cys	Ser	Gly
Vic/3/75	Pro	Val	**Val**	Asp	Ile	Asn	**Val**	**Lys**	Asp	Tyr	Ser	Ile	Asp	Ser	Ser	Tyr	Val	Cys	Ser	Gly
										310										320

RI/5^{-}/57	Leu	Val	Gly	Asp	Thr	Pro	Arg	Asn	Asp	Asp	Ser	Ser	Ser	Asn	Ser	Asn	Cys	Arg	Asp	Pro
Tokyo/67	Leu	Val	Gly	Asp	Thr	Pro	Arg	Asn	Asp	Asp	**Arg**	Ser	Ser	Asn	Ser	Asn	Cys	Arg	**Asn**	Pro
NT/60/68	Leu	Val	Gly	Asp	Thr	Pro	Arg	Asn	Asp	Asp	**Arg**	Ser	Ser	Asn	Ser	Asn	Cys	Arg	**Asn**	Pro
Udorn/72	Leu	Val	Gly	Asp	Thr	Pro	Arg	Asn	**Asn**	Asp	**Arg**	Ser	Ser	Asn	Ser	**Tyr**	Cys	Arg	**Asn**	Pro
Vic/3/75	Leu	Val	Gly	Asp	Thr	Pro	Arg	**Lys**	**Asn**	Asp	**Arg**	Ser	Ser	**Ser**	Ser	**Tyr**	Cys	Arg	**Asn**	Pro
										330										340

RI/5^{-}/57	Asn	Asn	Glu	Arg	Gly	Asn	Pro	Gly	Val	Lys	Gly	Trp	Ala	Phe	Asp	Asn	Gly	Asp	Asp	Val
Tokyo/67	Asn	Asn	Glu	Arg	Gly	**Thr**	**Gln**	Gly	Val	Lys	Gly	Trp	Ala	Phe	Asp	Asn	Gly	**Asn**	Asp	**Leu**
NT/60/68	Asn	Asn	Glu	Arg	Gly	Asn	**Gln**	Gly	Val	Lys	Gly	Trp	Ala	Phe	Asp	Asn	Gly	Asp	Asp	Val
Udorn/72	Asn	Asn	Glu	**Lys**	Gly	Asn	**His**	Gly	Val	Lys	Gly	Trp	Ala	Phe	Asp	**Asp**	Gly	**Asn**	Asp	Val
Vic/3/75	Asn	Asn	Glu	**Lys**	Gly	**Ile**	**His**	Gly	Val	Lys	Gly	Trp	Ala	Phe	Asp	**Asp**	Gly	**Asn**	Asp	Val
										350										360

RI/5^{-}/57	Trp	Met	Gly	Arg	Thr	Ile	Asn	Lys	Glu	Ser	Arg	Ser	Gly	Tyr	Glu	Thr	Phe	Lys	Val	Ile
Tokyo/67	Trp	Met	Gly	Arg	Thr	Ile	**Ser**	Lys	**Asp**	**Leu**	Arg	Ser	Gly	Tyr	Glu	Thr	Phe	Lys	Val	Ile
NT/60/68	Trp	Met	Gly	Arg	Thr	Ile	**Ser**	Lys	**Asp**	**Leu**	Arg	Ser	Gly	Tyr	Glu	Thr	Phe	Lys	Val	Ile
Udorn/72	Trp	Met	Gly	Arg	Thr	Ile	**Ser**	**Glu**	**Asp**	Ser	Arg	Ser	Gly	Tyr	Glu	Thr	Phe	Lys	Val	Ile
Vic/3/75	Trp	Met	Gly	Arg	Thr	Ile	**Ser**	**Glu**	**Asp**	Ser	Arg	Ser	Gly	Tyr	Glu	Thr	Phe	Lys	Val	Ile
										370										380

RI/5^{-}/57	Gly	Gly	Trp	Ser	Thr	Pro	Asn	Ser	Lys	Ser	Gln	Val	Asn	Arg	Gln	Val	Ile	Val	Asp	Asn
Tokyo/67	Gly	Gly	Trp	Ser	Thr	Pro	Asn	Ser	Lys	Ser	Gln	**Ile**	Asn	Arg	Gln	Val	Ile	Val	Asp	**Ser**
NT/60/68	Gly	Gly	Trp	Ser	Thr	Pro	Asn	Ser	Lys	Ser	Gln	**Ile**	Asn	Arg	Gln	Val	Ile	Val	Asp	**Ser**
Udorn/72	Gly	Gly	Trp	Ser	Thr	Pro	Asn	Ser	Lys	**Leu**	Gln	**Ile**	Asn	Arg	Gln	Val	Ile	Val	Asp	**Ser**
Vic/3/75	Gly	Gly	Trp	Ser	Thr	Pro	Asn	Ser	Lys	**Leu**	Gln	**Ile**	Asn	Arg	Gln	Val	Ile	Val	Asp	**Ser**
										390										400

RI/5^{-}/57	Asn	Asn	Trp	Ser	Gly	Tyr	Ser	Gly	Ile	Phe	Ser	Val	Glu	Gly	Lys	Ser	Cys	Ile	Asn	Arg
Tokyo/67	**Asp**	Asn	**Arg**	Ser	Gly	Tyr	Ser	Gly	Ile	Phe	Ser	Val	Glu	Gly	Lys	Ser	Cys	Ile	Asn	Arg
NT/60/68	**Asp**	Asn	**Arg**	Ser	Gly	Tyr	Ser	Gly	Ile	Phe	Ser	Val	Glu	Gly	Lys	Ser	Cys	Ile	Asn	Arg
Udorn/72	**Asp**	Asn	**Arg**	Ser	Gly	Tyr	Ser	Gly	Ile	Phe	Ser	Val	Glu	Gly	Lys	Ser	Cys	Ile	Asn	Arg
Vic/3/75	**Ala**	Asn	**Arg**	Ser	Gly	Tyr	Ser	Gly	Ile	Phe	Ser	Val	Glu	Gly	Lys	Ser	Cys	Ile	Asn	Arg
										410										420

RI/5^{-}/57	Cys	Phe	Tyr	Val	Glu	Leu	Ile	Arg	Gly	Arg	Pro	Gln	Glu	Thr	Arg	Val	Trp	Trp	Thr	Ser
Tokyo/67	Cys	Phe	Tyr	Val	Glu	Leu	Ile	Arg	Gly	Arg	**Lys**	Gln	Glu	Thr	Arg	Val	Trp	Trp	Thr	Ser
NT/60/68	Cys	Phe	Tyr	Val	Glu	Leu	Ile	Arg	Gly	Arg	**Lys**	Gln	Glu	**Ala**	Arg	Val	Trp	Trp	Thr	Ser
Udorn/72	Cys	Phe	Tyr	Val	Glu	Leu	Ile	Arg	Gly	Arg	**Glu**	Gln	Glu	Thr	Arg	Val	Trp	Trp	Thr	Ser
Vic/3/75	Cys	Phe	Tyr	Val	Glu	Leu	Ile	Arg	Gly	Arg	**Glu**	Gln	Glu	Thr	Arg	Val	Trp	Trp	Thr	Ser
										430										440

RI/5^{-}/57	Asn	Ser	Ile	Val	Val	Phe	Cys	Gly	Thr	Ser	Gly	Thr	Tyr	Gly	Thr	Gly	Ser	Trp	Pro	Asp
Tokyo/67	Asn	Ser	Ile	Val	Val	Phe	Cys	Gly	Thr	Ser	Gly	Thr	Tyr	Gly	Thr	Gly	Ser	Trp	Pro	Asp
NT/60/68	Asn	Ser	Ile	Val	Val	Phe	Cys	Gly	Thr	Ser	Gly	Thr	Tyr	Gly	Thr	Gly	Ser	Trp	Pro	Asp
Udorn/72	Asn	Ser	Ile	Val	Val	Phe	Cys	Gly	Thr	Ser	Gly	Thr	Tyr	Gly	Thr	Gly	Ser	Trp	Pro	Asp
Vic/3/75	Asn	Ser	Ile	Val	Val	Phe	Cys	Gly	Thr	Ser	Gly	Thr	Tyr	Gly	Thr	Gly	Ser	Trp	Pro	Asp
										450										460

RI/5^{-}/57	Gly	Ala	Asn	Ile	Asn	Phe	Met	Pro	Ile
Tokyo/67	Gly	Ala	Asn	Ile	Asn	Phe	Met	Pro	Ile
NT/60/68	Gly	Ala	Asn	Ile	Asn	Phe	Met	Pro	Ile
Udorn/72	Gly	Ala	**Asp**	Ile	Asn	**Leu**	Met	Pro	Ile
Vic/3/75	Gly	Ala	**Asp**	Ile	Asn	**Leu**	Met	Pro	Ile
									469

of A/Tokyo/3/67 influenza virus in the presence of monoclonal antibody to the neuraminidase.[12] These 'monoclonal variants', like those of influenza virus hemagglutinin previously described,[13] do not bind at all to the antibody used for their selection, and each shows a single change in the amino-acid sequence of the neuraminidase.

Three different neuraminidase variants, grown from Tokyo/3/67, have been partially sequenced, and they show sequence changes at residues 221, 344 and 368.[14] The latter two changes occurred in chain segments also showing high variation in field strain sequences. The other change, at residue 221, is in a segment of structure undergoing no change in the field strains. However, the three-dimensional structure shows residue 221 to be on an adjacent loop to 197-199 where variation is found in field strains.[15]

THREE DIMENSIONAL STRUCTURE OF N2 NEURAMINIDASE HEADS

Neuraminidase from two different field strains, A/Tokyo/3/67 and A/RI/5$^+$/57, hereafter referred to as Tokyo and RI/5$^+$ respectively, have been used in this study.[15,16]

Figure 4 shows the structure of influenza virus neuraminidase which provides the first known example of a tetrameric protein with circular four-fold symmetry. The polypeptide chain folds into six, topologically identical, four-stranded, antiparallel β-sheets which are themselves arranged like the blades of a propeller. Calcium ions are believed to be bound on the four-fold axis between a cluster of eight acidic groups. The most C-terminal of the five glycosylation sequences carries no carbohydrate.

The catalytic sites of influenza virus neuraminidase are located in a deep cleft on the upper corners of the box-shaped tetramer. These are flanked and surrounded by nine acidic residues, six basic residues and three hydrophobic residues (Figure 4) all of which are strictly conserved in all known influenza virus neuraminidase sequences.

Amino acids which change during antigenic drift are observed to cluster preferentially into the distal surface loops which connect the various strands of β sheets. Some of these variable residues are immediately adjacent to amino acids implicated in enzyme activity (Figure 4). The variable loops do not segregate into spatially distinct non-overlapping areas of variation. Rather they form a nearly continuous surface across the top of the subunit, encircling the catalytic site. It remains to be seen whether all of these variable loops can participate in forming complexes with immunoglobulins. Neuraminidase variants selected with monoclonal antibodies show that residues 221, 344 and 368 form part of

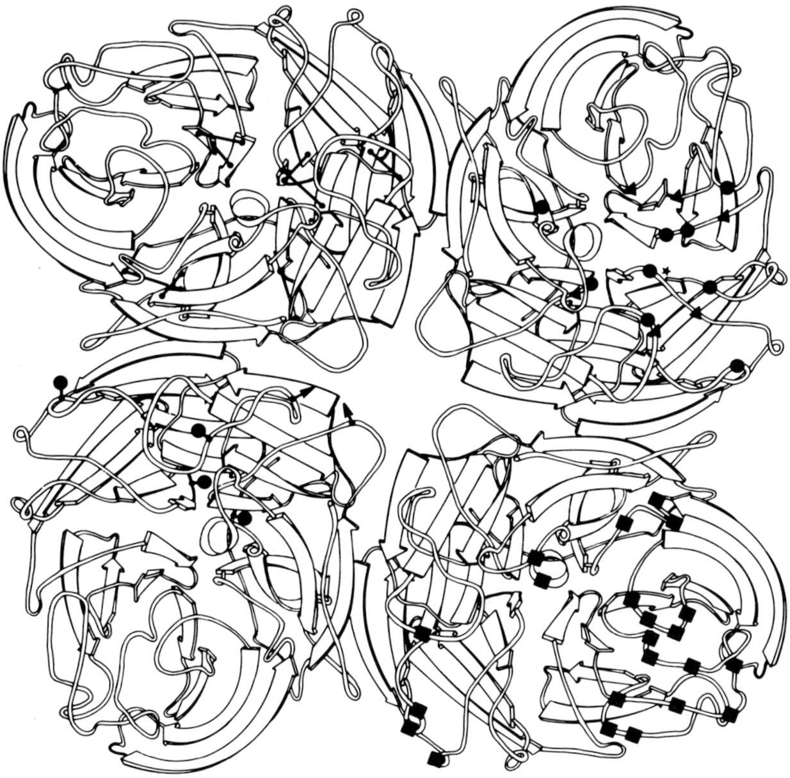

Figure 4. Schematic of the neuraminidase tetramer viewed from above down the symmetry axis. The four subunits highlight different features of the structure.
Top left: disulfide bonds.
Bottom left: carbohydrate attachment sites at 86,146, 200 and 234 and metal ligands ASP 113 and ASP 114
Bottom right: Residues which changed in field strains and in variants selected with monoclonal antibodies.
Top right: conserved acidic and basic residues in influenza A and B neuraminidase. *Marks the sialic acid binding site.

antigenic determinants and it is likely that when more mutants are characterised remaining regions of frequent field stain variation will also be seen to form sites for interaction with antibodies. Recently neuraminidase heads of subtype N9 have been crystallized (Figure 2). This neuraminidase possesses hemagglutinin activity and its structure is currently being determined.

REFERENCES

1 Laver, W.G. and Valentine, R.C. (1969). Virology 38, 105-119.

2 Wrigley, N.G., Skehel, J.J., Charlwood, P.A. and Brand, C.M. (1973). Virology 51, 525-529.

3 Wrigley, N.G. (1979). Brit. Med. Bull. 35, 35-38.

4 Blok, J., Air, G.M., Laver, W.G., Ward, C.W., Lilley, G.G., Woods, E.G., Roxburgh, C.M. and Inglis, A.S. (1982). Virology 118, 109-121.

5 Blok, J. and Air, G.M. (1982). Biochemistry 21, 4001-4007.

6 Elleman, T.C., Azad, A.A. and Ward C.W. (1982). Nucl. Acid Res. 10, No. 21, 7005-7015.

7 Ward, C.W., Elleman, T.C. and Azad, A.A. (1982). Biochem. J. 207, 91-95.

8 Bentley, D.R. and Brownlee, G.G. (1982). Nucl. Acids Res. 10, 5033-5042.

9 van Rompuy, L., Min-Jou, W., Huylebroeck, D. and Fiers, W. (1982). J. Mol. Biol. 161, 1-11.

10 Fields, S., Winter, G. and Brownlee, G.G. (1981). Nature 290, 213-217.

11 Hiti, A.L. and Nayak, D.P. (1982). J. of Virol. 41, 730-734.

12 Webster, R.G., Hinshaw, V.S. and Laver W.G. (1982). Virology 117, 93-104.

13 Laver, W.G., Air, G.M. and Webster, R.G. (1981). J. Mol. Biol. 145, 339-361.

14 Laver, W.G., Air, G.M., Webster, R.G. and Markoff, L.J. (1982). Virology 122, 450-460.

15 Varghese, J.N., Laver, W.G. and Colman, P.M. (1983). Nature 303, 35-40.

16 Colman, P.M., Varghese, J.N. and Laver, W.G. (1983). Nature 303, 41-44.

PLEIOTROPIC EFFECTS OF A SINGLE AMINO ACID CHANGE ON ANTIGENICITY AND BIOLOGIC FUNCTION OF SWINE INFLUENZA VIRUS HEMAGGLUTININ MUTANTS[1]

E. D. Kilbourne,[2] G. W. Both,[3] and W. Gerhard[4]

[2]Department of Microbiology
Mount Sinai School of Medicine
of the City University of New York
New York, New York 10029

[3]Molecular and Cellular Biology Unit, CSIRO
P.O. Box 184, North Ryde, NSW, Australia

[4]The Wistar Institute
Philadelphia, Pennsylvania 19104

Contemporary strains of swine influenza virus contain two mutant subpopulations which differ in their capacity to replicate in chick embryos, Madin-Darby Canine Kidney (MDCK) cells, and in swine (1,2,3). These mutants are also distinguishable with heterotypic antiserum to A/SW/CAM/39(H1N1) (SW/CAM) virus (Table 1). Previous studies have shown that the mutation responsible for the antigenic and biologic differences in these L and H mutants (1) is in the hemagglutinin (HA) gene. The studies reported here, a summary of recently completed work, will provide evidence that the mutation is at a site on the HA molecule of this H1 virus corresponding to the Sa antigenic site of A/PR/8/34(H1N1) (PR8) virus (4). Furthermore, nucleotide sequencing of cDNA copies of the swine influenza virus HA gene have further localized the site to amino acid 155 within that antigenic region (5).

MATERIALS AND METHODS

Materials and methods have been fully described in recent publications (4,5). The viruses employed include L

[1]This work was supported, in part, by United States Public Health Service Grant AI-09304.

Copyright © 1984 by Academic Press, Inc.
All rights of reproduction in any form reserved.
ISBN 0-12-183501-4

and H mutants of swine influenza virus A/NJ/11/76(H1N1) and various reassortants in which the swine influenza virus HA gene has been segregated through recombination with PR8 virus. Unless otherwise identified (Table 4), L and H reassortant viruses are reassortants X-53(cl)-PR8(2) and X-53(cl)-PR8(2)H[2], respectively (5). Nucleotide sequencing strategy and methods are presented elsewhere (5).

RESULTS AND DISCUSSION

CHARACTERIZATION OF THE L AND H MUTANT PHENOTYPES

L mutants are inhibited by high dilutions of SW/CAM heterotypic polyclonal antibody, are lower yielding in chick embryos, produce smaller plaques in MDCK cells, and are more infective for swine than H mutants (Table 1).

REACTION OF L AND H MUTANTS AND REASSORTANTS WITH MONOCLONAL ANTIBODIES TO PR8 HA

When reacted in RIA with a panel of 14 monoclonal antibodies which define the Sa, Ca and Cb regions of the PR8 HA (Table 2), the three antibodies reactive with the Sa antigenic site were found to distinguish the L and H mutants and

TABLE 1
DISTINGUISHING PROPERTIES OF L AND H MUTANTS AND THEIR REASSORTANT VIRUSES

Virus	HI with SW/CAM ab	Viral yield	Plaque size(mm)[c]	Swine ID_{50}[d]
A/NJ/11/76(L)	320[a]	16[b]	1-2.5	10^{2}
A/NJ/11/76(H)	20	128	2-4	10^{4}
Reassortant(L)	80	1024	1-3	$10^{4.4}$
Reassortant(H)	<10	8192	2-5	$>10^{5.6}$

[a]Reciprocal of antiserum dilution at endpoint.
[b]Reciprocal of virus dilution at endpoint of HA titration.
[c]In MDCK cells.
[d]Minimal egg infectious dose required to infect swine.
Data from Kilbourne _et al._, 1983 (4).

TABLE 2
CROSS REACTIONS IN RIA OF MONOCLONAL PR8 ANTIBODIES WITH L AND H MUTANTS AND REASSORTANTS

Monoclonal antibodies to HA site:	Virus				
	PR8	A/NJ(L)	A/NJ(H)	Rx(L)	Rx(H)
Sa	4390[a]	139	1470	94	1474
Ca	4657	1776	2115	1913	1837
Cb	4645	1383	2123	1692	1991

[a]Mean cpm above assay background.
Data summarized from Kilbourne _et al._, 1983 (4).
Rx - reassortant virus.

reassortants, by binding to significant levels with the H but not L viruses. (CPM in Table 2 are mean values respectively for assays with 3 Sa, 9 Ca and 2 Cb hybridomas.)

The Sa site reactive antibody Sa-13 was inhibitory to hemagglutination (HI) and plaque formation in MDCK cells (PI) of H but not L mutants. A monoclonal antibody (9C8) specifically inhibitory to L mutants was produced following inoculation of reassortant L into Balb C mice (4) (Table 3).

TABLE 3
INHIBITION OF BIOLOGIC ACTIVITY OF L AND H MUTANTS WITH MONOCLONAL ANTIBODIES

	Monoclonal antibody[a]			
	Sa-13		9C8	
Virus	HI[a]	PI[a]	HI	PI
A/NJ(L)	<10	40	320	25600
A/NJ(H)	640	1280	<10	<800
Rx(L)	<10	<40	320	25600
Rx(H)	320	1280	10	40

[a]See text.
[b]Reciprocal of antibody dilution at endpoint.

SELECTION AND PHENOTYPIC CHARACTERIZATION OF L AND H REVERTANT VIRUSES

Revertants of A/NJ(L) and reassortant (L) viruses were obtained by selection with 9C8 (L-specific) monoclonal antibody and were found to be changed both antigenically and biologically to H phenotype. Passage of reassortant (H) with Sa-5 (anti-H) monoclonal antibody yielded virus indistinguishable from reassortant (L). (It should be noted that reassortant H is itself a revertant of reassortant L not selected by antibody, but emergent as a large plaque variant.)

CORRELATION OF L AND H PHENOTYPES WITH HA NUCLEOTIDE SEQUENCE CHANGES (TABLE 4)

The complete nucleotide sequence was determined for virus 1 (Table 4), for 70% of the sequence of virus 2 and partial sequences were obtained through the Sa antigenic site for viruses 3-6 (5). Changes were found only for 4 nucleotides coding for amino acids 153, 155, 159 and 228. Changes at amino acid 155 (Gly to Glu) distinguish the L and H reassortants. Virus 3, not reactive with either Sa-13 or 9C8 antibody, was derived from an H reassortant. This virus has retained Glu at position 155 but has undergone mutation

TABLE 4
CORRELATION OF ANTIGENIC AND BIOLOGIC PHENOTYPE WITH MUTATION AT HA AMINO ACID 155

Virus	Phenotype ag.	Phenotype bio.	HA amino acid 153	HA amino acid 155[a]	HA amino acid 159	HA amino acid 228
1) X-53a(cl)	H	H	Lys	Glu	Pro	Ser
2) X-53(cl)-PR8	L	L	Glu	Gly	Pro	Asn
3) X-53a(cl)-PR8(H)	X[c]	H	Lys	Glu	Thr	Asn
4) X-53(cl)-Lp	L	L	Lys	Gly	Pro	Asn
5) X-53(cl)-PR8(2)	L	L	Lys	Gly	Pro	Asn
6) X-53(cl)-PR8(2)H[2]	H	H	Lys	Glu	Pro	Asn

[a]Equivalent to residue 158 in H3 model.

[b]Amino acid deduced from nucleotide sequence (data from Both et al., 1983) (5).

[c]Non-L, non-H.

from Pro to Thr at amino acid 159, a neighboring site where change could sterically affect antibody binding.

SUMMARY

Swine influenza virus L and H HA mutants differ in antigenicity and replication.

L and H mutants can be distinguished with monoclonal antibodies to the H1 (PR8) Sa antigenic site.

Nucleotide sequencing has identified differences of L and H mutants (Gly-Glu) at deduced amino acid 155 within the Sa site.

L to H and H to L revertants have simultaneous change in antigenic and biological phenotype.

An L to H revertant which has been sequenced has Gly to Glu change at amino acid 155.

Antigenic mutation may be the fortuitous concomitant of non-immunologic selection for replication characteristics of a virus.

REFERENCES

1. Kilbourne, E. D. (1978). Proc. Natl. Acad. Sci. USA, 75, 6258.
2. Kilbourne, E. D., McGregor, S., and Easterday, B. C. (1979). Infect. & Immun. 26, 197.
3. Kilbourne, E. D., McGregor, S., and Easterday, B. C. (1981). In "The Replication of Negative Strand Viruses" (D. H. Bishop and R. W. Compans, eds.), p. 449 Elsevier North Holland, Inc.
4. Kilbourne, E. D., Gerhard, W., and Whitaker, C. Proc. Natl. Acad. Sci. USA, in press.
5. Both, G. W., Shi, C. H., and Kilbourne, E. D. Proc. Natl. Acad. Sci. USA, in press.
6. Wiley, D. C., Wilson, I. A., and Skehel, J. J. (1981) Nature 289, 373.

SELECTION OF RECEPTOR VARIANTS FROM HUMAN AND AVIAN INFLUENZA ISOLATES WITH THE H3 HEMAGGLUTININ[1]

Gary N. Rogers, Xiao-fan Wang, Thomas J. Pritchett,
Lynne F. Haber, and James C. Paulson

Department of Biological Chemistry
UCLA School of Medicine
Los Angeles, CA 90024

Structural changes in the hemagglutinin molecule are responsible for the irregular occurance of influenza pandemics (1). Increasing evidence suggests that new pandemics may originate by transmission of influenza viruses from lower animals to man either by direct passage or by genetic recombination (2). At least one of the known human pandemic strains [A/Hong Kong/68 (H3N2)] appears to have arisen by such a recombination mechanism, the most likely source of the genetic information coding for the hemagglutinin of A/Hong Kong/68 being from a virus related to A/duck/Ukraine-/1/63 (2-4). Comparison of receptor binding, which is also a property of the hemagglutinin molecule, for several human and animal (avian and equine) influenza H3 isolates revealed a remarkable correlation between the ability to bind different sialyloligosaccharide structures and the host of origin (5). While human isolates of this subtype bind strongly to erythrocytes enzymatically modified to contain terminal SAα2,6Gal sequences, avian and equine isolates preferentially bind erythrocytes bearing SAα2,3Gal structures.[2] It has recently been shown that such differences in receptor specificity can arise as the result of a single amino acid substitution in the receptor binding site (6,7).

Since receptor specificity may be one factor relevant to adaptation of influenza viruses from one host to another, it became of interest to develop laboratory models which simulate potential host specified selective pressures that favor growth of one receptor type over another. The present study describes the isolation of receptor variants from human and avian influenza isolates bearing the H3 hemagglutinin by two separate approaches.

---

[1]This work was supported by USPHS research grant AI-16165.

[2]The abbreviations SAα2,6Gal and SAα2,3Gal refer to the terminal sialic acid (SA) linkage commonly found to terminate glycoprotein oligosaccharides linked to asparagine and to threonine or serine, respectively (8,9).

Copyright © 1984 by Academic Press, Inc.
All rights of reproduction in any form reserved.
ISBN 0-12-183501-4

TABLE 1

COMPARISON OF RECEPTOR SPECIFICITY FOR A/MEMPHIS/102/72 (M1) CLONES ISOLATED BY GROWTH IN THE PRESENCE OF HORSE SERUM

Virus[a]	Hemagglutination of Erythrocytes[b]				Adsorption[c]	
	Native	Asialo	SAα2,3Gal	SAα2,6Gal	SAα2,3Gal	SAα2,6Gal
	(HA Titer)				(%)	
Parent population	64	0	8	64	27.4	92
Insensitive clones						
1	64	0	64	0	114.6	0
2	64	0	0	32	31.8	26.8
3	128	0	64	0	91.3	2.9
4	64	0	64	0	101.4	0
5	64	0	64	0	173.0	0
6	64	0	64	0	119.8	0
7	64	0	64	0	129.5	0

[a]Virus preparations used in these experiments were MDCK cell lysates.

[b]Human erythrocytes were either unmodified (Native), treated with Vibrio cholera sialidase (Asialo), or sialidase treated cells resialylated with CMP-SA and purified sialyltransferases to contain the SAα2,3Gal or SAα2,6Gal sequences as reported earlier (5). Hemagglutination (HA) titers were determined by standard procedures and are expressed as the reciprocal of the greatest dilution of virus at which agglutination occurred.

[c]Adsorption of virus to modified erythrocytes [5%(v/v)] was performed as previously described (5). Results are expressed as the percentage of virus that bound native cells.

SELECTION OF RECEPTOR VARIANTS WITH GLYCOPROTEIN INHIBITORS OF INFECTION

The ability of glycoprotein inhibitors of infection, such as α_2 macroglobulin, present in non-immune horse serum to cause selection of receptor variants from human influenza isolates has been amply documented (6,10-12). these glycoproteins apparently act as receptor analogs, selectively suppressing the growth of particles that bind the SAα2,6Gal linkage, allowing growth to predominance of preexisting variants exhibiting the inhibitor insensitive, SAα2,3Gal specific phenotype. Homogeneous populations of receptor variants can be obtained by standard cloning procedures either during or subsequent to selection with horse serum. The results presented in Table 1 illustrate the power of this technique. Monolayers of Madin Darby canine kidney (MDCK) cells were inoculated, in the presence of 1% (^{v}/v) horse serum, with a laboratory adapted strain of the human isolate A/Memphis/102/72 (M1) and grown in the presence of medium containing 0.75% agar and 0.2% (^{v}/v) horse serum. While numerous small centers of infection could be identified under the microscope after a growth period of five days, only seven large, clearly visible plaques (of 2.5×10^4 plaque forming units that were applied) that exhibited typical morphology were observed. The progeny from each of these plaques were grown seperately in the absence of selective pressures and their receptor specificity was examined in both hemagglutination and direct adsorption assays. The results, presented in Table 1, show that while the parental population (M1) agglutinated and bound to erythrocytes containing either the SAα2,6Gal or the SAα2,3Gal linkage with a clear preference for the latter, six of the seven horse serum insensitive clones bound exclusively to erythrocytes modified to contain the SAα2,3Gal linkage. One of the insensitive variants agglutinated only SAα2,6Gal containing erythrocytes, but exhibited approximately equal binding to both the SAα2,3Gal and the SAα2,6Gal linkages when examined by adsorption to derivatized erythrocytes. The existance of influenza hemagglutinins exhibiting apparent dual specificity has been reported previously (5,6).

The selection procedures described above were also applied to two other seed stocks of A/Memphis/102/72 that exhibited very different receptor properties. The original isolate of A/Memphis/102/72 (M0) after a single passage in eggs, which agglutinates cells containing the SAα2,6Gal linkage but not cells containing SAα2,3Gal, did not yield any large plaques of SAα2,3Gal specific variants from an input of 5×10^4 PFU. In contrast, a second egg adapted laboratory strain (M2), which agglutinates only SAα2,3Gal derivatized cells, showed no inhibition by horse serum. The variations observed in the receptor specificity and horse serum senistivity of different preparations of A/Memphis/102/72 suggest that receptor variants are already present in the population. Selection of inhibitor insensitive variants by growth in the presence of horse serum is, therefore, simply a reflection of the heterogeneity found in the receptor binding properties of individual influenza isolates (10,11).

SELECTION OF RECEPTOR VARIANTS BY ADSORPTION TO DERIVATIZED ERYTHROCYTES

An extensive examination of potential glycoprotein inhibitors has failed to identify any inhibitor with properties analogous to horse serum but specific for virus that recognizes oligosaccharides terminating in SAα2,3Gal sequences. Therefore selection of SAα2,6Gal variants from a predominantly SAα2,3Gal specific population required a different approach. This was accomplished by selective adsorption and elution of A/duck/Ukraine/1/63 with erythrocytes resialylated to contain SAα2,6Gal sequences. Adsorption to the derivatized erythrocytes, under conditions of a large viral excess (approximately ten fold) was performed as previously described (5,12). The bound virus was eluted from red blood cells by addition of bacterial, Clostridium perfingens, sialidase (250 mU/ml) and incubation for 3 h at 37°C. The virus released from cells was then grown in MDCK cell culture to amplify. After a total of three such cycles the selected virus population exhibited a 15-20 fold increase in the level of detectable binding to SAα2,6Gal derivatized erythrocytes. A fourth adsorption was performed and the eluted virus was plaqued on MDCK cells without an intermediate amplification step. Initially sixty primary clones were picked and screened for growth on MDCK cells, in 96 well culture plates, in the presence and absence of 0.2% ($^{v}/v$) horse serum. The variants were classified into two categories depending on their ability to grow in the presence of horse serum. On the basis of previous work with human isolates (6,12), insensitive (HS^-) variants, i.e. those viruses that grew both in the presence and absence of horse serum, were expected to be SAα2,3Gal specific while senistive (HS^+) variants, those that grew in the absence but not in the presence of horse serum, were expected to be SAα2,6Gal specific. The receptor specificities of representative HS^+ and HS^- variants were examined directly by adsorption to derivatized erythrocytes. A summary of the results is shown in Table 2. As predicted all of the HS^- variants examined bound specifically to erythrocytes containing the SAα2,3Gal sequences. Of the senistive (HS^+)

TABLE 2
SURVEY OF RECEPTOR SPECIFICITY OF A/DUCK/UKRAINE/1/63 CLONES PICKED FOR SENSITIVITY AND INSENSITIVITY TO GROWTH IN THE PRESENCE OF HORSE SERUM

Type[a]	Receptor Specificity[b] (number of clones specific for linkage type/ number of clones tested)		
	SAα2,3Gal	SAα2,6Gal	SAα2,3Gal/ SAα2,6Gal
Insensitive (HS⁻)	5/5	0/5	0/5
Sensitive (HS⁺)	9/14	5/14	0/14

[a]Primary clones were classified inhibitor insensitive (HS⁻) or inhibitor sensitive (HS⁺) as described in the text.

[b]Receptor specificity of the primary clones is defined by adsorption of virus to derivatized human erythrocytes containing terminal SAα2,3Gal or SAα2,6Gal oligosaccharide sequences.

variants, five of the fourteen viruses examined were specific for red blood cells containing the SAα2,6Gal linkage while the remaining nine clones bound only SAα2,3Gal derivatized erythrocytes. These nine apparently inhibitor sensitive, SAα2,3Gal specific viruses are presumed to be false positives and not a new class of receptor type since upon reexamination, under more stringent conditions, growth in MDCK cell culture was not inhibited by the presence of horse serum.

Ten primary clones, five of each receptor type, were further purified by two additional plaque to plaque passes to ensure homogeneity of the populations. The receptor specificity for each of the ten variants was assessed by both hemagglutination and adsorption to derivatized erythrocytes. All of the variants processed in this manner retained the receptor properties of the primary clones. Adsorption of virus to different concentrations of enzymatically modified erythrocytes for UK 3, a representative SAα2,6Gal specific variant, is compared to the original parent population in Fig. 1. The cloned variant, UK 3, bound only erythrocytes containing the terminal SAα2,6Gal structure. In contrast, the parent strain bound strongly to SAα2,3Gal derivatized erythrocytes with no detectable binding to the

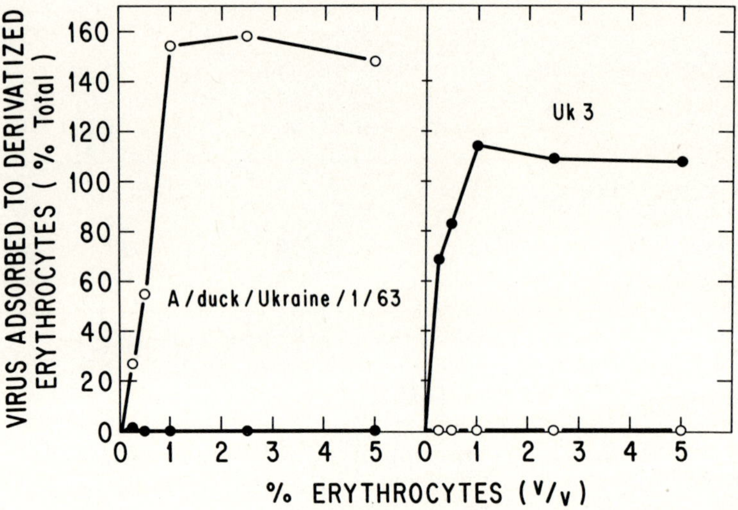

FIGURE 1. Comparison of A/duck/Ukraine/1/63 and an inhibitor sensitive variant by adsorption of virus to enzymatically modified erythrocytes. Preparations of A/duck/Ukraine/1/63 (right) and the cloned variant UK 3 (left) were incubated with increasing concentrations of derivatized erythrocytes containing the terminal SAα2,6Gal (●) or SAα2,3-Gal (○) linkages on cell surface sialyloligosaccharides. Viral adsorption was monitored by assaying washed, resuspended erythrocytes for the presence of viral neuraminidase as described (5). Results are expressed as percent of viral neuraminidase activity bound to 5% (V/v) native erythrocytes.

SAα2,6Gal containing erythrocytes. All of the cloned inhibitor insensitive, SAα2,3Gal specific variants exhibited binding properties similar to the parent strain while adsorption data for the other inhibitor sensitive, SAα2,6Gal specific variants closely resembled that shown for UK 3 (Fig. 1). Furthermore agglutination of native erythrocytes by the SAα2,6Gal specific variants was inhibited by purified equine α_2 macroglobulin whereas the SAα2,3Gal specific variants were not. Thus receptor variants from A/duck/Ukraine/1/63 exhibit all the properties previously associated with the corresponding receptor type isolated from the human H3 isolates A/Memphis/102/72 and X-31 (6,12).

DISCUSSION

A close association between receptor binding properties and host species of origin has been observed in field isolates of influenza virus bearing the H3 hemagglutinin (5). The data presented in this report illustrate two possible mechanisms by which host derived selective pressures could result in growth to predominance of a virus bearing a hemagglutinin with characteristic binding properties. The first is selective inhibition of infection by soluble glycoproteins. Horse serum glycoproteins strongly suppress infection of viruses that bind to cell surface oligosaccharides terminating in a SAα2,6Gal linkage allowing growth to predominance of variants that bind terminal SAα2,3Gal sequences (10-12). This is demonstrated by the ability to isolate <u>preexisting variants</u> with the inhibitor insensitive, SAα2,3-Gal specific phenotype from human H3 isolates, consisting of predominantly inhibitor sensitive, SAα2,6Gal specific particles (6,12). Such a mechanism would be equivalent to inhibition of viral infection by mucins or other soluble glycoproteins lining the respiratory tract.

Another potential means of host selection is at the level of the cell surface. In this case a host cell may only elaborate sialyloligosaccharides recognized as receptor determinants by one virus receptor type and not by others, thus lending a selective advantage to the virus that can adsorb to the cell surface. Using enzymatically modified erythrocytes containing the SAα2,6Gal linkage we have enriched a laboratory population of A/duck/Ukraine/1/63 for receptor variants that exhibit the inhibitor sensitive, SAα2,6Gal specific phenotype from a population comprised largely of inhibitor insensitive, SAα2,3Gal specific particles. Similar selection of receptor variants has been observed by mixed infection of influenza viruses with contrasting receptor specificities on enzymatically modified MDCK cells (S. Carroll and J. Paulson, unpublished observations).

The observations presented here clearly demonstrate the potential of sialyloligosaccharides elaborated by a host to exert selective pressures leading ultimately to the emergence of influenza virus receptor variants during host adaptation.

ACKNOWLEDGEMENTS

We are grateful to Dr. R.G. Webster for the generous gifts of seed stocks of A/Memphis/102/72 and A/duck/Ukraine /1/63, and for helpful discussion during the course of this work. We also wish to thank Jasminder Weinstein and Ursula de Souza-e-Silva for preparation of the sialyltransferases.

REFERENCES

1. Webster, R.G. and Laver, W.G. (1975). In "The Influenza Viruses and Influenza" (E.D. Kilbourne, ed.) p. 269. Academic Press, New York.
2. Laver, W.G. and Webster, R.G. (1979). Brit. Med. Bull. 35, 29.
3. Scholtissek, C., Rohde, W., von Hoyningen, V. and Rott, R. (1978). Virology 87, 13.
4. Fang, R., Min Jou, W. Huylebroeck, D., Devos, R. and Fiers, W. (1981). Cell 25, 315.
5. Rogers, G.N. and Paulson, J.C. (1983). Virology 127, 361.
6. Rogers, G.N., Paulson, J.C., Daniels, R.S., Skehel, J.J., Wilson, I.A. and Wiley, D.C. (1983). Nature 304, 76.
7. Wilson, I.A., Skehel, J.J. and Wiley, D.C. (1981) Nature, 289, 366.
8. Thomas, D.B. and Winzler, R.J. (1969). J. Biol. Chem. 244, 5943.
9. Yoshima, H., Furthmayr, H. and Kobata, A. (1980). J. Biol. Chem. 255, 9713.
10. Choppin, P.W. and Tamm, I. (1960). J. Exp. Med. 112, 895.
11. Choppin, P.W. and Tamm, I. (1960). J. Exp. Med. 112, 921.
12. Rogers, G.N., Pritchett, T.J., Lane, J.L. and Paulson, J.C., Virology, in press.

ANTIGENIC ALTERATIONS OF THE INFLUENZA VIRUS HEMAGGLUTININ DURING THE INFECTIOUS CYCLE[1]

Thomas Bächi,[2] Jonathan W. Yewdell,[3] and Walter Gerhard[3]

[2]Institute for Immunology and Virology
University of Zürich
8028 Zürich, Switzerland

[3]The Wistar Institute of Anatomy and Biology
36th Street at Spruce
Philadelphia, Pennsylvania 19104

ROLES OF THE HEMAGGLUTININ IN THE INFECTIOUS CYCLE

The hemagglutinin glycoprotein of influenza virus has important functions in the initiation of infection: it mediates adsorption of the virus particles to cell surface receptors and is responsible for the subsequent uncoating of the virus by a process of fusion between viral and host membranes. The latter event depends on the exposure of the cleaved hemagglutinin to acid pH and has been documented in detail by experiments on influenza virus-induced hemolysis and cell fusion (1-5). *In vivo*, the fusion step is part of the endocytic pathway of influenza virus infection and is triggered by the low pH of the endosomal compartment (6, 7). Exposure of virus or isolated hemagglutinin to acid pH *in vitro* results in conformational (8) and antigenic changes (9-11) in the hemagglutinin molecule. Recently, we have found that certain monoclonal anti-hemagglutinin antibodies detect antigenic alterations that coincide with the acid activation of this molecule. One of these antibodies has now been used to examine the replication of influenza virus by immunofluorescence techniques. By virtue of this antibody's ability to recognize acid-induced changes of the hemagglutinin molecule, it was possible to directly visualize *in situ* the interaction of the virus with the endosomal compartment.

[1]This work was supported by Public Health Service grants AI-13989 and CA-10815 from the National Institutes of Health.

Copyright © 1984 by Academic Press, Inc.
All rights of reproduction in any form reserved.
ISBN 0-12-183501-4

IMMUNOFLUORESCENCE OF INFLUENZA VIRUS HEMAGGLUTININ IN MDCK CELLS

Monoclonal Anti-Hemagglutinin Antibodies. The antigenicity of the hemagglutinin of influenza virus A/PR/8/34 has been studied by antigenic (12) and structural (13) analysis of virus mutants selected in the presence of monoclonal anti-hemagglutinin antibodies. These studies indicated the presence of four major antigenic regions (13). More recently, it was found that many monoclonal anti-hemagglutinin antibodies could clearly differentiate between untreated and pH 5-treated virus in hemagglutination inhibition tests and radioimmunoassays (11). In the present study, two antibodies were employed, one, H28-E23, is equally reactive with the hemagglutinin of untreated and pH 5-treated virus and the other, Y8-10C2, reacts strongly with the hemagglutinin on pH 5-treated virus and reacts poorly, if at all, with the hemagglutinin on untreated virus. The pattern of reactivity of these two antibodies at different stages of virus infection is summarized in Table 1. The intensity of the reaction of H28-E23 with infected cells seems to be solely influenced by the amount of hemagglutinin. In contrast, the reactivity of Y8-10C2 with the hemagglutinin is dependent upon the stage of the hemagglutinin in the infectious cycle as further described below.

Infection of Cells by Influenza Virus. Influenza virus replicates well in Madin-Darby canine kidney (MDCK) cells, but the inefficient cleavage of the hemagglutinin by these cells results in the production of largely non-infectious progeny. This provides a good system for studying viral replication under single cycle growth conditions. For the present experiments, MDCK cells were seeded onto microscope glass coverslips and were infected 1 or 2 days later with stock virus of the A/PR/8/34 strain grown in the allantoic cavity of chicken eggs. In experiments designed to study the mechanism of virus penetration and uncoating, it was necessary to use a concentrated virus inoculum prepared from virus pelleted from allantoic fluid. For the study of late stages of infection, the cells were infected with a low multiplicity (approximately 1). In all cases, infections were prepared by incubating cells with virus for 30 min at 0°C, washing away unadsorbed virus and incubating the cultures at 37°C with culture medium containing 2% fetal calf serum.

Localization of Hemagglutinin by Immunofluorescence. Cells were prepared for immunofluorescence studies at various stages of infection (Table 1). Immunoreactions were performed on unfixed cells at 0°C (infection stages I and IV) for detection of hemagglutinin present on the cell surface. Hemagglutinin present internally was detected after fixation of the cells with formaldehyde and subsequent permeabilization with Triton X-100. Bound monoclonal antibodies were revealed by reaction with goat anti-mouse IgG-rhodamine conjugate (Cappel).

In Figure 1, micrographs of four representative stages of a single cycle infection of MDCK cells with influenza virus A/PR/8/34 are shown (compare also to Table 1). In an attempt to localize the immunofluorescence (Fig. 1a,c,e,g), the cells were further examined by Nomarski interference contrast optics (Fig. 1b,d,f,h). Following adsorption, a large amount of surface-bound virus could be detected using antibody H28-E23, which also detected viral penetration (stage II) (data not shown). This constrasted markedly with the inability of antibody Y8-10C2 to detect surface-bound

TABLE 1
IMMUNOFLUORESCENCE OF HEMAGGLUTININ IN INFECTED CELLS

Infection stage	Time	Antibodies H28-E23[a]	Antibodies Y8-10C2[b]	Localization of hemagglutinin
I Adsorption[c]	0 min	+++	-[g]	Cell surface
II Uncoating[c]	10 min[e]	+	+[f]	Endosome
II Uncoating[c]	30 min	++	++[f]	Endosome
III Synthesis[d]	90 min	+	+	Nuclear membrane and RER
IV Assembly[d]	180 min	+++	-[g]	Cell surface

[a]Anti-hemagglutinin antibody reacting with native and pH 5-treated virus.

[b]Antibody reacting with pH 5 but not with native virus.

[c]Experiments using high multiplicity of infection. Virus was allowed to adsorb to cells for 30 min at 0°C.

[d]Experiments using low multiplicity of infection.

[e]Time indicates time of incubation of cells at 37°C after adsorption stage.

[f]Antigen not detectable when cells are infected in presence of ammonium chloride.

[g]Antibody binding can be achieved only by treatment of cells for 5 min at 37°C at pH 5.

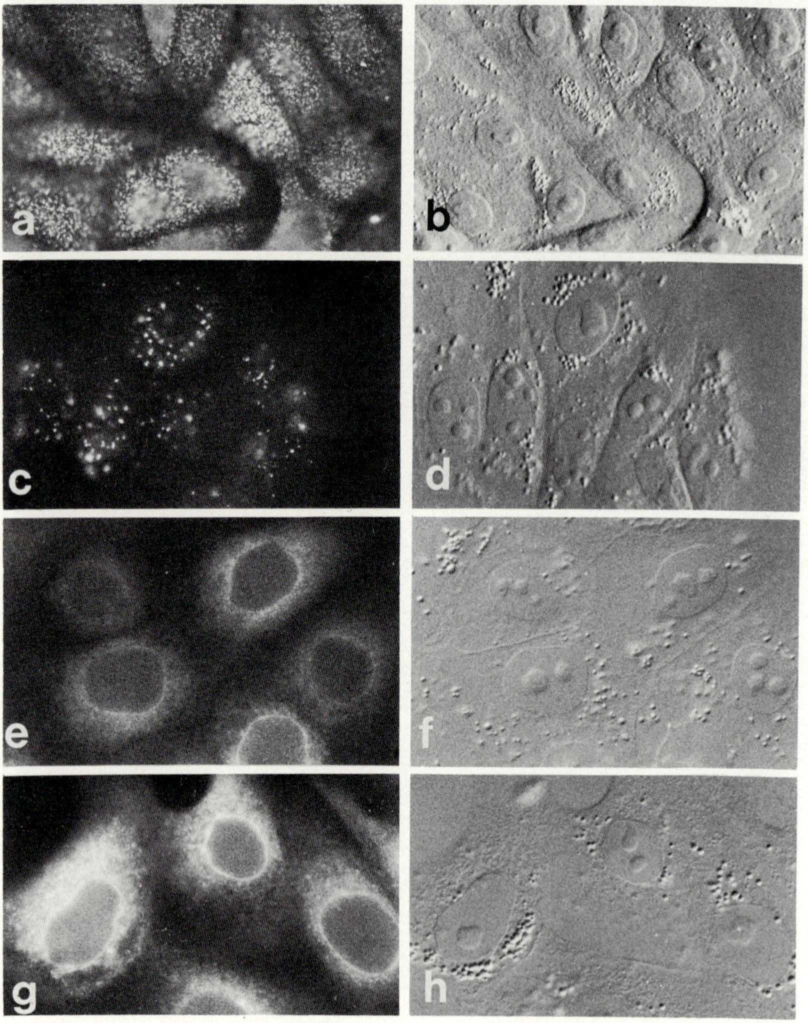

FIGURE 1. Immunofluorescence of MDCK cells after infection with influenza virus (a,c,e,g) and the corresponding pictures taken with Nomarski optics (b,d,f,h). Following incubation of cells and virus for 30 min at 0°C, antibody H28-E23 reveals virus adsorbed onto the cell surface (a). After incubation of cells for 10 min at 37°C, antibody Y8-10C2 becomes reactive with hemagglutinin of virus exposed to the low pH of the endosomal compartment (c). The biosynthesis of the hemagglutinin is first seen after 90 min of infection (e) and reaches maximal fluorescence after 180 min (g).

virus. On the other hand, Y8-10C2 proved to be a sensitive probe for internalized virus. Ten minutes after shifting the cultures to 37°C following viral adsorption at 0°C, a granular fluorescence was observed which almost certainly represents the hemagglutinin in the endosomal compartment (Fig. 1c). This pattern of fluorescence steadily increased during the first 30 min following the temperature shift, and decreased thereafter.

The fate of the hemagglutinin during penetration was further investigated by infecting cells in the presence of ammonium chloride (which has been shown to neutralize the pH of the endosomal compartment). This treatment resulted in the accumulation of virus in the cytoplasm which remained nonreactive with antibody Y8-10C2. This observation agrees with previous studies reporting the inhibition of uncoating of influenza virus by cells treated with ammonium chloride (6). The reactivity of penetrated virus with antibody Y8-10C2 is therefore likely to reflect an antigenic change of the hemagglutinin in the cell which is similar, if not identical, to the antigenic conversion of this molecule *in vitro* by treatment at pH 5.

The appearance of newly synthesized hemagglutinin was signaled 90 min after viral penetration by a distinct perinuclear fluorescence and a gradual increase in cytoplasmic fluorescence. Identical patterns were observed using either antibody H28-E23 (data not shown) or Y8-10C2 (Fig. 1e,g). In contrast, only H28-E23 was reactive with hemagglutinin transported to the cell surface. Antibody Y8-10C2 was reactive with cell surface hemagglutinin only after the cells were briefly exposed to acid media (pH 5 for 5 min at 37°C) prior to immunoreaction.

Alterations in Hemagglutinin Structure During the Infectious Cycle Revealed by Antibody Y8-10C2. Using antibody Y8-10C2 as a probe of hemagglutinin structure, we have demonstrated that shortly after internalization the hemagglutinin undergoes the conformational alterations which we have previously shown to coincide with acid-induced activation of viral fusion activity. This provides additional direct evidence for previous reports that fusion of viral and cellular envelopes occur only after internalization (6, 7). We have also shown that the acid-induced determinant recognized by antibody Y8-10C2 is fully accessible to antibody on newly synthesized hemagglutinin present intracellularly but not on hemagglutinin which has been transported to the cell surface. The identical patterns of intracellular fluorescence observed at later times in the infectious cycle with antibodies H28-E23 and Y8-10C2 suggest that the masking of the epitope occurs after the exit of the hemagglutinin from the Golgi complex. In preliminary experiments using the potent inhibitor of glycolysis tunicamycin, we have failed to observe alterations in the binding pattern of these antibodies. This suggests that glycosylation is not responsible for the biosynthetic alterations in the hemagglutinin detected by antibody Y8-10C2. Instead, this dramatic effect may be due to either a) other modifications of the hemagglutinin monomer, b) association of hemagglutinin monomers into the trimeric form found on cell surfaces and virions, or c) association of the hemagglutinin with other viral components. Experiments are currently in progress to distinguish between these possibilities.

ACKNOWLEDGMENTS

This work was done with the excellent technical assistance of Miss R. Keller.

REFERENCES

1. Maeda, T., and Ohnishi, S. (1980). FEBS Lett. 122, 283.
2. Lenard J., and Miller, D. K. (1981). Virology 110, 479.
3. Maeda, T., Kawasaki, K., and Ohnishi, S. (1981). Proc. Natl. Acad. Sci. U.S.A. 78, 4133.
4. White, J., Kartenbeck, J., and Helenius, A. (1982). EMBO J. 1, 217.
5. White, J., Matlin, K., and Helenius, A. (1981). J. Cell Biol. 89, 674.
6. Matlin, K. S., Reggio, H., Helenius, A., and Simons, K. (1981). J. Cell Biol. 91, 601.
7. Yoshimura, A., Kuroda, K., Kawasaki, K., Yamashina, S., Maeda, T., and Ohnishi, S. (1982). J. Virol. 43, 284.
8. Skehel, J. J., Bayley, P. M., Brown, E. B., Martin, S. R., White, J. M., Waterfield, M. D., Wilson, I. A., and Willey, D. C. (1982). Proc. Natl. Acad. Sci. U.S.A. 79, 968.
9. Webster, R. G., Brown, L. E., and Jackson, D. C. (1983). Virology 126, 587.
10. Daniels, R. S., Douglas, A. R., Skehel, J. J., and Wiley, D. C. (1983). J. Gen. Virol. 64, 1657.
11. Yewdell, J. W., Gerhard, W., and Bächi, T. (1983). J. Virol., in press.
12. Gerhard, W., Yewdell, J., Frankel, M. E., and Webster, R. (1981). Nature 290, 713.
13. Caton, A. J., Brownlee, G. G., Yewdell, J., and Gerhard, W. (1982). Cell 31, 417.

ANTIGENIC DRIFT OF THE H1 SUBTYPE HAEMAGGLUTININ FROM 1977 TO 1980

F Lucy Raymond, Andrew J Caton, George G Brownlee

Sir William Dunn School of Pathology
University of Oxford, Oxford, UK

Nancy J Cox, Alan P Kendal

Center for Disease Control
Atlanta, Georgia 30333, USA

Previous analysis of an extensive set of laboratory-selected antigenic mutants of influenza virus A/PR/8/34 (H1N1 subtype) defined the location of 4 antigenic sites on the virus haemagglutinin (1,2). In order to estimate the relevance of these antigenic sites to the nature of antigenic variation which occurs during evolution of the virus in man, we have performed sequence analysis of the antigenic domain of a series of recent H1 subtype strains (3). These strains include the re-emerged H1N1 'Russian flu' A/USSR/90/77 and 4 subsequent antigenically drifted variants. From the results we derive an evolutionary pathway and discuss the observed amino acid substitutions with reference to the antigenic sites characterised previously in A/PR/8/34.

EVOLUTION OF H1N1 SUBTYPE FIELD STRAINS

The amino acid sequence of the HA1 subunit of influenza virus strains A/USSR/90/77, A/Brazil/11/78, A/Lackland/3/78, A/England/333/80 and A/India/6263/80 was derived by nucleotide sequence analysis of virion RNA (3,4).

The deduced amino acid substitutions observed in the five strains were compared to the definitive HA1 domain of the haemagglutinin of A/PR/8/34 (Table I)(5). Forty-two amino acid substitutions occur between A/PR/8/34 and A/USSR/90/77 but no insertions or deletions were observed. This degree of amino acid substitution constitutes an evolutionary drift of 13%. As A/USSR/90/77 is closely related to the strain A/FW/1/50 and probably represents the re-emergence of a 1950 virus strain (6), this corresponds to a rate of antigenic drift of

Copyright © 1984 by Academic Press, Inc.
All rights of reproduction in any form reserved.
ISBN 0-12-183501-4

0.8% per year between 1934 and 1950. This is a similar rate to that observed for H3 subtype strains (7), suggesting that influenza viruses of different subtype probably undergo relatively constant rates of evolution during circulation in man.

Amino acid substitution at residues 96, 158 and 163 cause A/USSR/90/77 to gain 3 potential glycosylation sites (i.e. AsnX $\frac{\text{Ser}}{\text{Thr}}$ (8)) compared to A/PR/8/34. These extra glycosylation sites are retained in the subsequent strains with the exception of the site at residue 163 which is lost again in A/England/333/80.

Fifteen amino acid substitutions have occurred during the evolution of recent H1N1 subtype strains between 1977 and 1980 (Table I). On the basis of the individual amino acid substitutions an evolutionary pathway for the recent H1N1 subtype strains can be postulated (Figure 1). The epidemic strains are all shown as dead ends, the mainline being maintained through the postulated intermediates X_0, X_1, X_2, X_3. Altered amino acid residues are noted by number and where a residue is part of an antigenic site the residue number is underlined.

ANTIGENIC FEATURES OF THE H1 SUBTYPE

Table I identifies the positions of changed amino acid residues with the location of the previously established antigenic sites, Sa, Sb, Ca and Cb of A/PR/8/34 (1). Assuming that the three-dimensional structure of A/USSR/90/77 and A/PR/8/34 are similar, 69% of the changed surface residues in A/USSR/90/77 map to antigenic sites. It seems that during the evolution to A/USSR/90/77 mutations on the surface of the molecule occur mainly but not exclusively within dominant antigenic sites and that point mutations have occurred at all four sites.

TABLE I. The location of amino acid changes occurring in the recent H1 strains compared to A/PR/8/34 (5). Only residues differing from that of A/PR/8/34 are described.
a) The residues are numbered to facilitate alignment with the sequence of A/Aichi/2/68 and the location of amino acids on the three dimensional structure of the haemagglutinin (11).
b) The one letter amino acid code indicates the residue for each virus strain studied. Residue 65 for A/Lackland/3/78 was not determined.
c) This identifies antigenic residues by the sites (1). A (?) indicates residues which lie immediately adjacent to antigenic residues in the three dimensional structure and are tentatively assigned to the sites shown.

Amino acid[a] Position	A/PR/8/34	A/USSR/90/77	A/Lackland/3/78	A/Brazil/11/78	A/England/333/80[b]	A/India/6263/80	Antigenic site[c]	Location[d] on molecule
65	N	N		S	S	S	-	Buried
70	D	E	E	E	E	E	?Cb	Surface
77a	P	S	S	S	S	S	?Cb	Surface
79	L	F	F	F	F	F	Cb	Surface
80	P	S	S	S	S	S	?Cb	Surface
81	V	K	K	K	K	K	Cb	Surface
82	R	K	K	K	K	K	Cb	Surface
88	V	A	A	A	A	A	-	Buried
96	I	T	T	T	T	T	-	Surface
101	D	Y	Y	Y	Y	Y	-	Buried
103	I	A	A	A	A	A	-	Buried
125c	S	R	R	R	S	S	?Sa	Surface
129	N	K	K	K	K	K	Sa	Surface
132	T	V	V	I	V	V	-	Surface
134	K	R	R	R	R	R	-	Surface
138	A	S	S	S	S	S	-	Buried
142	A	K	K	K	K	K	?Ca	Surface
157	K	E	E	E	K	K	?Sa/Sb	Surface
158	E	N	N	N	N	N	Sa	Surface
163	K	N	N	N	S	N	Sa	Surface
165	K	S	S	S	S	S	Sa	Surface
166	N	K	K	K	K	K	Sa	Surface
171	K	N	N	N	N	N	?Ca	Surface
173	G	E	E	E	E	E	Ca	Surface
182	I	V	V	V	V	V	-	Buried
188	S	I	I	I	I	I	-	Surface
189	K	E	K	E	E	E	?Sb	Surface
190	D	D	D	D	D	N	-	?Surface
192	Q	K	K	K	K	K	Sb	Surface
193	N	T	T	T	T	T	Sb	Surface
196	Q	R	R	R	R	R	Sb	Surface
197	N	K	K	K	K	K	?Sb	Surface
206	T	S	S	S	S	S	-	Buried
208	N	N	N	N	H	H	?Ca	Surface
219	E	E	E	K	K	K	?Ca	Surface
225	D	G	D	G	G	D	Ca	Surface
227	A	A	A	E	E	E	?Ca	Surface
230	M	I	I	I	I	I	-	Buried
238	K	E	E	E	E	E	-	Surface
255	R	W	W	W	W	W	-	?Surface
256	Y	H	Y	Y	Y	Y	-	Buried
260a	S	N	S	S	S	S	-	Buried
275	H	D	D	D	D	D	-	Surface
278	N	D	D	D	D	D	-	Surface
279	T	T	T	T	A	A	-	Surface
285	L	Q	Q	Q	Q	Q	-	Surface
297	I	I	V	V	V	V	-	Buried
312	A	T	T	T	T	T	-	Surface

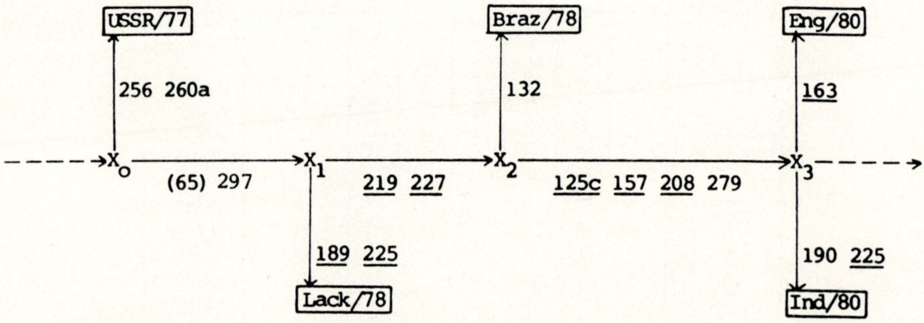

FIGURE 1. The evolution of recent epidemic H1 subtype virus strains. X_0, X_1, X_2 and X_3 are proposed intermediates during the evolutionary pathway. Since residue 65 is unknown for A/Lackland/3/78, it is bracketed. Antigenic residues are underlined (see also Table I).

Similarly, each of the subsequent virus strains emerging after A/USSR/90/77 have acquired a point mutation in at least one of the antigenic sites defined for A/PR/8/34 (Figure 1 and Table I). These results support the previous identification of the 4 antigenic sites for A/PR/8/34 and extend it to other H1N1 subtype strains. Furthermore, it suggests that the immune recognition of the haemagglutinin in man is similar to that of the mouse system used in the model experiments and supports the view that antibody selection plays a crucial role during evolution of new virus strains in the field.

A/England/333/80 vs. A/India/6263/80. Although the epidemic properties of a virus are multifactorial it is widely accepted that structural changes in the haemagglutinin lead to epidemic outbreaks of influenza in man (9). A/England/333/80 is epidemiologically more successful than A/India/6263/80 (10) and we tentatively suggest that this may result from one or both of the following structural differences. Firstly, A/India/6263/80 has a changed residue at 190. Wilson et al (11) have proposed that residue 190 constitutes part of the virus-host cell receptor binding site. A change from a charged asp residue (A/England/333/80) to an uncharged asn residue (A/India/6263/80) could affect virus receptor binding detrimentally. Secondly, it is possible that the loss of a carbohydrate attachment site at residue 163 during the evolution to A/England/333/80 caused a major antigenic change which conferred a selective advantage for this emerging strain over the other circulating virus strain.

SUMMARY

Five H1N1 subtype strains isolated between 1977 and 1980 share a minimum of 85% amino acid sequence homology with A/PR/8/34. Most of the surface amino acid substitutions which occurred during the evolution of A/PR/8/34 to A/USSR/90/77 and subsequently in the 1977-1980 strains are located in the 4 antigenic sites previously defined by an analysis of laboratory-selected mutants of A/PR/8/34. Identification of changed amino acid residues allowed an evolutionary pathway for the strains to be established, and may indicate that the different epidemic properties of individual strains are a consequence of only a few amino acid changes.

ACKNOWLEDGEMENTS

We thank Keith Gould for oligonucleotide synthesis facilities and Don Wiley for advice on which are surface residues on the three-dimensional structure, and the MRC for a programme grant to GGB supporting this work. FLR thanks Lady Margaret Hall for financial support early in this project and the MRC for a Research Studentship.

REFERENCES

1. Caton, A.J., Brownlee, G.G., Yewdell, J.W. and Gerhard, W. (1982). Cell 31, 417-427.
2. Caton, A.J., Raymond, F.L., Brownlee, G.G., Yewdell, J.W. and Gerhard, W. (1983). Biochem.Soc.Transactions 11, 435-441.
3. Raymond, F.L., Caton, A.J., Cox, N.J., Kendal, A.P. and Brownlee, G.G. (submitted for publication).
4. Sanger, F., Nicklen, S. and Coulson, A.R. (1977). P.N.A.S. 74, 5463-5467.
5. Winter, G., Fields, S. and Brownlee, G.G. (1981). Nature 292, 72-75.
6. Kendal, A.P., Noble, G.R., Skehel, J.J. and Dowdle, W.R. (1978). Virology 89, 632-636.
7. Sleigh, M.J. and Both, G.W. (1981). In: "Genetic Variation among Influenza Viruses" (D.P. Nayak, ed.), pp. 341-353, Academic Press, New York.
8. Neuberger, A., Gottschalk, A., Marshall, R.D. and Spiro, R.G. (1972). In: "Glycoproteins: their composition, structure and function" (A. Gottschalk, ed.) pp. 450-490, Elsevier, Amsterdam.
9. Stuart-Harris, C. (1979). Brit.Med.Bull. 35, 3-8.
10. Cox, N.J., Bai, Z.S. and Kendal, A.P. (1983). Bulletin WHO, 61, 143-152.

11. Wilson, I.A., Skehel, J.J. and Wiley, D.C. (1981). Nature 289, 366-373.

THE BINDING SITES TO MONOCLONAL ANTIBODIES ON A/USSR/90/77(H1N1) HEMAGGLUTININ

S. Nakajima, K. Nakajima, and A. P. Kendal

The Institute of Public Health
6-1, Shirokanedai 4 chome, Minato-ku,
Tokyo 108, Japan

and

Centers for Disease Control
Atlanta, Georgia 30333, USA

ANTIGENIC VARIANTS SELECTED IN VITRO

Antigenic drift of the hemagglutinin (HA) molecule of type A influenza viruses has been thought to occur by the accumulation of a series of point mutations in the antigenically important regions of the molecule. In order to study the evolution of antigenic variants of influenza viruses, we have selected antigenic variants of influenza A/USSR/90/77 (H1N1) strain with six monoclonal antibodies to the virus HA (1). Among the variants, the clones which had changed reactivity to monoclonal antibody 264 showed antigenic drift detectable by post-infection ferret sera (1). As most of the natural variants of H1N1 influenza viruses isolated since 1977 also had the change in this epitope (2,3), the epitope recognized by antibody 264 on the HA molecule is thought to play an important role in the antigenic drift of H1N1 influenza viruses. In the present study, we have newly isolated several variant clones selected by antibody 264.

CLONING OF THE RECOMBINANT DNA

The HA gene of B-1-23, a variant clone of A/USSR/90/77 virus selected by antibody 264, was cloned in bacteria as a double-stranded cDNA (dsDNA) copy in the Pst I site of plas-

[1]This work was supported by a research grant from the Science and Technology Agency and a Scientific Research Grant from the Ministry of Education in Japan.

Copyright © 1984 by Academic Press, Inc.
All rights of reproduction in any form reserved.
ISBN 0-12-183501-4

mid pBR322 by the method described (4). Total viral RNA was transcribed into a plus-stranded DNA (cDNA) using 5'-specific dodecanucleotide primer. dsDNA was synthesized from cDNA using 3'terminal loopback of the cDNA as a self primer. Hairpin product of dsDNA was cut with nuclease S_1 and then poly (dC) was added to the S_1-treated dsDNA. Annealing of Pst I-cleaved pBR322 tailed with poly (dG) and dsDNA tailed with poly (dC) was carried out and then E. coli K12 C600 strain was transformed with the annealed DNA. Tetracyclin-resistant, ampicillin-sensitive colonies were picked and the recombinant plasmids containing HA gene were selected by colony hybridization using ^{32}P-labeled recombinant plasmid HA DNA (5).

BIOHAZARD CONSIDERATIONS

To guard against potential biohazards, the transformation and all subsequent manipulations with live bacteria which might carry recombinant plasmids were carried out in P2 physical containment.

NUCLEOTIDE SEQUENCE OF THE CLONED DNA

Among the DNAs containing HA gene copy of B-1-23 variant clone, the pp392 DNA (the longest insert DNA) was sequenced by Maxam-Gilbert method (6). The comparison with the published nucleotide sequence of WSN HA gene showed that the pp392 DNA corresponded to the region coding for nearly entire HA1 polypeptide (7).

NUCLEOTIDE SEQUENCE OF THE HA1 GENE OF THE MONOCLONAL VARIANTS

Nucleotide sequences of the HA gene (HA1 portion) of the parental A/USSR/90/77 virus and monoclonal variant clones were obtained using appropriate ^{32}P-labeled restriction fragment DNAs of the pp392 as primers for sequencing the vRNA by dideoxy chain termination method (8). In B-1-23 variant clone, the nucleotide sequence determined from the cloned DNA (pp392) and that from vRNA differed in one nucleotide base, i. e. at position 599, T read from cloned DNA while C from vRNA, resulting in the amino acid difference of phenylalanine vs serine. The result obtained from vRNA was adopted in the present study.

AMINO ACID SUBSTITUTION IN VARIANT CLONES

Amino acid sequences of the HA1 polypeptides of A/USSR/90/77 virus and monoclonal variants were deduced from the

```
                               1        10        20        30        40        50
Aichi/68(H3)    MKTIIALSYIFCLALGQDLPGNDNSTATLCLGHHAVPNGTLVKTITDDQIEVTNATELVQSSSTGKICN-NPHRI
USSR/77(H1)     MKAKLLVLLCALSATDA---------DTICIGYHANNSTDTVDTVLEKNVTVTHSVNLLEDSHNGKLCRLKGIAP
B-1-23(264)

                 60        70        80         90         100       110       120
Aichi/68        LDGIDCTLIDALLGDPHCDVFQNETWDLFV-ERSKA-FSNCYPYDVPDYASLRSLVASSDTLEFITEGF---TWT
USSR/77         LQLGKCNIAGWILGNPECESLFSKKSWSYIAETPNSENGTCYPGYFADYEELREQLSSVSSFERFEIFPKERSWP
B-1-23

                 130        140       150       160       170       180       190       200
Aichi/68        GV-TQNGGSNACKRGPGSGFFSRLNWLTKSGSTYPVLNVTMPNNDNFDKLYIWGIHHPSTNQEQTSLYVQASGRV
USSR/77         KHNVTRGVPASCSHKGKSSFYRNLLWLTEKNGSYPNLSKSYVNNKEKEVLVLWGVHHPSNIEDQKTIYRKENAYV
B-1-23                                                                         N

                      210       220       230       240       250       260        270
Aichi/68        TVSTRRSQQTIIPNIGSRPWVRGLSSRISIYWTIVKPGDVLVINSNGNLIAPRGYFKMRTGKSS-IMRSDAPIDT
USSR/77         SVVSSNYNRRFTPEIAERPKVRGQAGRINYYWTLLEPGDTIIFEANGNLIAPWHAFALNRGFGSGIITSNASMDE
B-1-23

                   280       290      300        310       320    328
Aichi/68        CISECITPNGSIPNDKPFQNVNKITYGACPKYVKQNTLKLATGMRNVPEKQT
USSR/77         CDTKCQTPQGAINSSLPFQNIHPVTIGECPKYVRSTKLRNVTGLRNIPSIQS
B-1-23
```

FIGURE 1. Amino acid sequences of the HA1 polypeptide deduced from the nucleotide sequences in A/Aichi/2/68(H3N2), A/USSR/90/77(H1N1), and B-1-23 variant. B-1-23 is a variant clone of A/USSR/90/77 selected by monoclonal antibody 264.

TABLE 1

CHANGED AMINO ACID RESIDUES IN VARIANT CLONES

Variant clone	Selecting antibody	Altered reactivity in HI test with monoclonal antibody[a]	Changed amino acid residue
B-1-23[b]	264	264	190
B-1-10	264	264	190
B-1-33	264	264	190
B-1-63	264	264	190
B-1-101	264	264	189
B-1-201	264	264	190
D-1-14	110	110	125C[c]
D-1-14/264	110, 264	110, 264	125C[c], 187
F-1-1	22	22, 70, 385	90, 129, 132
A-1-42	70	70, 22	157
A-1-42/264	70, 264	70, 22, 264	130A, 187
C-1-1	W18	W18, 264, 385	125C, 189
C-2-14	W18	W18, 264	125C, 189
E-1-1	385	385, 264	129, 189, 219
E-2-1	385	385	159

[a] Antibody which had HI titer ≥ 16-fold lower than that obtained with A/USSR/90/77 virus is shown.

[b] B-1-101, B-1-201, and the other four variant clones are obtained from independent experiments.

[c] The changed amino acid is different from that seen in C-1-1 and C-2-14.

nucleotide sequences of vRNAs and numbered according to the sequence of Hong Kong (H3N2) HA1 polypeptide (9) (Fig. 1). Amino acids which were absent in H3 subtype HAs were numbered using alphabetic suffixes (e. g. 125A). Substitutions in amino acids in 15 variant clones examined in the present study compared with parental A/USSR/90/77 virus were detected at residues 90, 125C, 129, 130A, 132, 157, 159, 187, 189, and 190 (Table 1). All amino acid changes except in residue 90 were located at the upper external surface of the globular head of the HA1 molecule shown by Wilson et al. (10).

AMINO ACID CHANGES ASSOCIATED WITH EPITOPE 264

Six B-1 group variants selected by antibody 264 and obtained from three independent experiments were studied for their changed amino acids. Two variant clones (D-1-14 and A-1-42) were further neutralized with antibody 264 and the variant clones D-1-14/264 and A-1-42/264 were obtained. The latter two variant clones have lost reactivity in HI test with antibody 264 in addition to their original one. Substitution in amino acid in five B-1 group variants was in residue 190 from aspartic acid (D) to asparagine (N) and in one variant in residue 189 from glutamic acid (E) to lysine (K). Two variant clones newly selected by antibody 264 (A-1-42/264 and D-1-14/264) had the same amino acid substitution in residue 187 from asparagine (N) to lysine (K).

AMINO ACID CHANGES ASSOCIATED WITH OTHER EPITOPES

Epitope W18. The two variant clones selected by antibody W18 (C-1-1 and C-2-14) had reduced reactivity with antibodies W18 and 264. Both had substitutions at two residues, namely 125C and 189, from arginine (R) to glycine (G) and glutamic acid (E) to lysine (K), respectively. The difference in the reactivity in HI test with antibody 385 between two variants was not clarified.

Epitopes 22 and 70. The previous studies had shown the overlapping of these two epitopes (1). Variant F-1-1 selected by antibody 22 contained substitutions in amino acid residues 90, 129, and 132 from threonine (T) to alanine (A), lysine (K) to glutamic acid (E), and threonine (T) to alanine (A), respectively. Considering the location of residue 90 on the 3-dimensional structure of the HA molecule (10), the substitution in this residue might be coincidental and antigenically irrelevant. Variant A-1-42, selected by antibody 70, contained a single substitution at residue 157 from lysine (K) to glutamic acid (E). Variant A-1-42/264, which was further selected by antibody 264, had changes in residues 130A and 187 from asparagine (N) to aspartic acid (D) and asparagine (N) to lysine (K), respectively. The original amino acid change which variant A-1-42 had, was lost in this variant. However, the variant A-1-42/264 showed altered reactivity with antibody 22 and 70. The most probable explanation is that A-1-42 clone contains heterogeneous populations. The same amino acid change in residue 187 is also detected in variant D-1-14/264, which was doubly selected by antibodies 110 and 264. The change in this residue might be assigned to epitope 264.

Epitope 110. Variant D-1-14 selected by antibody 110 had a single amino acid substitution at residue 125C from arginine (R) to lysine (K). The amino acid residue 125C has also been assigned to epitope W18, but the substituted amino acid was different

Epitope 385. Variants E-1-1 and E-2-1 selected by antibody 385 possessed four amino acid substitutions in total at residues 129, 159, 189, and 219 from lysine (K) to glutamic acid (E), glycine (G) to aspartic acid (D), and in the latter two residues, glutamic acid (E) to lysine (K).

BINDING SITES TO MONOCLONAL ANTIBODIES ON A/USSR/90/77 HA

Most amino acid changes found by us in USSR monoclonal variants fell on the antigenic sites proposed for H1 PR8 and H3 subtype HAs (11, 12, 13). Epitope 264 whose change is thought to play an important role in the antigenic drift of H1N1 influenza viruses is the area including amino acid residues 187-190 and probably corresponds to Sb site in PR8 HA or site B in H3 subtype HAs. In two natural variants studied (A/Brazil/11/78 and A/Lackland/3/78), which had altered reactivity with antibody 264, A/Lackland/3/78 had the change in residue 189, but A/Brazil/11/78 did not have any change in residues 187-190 (unpublished data). This indicates either that the amino acids locating closely to residues 187-190 in the 3-D structure is involved in epitope 264 or that the amino acid changes occurred in other residues may cause a conformational change in the epitope 264. Epitope W18 is the area including amino acid residue 125C and probably corresponds to Sa site of PR8 HA. The change in residue 189 associated with epitope W18 may be explained as follows. One possibility is that the structural and/or functional disturbance caused by the change at one residue had to be compensated by the second change at another residue. The other possibility is that one antibody reacted with multiple amino acids at the different positions and was responsible for the multiple changes. The amino acid substitution in residue 125C with charge difference may accompany the change in amino acid residue 189. Epitopes 22/70 are in the area including amino acid residues 129, 130A, 132, and 157 and are probably located on the area between Sa and Sb sites. Epitope 110, including amino acid residue 125C overlaps with epitope W18. Epitope 385 is the area including at least amino acid residues 129, 159, and 189, which are located on the area stradding Sa and Sb sites. The amino acid changes corresponding to Cal, and Ca2, and Cb sites in PR8 HA or site C in H3 subtype HAs were not found, probably because of the small panel of monoclonal antibodies used for the selection of monoclonal variants.

ACKNOWLEDGMENT

We thank Dr. Y. Kuchino for helpful technical advices Dr. R. G. Webster for providing A/USSR/90/77 specific monoclonal antibodies, and Dr. A. Sugiura for reading the manuscript.

REFERENCES

1. Nakajima, S. and Kendal, A. P. (1981). Virology 113, 656.
2. Webster, R. G., Kendal, A. P., and Gerhard, W. (1979).
3. Nakajima, S., Cox, N. J., and Kendal, A. P. (1981). Infec. Immun. 32, 287.
4. Nakajima, S., Nakajima, K., and Kendal, A. P. Virology,in press.
5. Davis, A. R., Hiti, A. L., and Nayak, D. P. (1980). Gene 10, 205.
6. Hiti, A. L., Davis, A. R., and Nayak, D. P. (1981). Virology 111, 113.
7. Maxam, A. M. and Gilbert, W. (1977). Proc. Nat. Acad. Sci. USA 74, 560.
8. Sanger, F., Nicklen, S., and Coulson, A..R. (1977). Proc. Nat. Acad. Sci. USA 74. 5463.
9. Verhoeyen, M., Fang, R., Min Jou, W., Devos, R., Huylebroeck, D., Samen, E., and Fiers, W. (1980). Nature (London) 286, 771.
10. Wilson, I. A., Akehel, J. J., and Wiley, D. C. (1981). Nature (London) 289, 373.
11. Wiley, D. C., Wilson, I. A., and Skehel, J. J. (1981). Nature (London) 289, 373.
12. Caton, A. J., Brownlee, G. G., Yewdell, J. W., and

Gerhard, W. (1982). Cell 31, 417.
13. Underwood, P. A. (1982). J. gen. Virol. 62, 153.

SEQUENCE CHANGES IN THE HEMAGGLUTININ OF AN ENTEROTROPIC H3 INFLUENZA VIRUS

C.W. Naeve, V.S. Hinshaw, and R.G. Webster

Division of Virology and Molecular Biology
St. Jude Children's Research Hospital
Memphis, Tennessee 38101

The hemagglutinin glycoprotein of influenza virus interacts with sialic-acid-containing cell membrane receptors (1). Recent studies have demonstrated that antigenically similar hemagglutinins have different receptor specificities (2), that these differences may correlate with species of origin (3), and that amino and substitutions in the hemagglutinin can alter receptor-binding specificity (4). These substitutions occur at HA1 residue 226, a part of the putative receptor binding site. Thus, it is likely the hemagglutinin and its interactions with host cell receptors will play a key role in the important biological property of tissue tropism.

As an initial step toward understanding the role of the hemagglutinin in tissue tropism, we prepared reassortant viruses between an influenza strain A/Mal/NY/6750/72 (H2N2 [Mal/NY] which replicates in the intestinal tract of ducks and a human strain, A/Udorn/307/72 (H3N2) [Udorn] which does not (5). The reassortants were produced at 42°C in the presence of monoclonal antibodies to select for viruses with the avian strain NA and internal genes and the human strain hemagglutinin. Thirty-four isolates were obtained and analyzed for genetic content.

GENETIC ANALYSIS

The avian-human reassortants, designated R1-R34, were genotyped by a sensitive hybridization - S1 nuclease assay in which ^{32}P-labeled cDNA probes complementary to both parental type virus RNAs are hybridized to the reassortant

[1]This work was supported by Public Health Service research grants AI-02649 and AI-08831 from the National Institute of Allergy and Infectious Diseases, by CORE grant CA-21765 from the National Cancer Institute, and by ALSAC.

Copyright © 1984 by Academic Press, Inc.
All rights of reproduction in any form reserved.
ISBN 0-12-183501-4

virion RNA, treated with S1 nuclease and the surviving duplexes denatured and resolved on agarose gels as described (5). Eight isolates were shown to have the same gene constellation, having received hemagglutinin gene segment 4 from the human strain, Udorn, and the remaining seven gene segments from Mal/NY (Table 1).

BIOLOGICAL ANALYSIS

Since mammalian influenza viruses do not replicate in the intestinal tract of ducks, we examined the effect of the human HA on the enterotropism of the eight reassortants having the Udorn HA as the only mammalian gene. The results of this study (Table 2) show that the Udorn HA restricts replication in the intestinal tract of Peking ducks in 7 of the 8 reassortants with the same genotype. The exception, reassortant R2, is biologically distinct from the other reassortants. R2, like the parental Mal/NY, replicated to high titers in the intestinal tract. These results suggested that a mutation had occurred in the R2 genome, perhaps in the HA, altering the tissue tropism of the virus. To determine if the Udorn HA on the reassortant R2 had changed, R2 was analysed for antigenic differences.

ANTIGENIC ANALYSIS

Antigenic characterization of the reassortants with polyclonal antisera and monoclonal antibodies indicated that R2 was an antigenic variant. To characterize this antigenic

TABLE 1
AVIAN-HUMAN REASSORTANT GENOTYPES

		1-6	8-9	13	15	16	24
PB2, PB1	1,2	D	D	U	U	U	U
PA	3	D	D	U	U	U	U
HA	4	U	U	U	D	D	D
NP	5	D	D	U	U	U	U
NA	6	D	D	U	D	U	D
NS	7	D	D	U	D	D	U
M	8	D	D	U	U	U	U

[a]U= A/Udorn/307/72; D = A/Mal/NY/6750/72.

TABLE 2
REPLICATION OF PARENTAL AND REASSORTANT VIRUSES IN THE INTESTINAL TRACT OF DUCKS

Virus	No. of animals infected/No. of animals inoculated	Peak virus titer ($log_{10}EID_{50}/G$) Feces	Rectum
Parental			
A/Udorn/307/72	0/4	0	0
A/Mallard/NY/6870/78	4/4	7.0	5.5
Reassortants			
R1, R3-R6, R8, R9	0/21	0	0
R2	6/7	6.0	5.8

[a]Pekin ducks were inoculated orally with 10^7 EID_{50} of virus. Tissues were collected at days postinfection and titrated for infectious virus (Webster et al., 1978).

[b]Number of animals infected is based on virus recovery from cloaca, bursa, rectum, or feces from inoculated birds.

change, the reassortants were tested with polyclonal and monoclonal antibodies to H3 hemagglutinin (Table 3). Results of HI tests with antisera specific to the hemagglutinin of Aichi/2/68 or England/42/72 indicated that the reactivity of R2 virus was at least 10-fold different from the parental virus. These results indicate that the hemagglutinin of R2 virus is still of the H3 subtype but is antigenically different from the parental Udorn strain. Monoclonal and polyclonal antibodies to the H2 hemagglutinin of Mal/NY did not react with the R2 virus in either HI or ELISA assays (results not shown).

Monoclonal antibodies specific for the different antigenic regions on the hemagglutinin of Mem/1/71 (H3) have been previously characterized (6). These monoclonal antibodies were tested in HI assays with the reassortant viruses in an attempt to determine which antigenic region of the HA molecule on R2 was altered. Monoclonal antibody to site A (138/1) did not react with either the parental virus Udorn HA or with the reassortants so that no information could be obtained about this region (Table 3). Antibody to site B (4/1) reacted strongly to Udorn and reassortants R1 and R3 to R6 but did not react with R2. Monoclonal antibodies to site C (H14/A2) reacted with the parental Udorn strain and reassortants R3 to R6, but only reacted with the R2 strain after the virus had been disrupted (data not shown), indicating that this antigenic site is probably present but is masked in some way in the intact R2 virion. These results indicate than an antigenic change has occurred in site B of the R2 virus, that site C is still present but perhaps masked, and that site A in the parental virus is changed so that no information could be obtained about this region. Thus, it appeared the R2 HA had acquired a mutation(s) altering its antigenic characteristics and the tissue tropism of the virus.

SEQUENCE ANALYSIS

To determine the location of mutations in the R2 hemagglutinin, the RNA genes coding for the HA1 of both the parental Udorn and the tissue tropism mutant R2 were sequenced. These were obtained by making synthetic oligodeoxynucleotides and using them to prime dideoxynucleotide chain termination sequencing reactions containing total virus RNA, reverse transcriptase and ^{32}P-labeled deoxy-

TABLE 3
ANTIGENIC ANALYSIS OF REASSORTANT INFLUENZA VIRUSES

	HI titers to the following influenza virus antigens		
	Udorn/307/72 Virus	R2 Virus	R1 and R3-R6 Virus
Polyclonal antisera to			
Aichi/2/68 (H3N2)	8,100	140	8,100
Eng/42/72 (H3N2)	24,000	4,000	24,000
R2	10,000	1,500	10,000
Monoclonal antibodies to			
Mem/1/71 (H3)			
138/1 (Site A)	<	<	<
4/1 (Site B)	50,000	<	50,000
H14/A2 (Site C)	900	<	900

The values give the reciprocal of the dilution inhibiting 4 HA units of virus.

< = Less than 50.

nucleotides (7). The deduced HA1 amino acid sequences are presented in Table 4. Only two amino acid changes are present in the R2 HA molecule at residues 226 and 228 both in the receptor binding site. The leucine at residue 226 in Udorn was changed to glutamine in R2 as a result of a codon change from CTG to CAG, while the serine at 228 in Udorn was changed to glycine in R2 as a result of a codon change from AGT to GGT.

Studies by Rogers et al. (4) demonstrated that the change at residue 226 from leucine to glutamine correlated with sensitivity to neutralization by an inhibitor present in horse serum and furthermore, this change correlated with altered sialic-acid binding specificities, i.e., those HAs with leucine at 226 bind SAα2-6Gal linkages while those with glutamine bind SAα2-3Gal linkages. To determine if only the change at 226 was required for the altered tissue tropism of R2, additional reassortants with the same genotype were sequenced through the receptor binding site and were tested for sensitivity to the horse serum inhibitor (data not shown). Those reassortants with leucine at 226 are sensitive to the inhibitor, those with glutamine are

TABLE 4

AMINO SEQUENCES OF THE UDORN (Ud) AND REASSORTANT R2 HA 1

	-18	1	14	24	34	44
Ud	MKTIIALSYI	FCLVLGQDFP	GNDNSTATLC	LGHHAVPNGT	LVKTITNDQI	EVTNATELVQ
R2	MKTIIALSYI	FCLVLGQDFP	GNDNSTATLC	LGHHAVPNGT	LVKTITNDQI	EVTNATELVQ
	54	64	74	84	94	104
Ud	SSSTGKICNN	PHRILDGIDC	TLIDALLGDP	HCDGFQNETW	DLFVERSKAF	SNCYPYDVPD
R2	SSSTGKICNN	PHRILDGIDC	TLIDALLGDP	HCDGFQNETW	DLFVERSKAF	SNCYPYDVPD
	114	124	134	144	154	164
Ud	YASLRSLVAS	SGTLEFISEG	FTWTGVTQNG	GSNACKRGPD	SGFFSRLNWL	YKSGSTYPVL
R2	YASLRSLVAS	SGTLEFISEG	FTWTGVIQNG	GSNACKRGPD	SGFFSRLNWL	YKSGSTYPVL
	174	184	194	204	214	224
Ud	NVTMPNNDNF	DKLYIWGVHH	PSTDQEQTSL	YVQASGRVTV	STKRSQQTII	PNIGSRPWVR
R2	NVTMPNNDNF	DKLYIWGVHH	PSTDQEQTSL	YVQASGRVTV	STKRSQQTII	PNIGSRPWVR
	234	244	254	264	274	284
Ud	GLSSRISIYW	TIVKPGDILV	INSNGNLIAP	RGYFKMRTGK	SSIMRSDAPI	GTCISECITP
	* *					
R2	GQSGRISIYW	TIVKPGDILV	INSNGNLIAP	RGYFKMRTGK	SSIMRSKAPI	GTCISECITP
	294	304	314	324		
Ud	NGSIPNDKPF	QNVNKITYGA	CPKYVKQNTL	KLATGMRNVP	EKQT	
R2	NGSIPNDKPF	QNVNKITYGA	CPKYVKQNTL	KLATGMRNVP	EKQT	

insensitive in agreement with Rogers et al. (4). However, the change at 226 did not correlate with tissue tropism since examples of reassortants with either leucine or glutamine at 226 but with serine at 228 did not replicate in the intestinal tract of ducks. Only R2 with the serine to glycine mutation at 228 in addition to the change at 226 exhibited avian enterotropism.

Examination of the published sequences of 28 HA1 molecules reveals only four that can be considered avian strains: A/Duck/Ukraine/1/63 (8), fowl plague virus (9), A/Seal/Mass/1/80 (10), and now reassortant R2 (this report). All have the same amino acid sequence in the receptor binding site - gly-gln-ser-gly-arg-ile-.

As a result of these studies, we suggest that i) mutations in the influenza hemagglutinin receptor binding site can alter the important biological property of tissue tropism, ii) the change at 226 correlates with horse serum inhibitor sensitivity and sialic-acid binding specificity but an additional change at 228 is required for altered tissue tropism, and iii) all avian strain receptor binding sites are likely to have the same amino acid sequence.

ACKNOWLEDGMENTS

We acknowledge the excellent technical assistance of Jim Boyle, Deanna Williams, Mary Ann Bigelow and Kenneth Cox.

REFERENCES

1. Hoyle, L. (1968). Virol. Monogr. 4,1.
2. Carroll, S.M., Higa, H.H., and Paulson, J.C. (1981). J. Biol. Chem. 256, 8357.
3. Rogers, G.N. and Paulson, J.C. (1983). Virology 127, 361.
4. Rogers, G.N., Paulson, J.C., Daniels, R.S., Skehel, J.J., Wilson, I.A., and Wiley, D.C. (1983). Nature 304, 76.
5. Naeve, C.W., Hinshaw, V.S., and Webster, R.G. (1983). Virology 128, 331.
6. Webster, R.G. and Laver, W.G. (1980). Virology 104, 139.
7. Air, G.M. (1979). Virology 97, 468.
8. Ward, C. and Dopheide, T. (1982). ICN-UCLA Symp. Molec. Cell Biol. 22, 331.
9. Porter, A.G., Barber, C., Carey, N.H., Hallewell, R.A., Threlfall, G., and Emtage, J.S. (1979). Nature 282, 471.
10. Naeve, C.W. and Webster, R.G. (1983). Virology (in press).

STRUCTURE OF THE CLEAVAGE SITE OF HEMAGGLUTININS OF PATHOGENIC AND NONPATHOGENIC H7 INFLUENZA VIRUSES*

F. X. Bosch, and R. Rott

Institut für Virologie, Justus-Liebig-Universität Giessen, D-6300 Giessen, West Germany

BIOLOGICAL ACTIVATION OF HEMAGGLUTININS

The proteolytical cleavability of the hemagglutinin (HA) in a wide range of host cells correlates with the pathogenic properties of natural avian influenza virus isolates (1). Our current understanding of the biological activation of "cleavable" HAs (of pathogenic avian strains) and of the *in vitro* activation by trypsin of "noncleavable" HAs (of nonpathogenic avian and other influenza viruses) is as follows: 1. The primary amino acid sequence of the connecting peptide region directly determines the cleavability of the HA and thereby biological properties of the virus particle. Cleavable HAs of pathogenic H7 and H5 strains have more than one basic amino acid in the connecting peptide, noncleavable HAs of nonpathogenic H7 and H5 strains and of all human influenza viruses have a single basic amino acid as 'connecting' peptide (2, unpublished results).

2. The initial cleavage between arginine and glycine, performed either by a cellular, trypsin-like enzyme or *in vitro* with trypsin, is followed by the elimination of arginine or arginine plus additional basic amino acids. The elimination of the connecting peptide is performed by a virus associated carboxypeptidase B-type activity, since this reaction is observed, when purified virions containing uncleaved HA are treated with trypsin, but not with nonactivating proteases (3).

Here we have addressed ourselves to the question of the structural requirements on the HA for the elimination reaction by analyzing the effects of inhibitors of glycoprotein processing.

Tunicamycin (TM) and 2-deoxyglucose (2-dg) both inhibit the synthesis of the lipid-linked oligosaccharide chain and thereby block N-glycosidic glycosylation of proteins (4).

*This work was supported by the Deutsche Forschungsgemeinschaft (Sonderforschungsbereich 47).

Copyright © 1984 by Academic Press, Inc.
All rights of reproduction in any form reserved.

ISBN 0-12-183501-4

Firstly we have shown, that the trypsin-like cleavage of all cleavable HAs is retained in the presence of these inhibitors, as reported previously for fowl plague virus (5). The effects of TM and 2-dg on the elimination of the connecting peptide were then analyzed using the technique of nonreducing isoelectric fucusing of infected cell extracts followed by SDS-PAGE under reducing conditions (nr-2d-analysis). This technique enables us to determine if, subsequent to trypsin-like cleavage, the connecting peptide is removed or not (2,3).

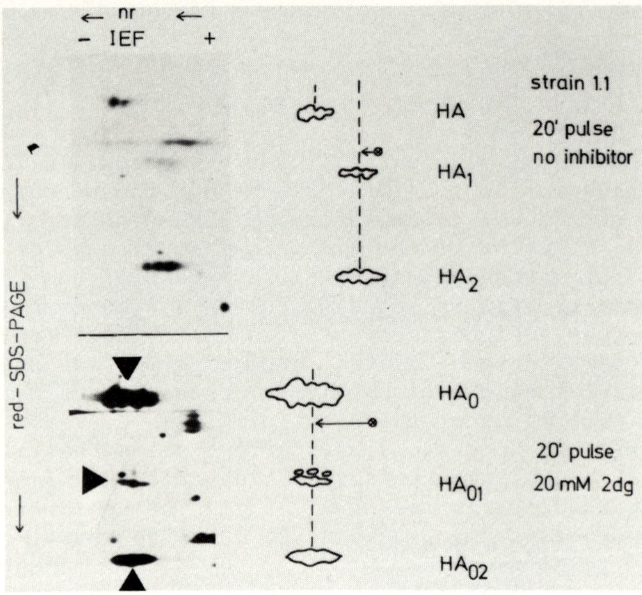

FIGURE 1. Inhibition of the elimination of the connecting peptide by 2-deoxyglucose (2-dg); details of nr-2d-analyses. Cells were infected with the pathogenic strain 1.1 and treated or not treated with 20 mM 2-dg, added 30 min p.i.. Cells were pulse-labeled with 50 µCi/ml ^{35}S-methionine for 20 min at 5 h p.i.. The solubilization in O'Farrell's lysis buffer which included 5% trasylol but lacked ß-mercaptoethanol was for 60 min. The schematic presentation of HA spots to the right illustrates the absence of charge shift in the treated cells (lower panel). Other procedures were as described previously (2)

▼ HA_0; ▶ HA_{01}, ▲ HA_{02}, (x) position of NA and NA_0, respectively.

Fig. 1 shows the result of a nr-2d-analysis from 2-dg treated or untreated cells, infected with the pathogenic H7 strain 1.1 (A/FPV/Rostock/34). In the untreated cells there is a significant charge shift towards the acidic side of the pH gradient of the HA subunits as compared to the uncleaved precursor. In the presence of 2-dg the first cleavage reaction still takes place, the cleavage products HA_{01} (▶) and HA_{02} (▲), however, migrate to the same pI-position as the uncleaved, nonglycosylated precursor (▼), i.e. the elimination of the connecting peptide has not occurred. The same result was obtained with other H7 strains and with TM as glycosylation inhibitor (not shown). Thus, glycosylation of HA is not a prerequisite for the trypsin-like cleavage, but is a structural requirement for the elimination of the connecting peptide to occur.

In contrast to TM and 2-dg, the α-glucosidase inhibitors bromoconduritol (BC) and N-Methyl-deoxynojirimycin (MdN) do not block formation and transfer to protein of lipid-linked oligosaccharides, but interfere with their subsequent processing. BC inhibits trimming of the innermost glucose from protein-linked oligosaccharides and thereby the formation of both 'high-mannose' and complex carbohydrate side chains found in mature glycoproteins (6). MdN inhibits the trimming off of all three glucoses; the oligosaccharide structure found with MdN is $Glc_3Man_x(GlcNAc)_2$ (x = 7,8,9) (7).

As to be expected, BC and MdN like the other inhibitors employed, did not inhibit the first activation reaction, trypsin-like cleavage, performed by a cellular enzyme. However, in contrast to TM and 2-dg the elimination of the connecting peptide is also not inhibited by BC or MdN. This is exemplified in Fig. 2 with the inhibitor MdN including both a cleavable (A) and a noncleavable HA (B) in the nr-2d-analysis. Both cleavage products HA_1 and HA_2 as well as the HA precursors show a slightly reduced electrophoretic mobility in MdN-treated cells, due to the inhibition of glucose trimming. The cleaved HAs, however, irrespective of whether cleaved in the cell (A) or <u>in vitro</u> with trypsin, show the same shift in charge as do the HAs in the untreated cells (2,3), indicating the removal of a basic connecting peptide (A) or a basic amino acid (B).

These results show that the structural requirements on the HA for the two reactions comprising biological activation are different. Cleavage of the peptide bond between HA_1 and HA_2 is not dependent on the attachment of carbohydrate side chains, but is determined solely by the amino acid sequence of the cleavage site. The elimination of the connecting peptide between HA_1 and HA_2, on the other hand, is blocked when transfer of oligosaccharide side chains onto the HA polypeptide is inhibited by tunicamycin or 2-deoxyglucose. The nonglycosylated cleavage products migrate to the same position in the pH-gradient as the HA_0 precursor (Fig. 1).

N-methyl-deoxynojirimycin and bromoconduritol do not inhibit the transfer of lipid-linked oligosaccharide at all, but inhibit the initial trimming of the glucose residues, on the protein bound oligosaccharide (6,7). In the presence of these inhibitors the elimination of the connecting peptide occurs (Fig. 2). Thus, transfer of the oligosaccharide moiety to the HA protein without processing to the mature 'high mannose' and complex type of oligosaccharide, is sufficient to restore the structural basis for the elimination reaction.

Based on the assumption that in the presence of TM or 2-dg only the substrate of the elimination-reaction, the HA, is altered, the results described would imply that the nonglycosylated HA, which has already been cleaved by a trypsin-like enzyme, adopts a conformation which precludes its functioning as a substrate for the subsequent exopeptidase. However, it is also possible that the eliminating enzyme, which is tightly associated with purified virus, itself is a glycoprotein inactive in the nonglycosylated form. This

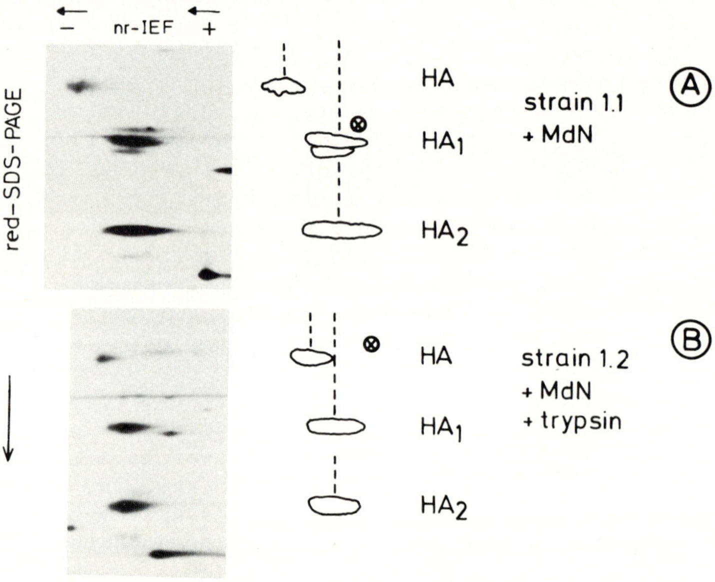

FIGURE 2. The elimination of the connecting peptide is not inhibited by N-methyl-deoxynojirimycin (MdN); details of nr-2d-analyses. Cells were infected with the pathogenic strain 1.1 (A) or the nonpathogenic strain (1.2). They were treated with 2 mM MdN at 3 h p.i. and labeled at 5 h p.i. for 30 min (A) or for 60 min (B). Trypsin treatment in B was as described (2). Other procedures were as in Fig. 1 and described previously (2).

view is supported by the finding, that the only potent inhibitor of the elimination of the connective peptide found so far, is 2-mercaptomethyl-3-guanidinoethylthiopropanoic acid, which very efficiently inhibits carboxypeptidase N (8). This enzyme is a glycoprotein.

During the course of these studies we noted a striking difference in the metabolic stability between cleavable and "noncleavable" HAs in the nonglycosylated form, in particular when TM was used as glycosylation inhibitor. All cleavable HAs were proteolytically degraded in glycosylation inhibited cells and could not be detected 3 hr after a pulse label, whereas the noncleavable HAs were readily detected in such cells for several hours. This indicates, that the nonglycosylated HAs after initial specific cleavage become susceptible to further proteolytic attack.

NUCLEOTIDE SEQUENCING OF THE CONNECTING PEPTIDE REGION OF H7 and H5 HEMAGGLUTININS

Differences in the proteolytic cleavability of the HA determining different pathogenic properties of the virus have been found only within the avian subtypes H7 and H5, isolates from other known subtypes are all nonpathogenic (1). To determine the genetic basis for the difference in proteolytic cleavability of H7-HAs, we have initiated sequencing experiments using synthetic oligonucleotide primers and the dideoxy technique (9). Table 1 shows the region on the HA gene (vRNA) of fowl plague virus, strain Rostock, used as priming site (______) and the connecting peptide region x (═══, taken from 10). The region underlined as priming site was found to be conserved among several HAs of human viruses except the A residue marked by an asterix which was found to be replaced by T on some other HAs. Two primers were therefore synthesized complementary to this region, given at the bottom of Table 1. To our surprise, we could prime only the HAs of two different H7 strains successfully with primer F4, those of the Rostock strain of fowl plague virus of our Institute and another pathogenic H7N1 strain (strain 1.8), isolated in Germany 1972. Both strains showed homology to the published HA sequence of FPV (10) in the region which we could readily sequence with one significant exception: On both HAs instead of CUU we found the C replaced by U, changing GAA in the coding strand to AAA. Thus, the connecting peptide region of the two cleavable HAs sequenced with this primer consist exclusively of basic amino acids, whereas the connecting peptide of FPV as reported (10), contains a Glu residue (see boxed letters).

TABLE 1

NUCLEOTIDE SEQUENCE OF THE PROTEOLYTIC CLEAVAGE SITE OF FPV-HA AND OF THE REGION USED AS PRIMING SITE

FPV-HA (10)	5'v RNA									
	AAU	AAC	CGU	UGA*	CCC	UAC	UUC	UUG	CAA	GGG
	CUU	GGA	AGG	UUU	UUU	UCC	CUU	UUU	UCU	CCG
			Ser	Lys	Lys	Arg	Glu	Lys	Arg	Gly
FPV-HA-Giessen (2) strain 1.1			UCC	UUU	UUU	UCC	UUU	UUU	UCU	CCG
			Ser	Lys	Lys	Arg	Lys	Lys	Arg	Gly
A/Carduelis/Germany/72 (2) strain 1.8			UCC	UUU	UUU	UCC	UUU	UUU	UCU	CCG
			Ser	Lys	Lys	Arg	Lys	Lys	Arg	Gly
A/seal/Mass/1/80-HA (11)			CCA				UUC	UGG	UCU	CCU
			Pro				Lys	Thr	Arg	Gly

——— priming site

═══ connecting peptide

*substituted by T on other HAs

Primers synthesized: F4 5'GGCAACTGGG 3'

F3 5' GCAACAGGGA 3'

Corresponding seal HA sequence: 5'GGCTACAGGAA 3'

▲ ▲ ▲

During the preparation of this manuscript the complete nucleotide sequence of the A/seal/Mass/1/80-HA (11) was made available to us and we have included the relevant regions in Table 1. A primer homologous to seal HA would require 3 changes to primer F4, pointed out by arrows. Thus, our failure to obtain sequences of the HA cleavage sites of the other H7 and the H5 strains, was most likely due to unexpected changes within this region on these HA genes. The connecting peptide region of this HA, which biologically and in the nr-2d-analysis behaves like a 'noncleavable' HA, displays a single tryptic cleavage site. However, there are two more amino acids within this region with a Lys-residue separated from Arg by Thr. Two point mutations within the Thr-codon would convert this HA into a cleavable one, or vice versa, the same point mutations might have converted this HA from a originally cleavable HA into a 'noncleavable' one. Experiments to determine the genetic structure of the cleavage sites of other cleavable and noncleavable H7 and H5 HAs are under way.

ACKNOWLEDGMENTS

We are greatly indebted to Dr. J. J. Skehel for the synthesis of the oligonucleotide primers and to Dr. Rod Daniels for introducing us to the dideoxy sequencing technique.

REFERENCES

1. Bosch, F. X., Orlich, M., Klenk, H.-D., and Rott, R. (1979). Virology 95, 197-207.

2. Bosch, F. X., Garten, W., Klenk, H.-D., and Rott, R. (1981). Virology 113, 725-735.

3. Garten, W., Bosch, F. X., Linder, D., Rott, R., and Klenk, H.-D. (1981). Virology 115, 361-374.

4. Schwarz, R. T., and Datema, R. (1980). Trends Biochem. Sci. 5, 65-67.

5. Klenk, H.-D., Wöllert, W., Rott, R., and Scholtissek, C. (1974). Virology 57, 28-41.

6. Datema, R., Romero, P. A., Legler, G., and Schwarz, R. T. (1982). Proc. Natl. Acad. Sci. USA 79, 6787-6791.

7. Romero, P. A., Datema, R., and Schwarz, R. T., Virology, submitted.

8. Garten, W., and Klenk, H.-D. (1983). J. gen. Virol., in the press.

9. Sanger, F., Nicklen, S., and Coulson, A. R. (1977). Proc. Natl. Acad. Sci. USA 74, 5463-5467.

10. Porter, A. G., Barber, C., Carey, N. H., Hallewell, R. A., Threlfall, G., and Emtage, J. S. (1979). Nature 282, 471-477.

11. Naeve, C. W., and Webster, R. G. (1983). Virology 129, in press.

AN INFLUENZA VIRUS MUTANT WHOSE LESION RESULTS IN TEMPERATURE SENSITIVE HAEMAGGLUTINATION[1]

John W. McCauley

Division of Virology
Dept. of Pathology
University of Cambridge,
Laboratories Block,New Addenbroke's Hospital,
Hills Road, Cambridge.

The influenza virus haemagglutinin is responsible for binding to and fusion with the cell membrane.Until recently the region of the molecule responsible for binding to sialic acid residues was identified by the structural similarity to the sialic acid bnding site of wheat germ agglutinin(1). Analysis of the HA sequences of viruses which show differing sensitivities to the gamma inhibitors present in horse or guinea pig serum indicate that a single amino acid change at residue 226 located in the proposed sialic acid binding site is responsible for receptor specificity(2).

This communication describes a mutant of fowl plague virus with a ts lesion in the HA which results in 1)temperature sensitive agglutination of chick red cells,2) an inability to plaque in horse serum at the non-permissive temperature and reports analysis of the nucleotide sequence of regions encoding the HA1 portion of this virus.

[1]This work was supported by an MRC program grant PG 979/140

Copyright © 1984 by Academic Press, Inc.
All rights of reproduction in any form reserved.
ISBN 0-12-183501-4

SENSITIVITY TO GROWTH IN HORSE SERUM

The mutant virus ts C US4 was isolated in Cambridge by Almond (3) from a mutagenised stock of avian (fowl plague/Rostock/34) influenza. Recombination analysis with the Cambridge ts mutants indicated that it contained a lesion in the HA gene and its phenotype showed temperature labile agglutination of chick red cells(3,4) and is shown in figure 1.

Assay of plaque formation in chick embryo fibroblasts grown in horse serum shows a reduction in titre of 2-3 X 10^2 when grown at 40.5^{o}. In order to discover whether the ts lesion was receptor specific or a general sialic acid binding property of the virus assay of temperature sensiivity was performed in CEF cells grown and maintained in 2% horse serum. The results which are shown in table 1 indicate that whilst the parental virus plaques less efficintly in horse serum at 40.5^{o} by approximately 10 fold the mutant shows no plaques at the higher tempeature, a ts reduction in plaquing efficiency of $>10^6$ (table 1).

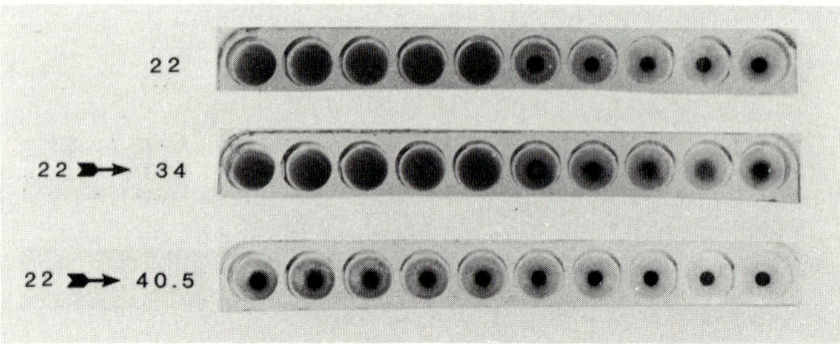

Figure 1. Haemagglutination assays of ts C US4 infected alantoic fluid were carried out at room temperature and subsequently raised to 34 or 40.5^{o}.

TABLE 1

	PLAQUE TITRATION OF VIRUS		
	CALF SERUM	HORSE SERUM	TEMP.
ts C US4	$2.5X10^7$	$1.2X10^7$	34
	$1.5X10^5$	<10	40.5
Parent	$3X10^8$	$4.7X10^8$	34
	$3X10^7$	$5X10^6$	40.5

The results from these inhibition studies suggest one of two possibilities. Since inhibition by horse serum is implicated in virus binding to N-linked sugars(2), the ts lesion is probably due to either a reduced binding to O-linked sugars at the non-permissive temperature or a temperature dependant binding to N-linked sugars, both of which would result in an increased sensitivity to horse serum at the non-permissive temperature.At present no evidence can be presented to distinguish between these two possibilities.

SEQUENCE ANALYSIS OF GENE 4 OF TS C US4

Oligonucleotides were synthesised using a phosphotriester method on a solid phase support(5) and purified by HPLC. Sequencing of vRNA was performed using reverse transcriptase and chain terminating inhibitors with the label as 5' polynucleotide kinase labeled primer or as an internal label using a single alpha-^{32}P labelled nucleotide triphosphate. Sequence analysis of 800 nucleotides from the 3' terminus of

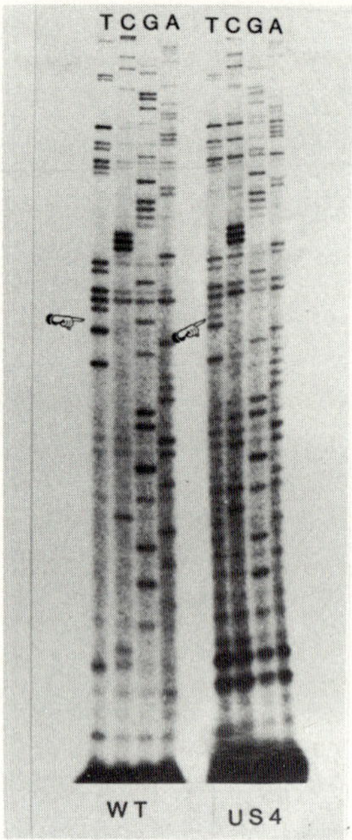

Figure 2.A synthetic oligonucleotide primer (5) corresponding in sequence to bases 292 to 303 of RNA segment 4 of FPV (9) was used to prime reverse transcription of purified vRNA of wild type and mutant virus in the prescence of dideoxynucleotide triphosphate chain terminators. The primer was labeled at the 5' end with ^{32}P-ATP and polynucleotide kinase and subsequently purified by polyacrylamide gel electrophoresis.Primer extension was carried out in 50 mM TrisHCl ,10mM $MgCl_2$,20mM DTT,and 0.4mM deoxynucleotide triphosphates and 0.25mM dideoxynucleotides with 5 units of reverse transcriptase(Life Sciences) for 120 minutes at 40°. The products were ethanol precipitated and dissolved in 2.5 microlitres of 8M urea,10mM TrisHCl pH 7.4 and separated on a 12% polyacrylamide gel.The nucleotide change between wild type and mutant virus is indicated.

gene 4 has shown a single base change at nucleotide 331 which results in a change from Val to Phe in ts C US4 at position 94 in HA_1,adjacent to the sequence Cys-Tyr.The sequence of this region has been shown to be conserved in all analysed sub-types,consistant with its location in the sialic acid binding site(1).

Sub-Type	Amino Acid around Cys(95) References 6-10.
H1	I C Y
H2	L C Y
H3	N C Y
H5	L C Y
H7	V C Y
H7 (ts)	F C Y
H10	Y C Y

ANALYSIS OF RECOMBINANTS

Recombination of ts C US4 with A/Chick/Germany/42 showed that only viruses which contained the US4 haemagglutanin formed plaques,all of which were temperature sensitive for growth in both calf and horse serum.These results suggest that the haemagglutinin alone is responsible for the temperature sensitive inhibition of growth by gamma inhibitors.

Plaque morphology of the temperature sensitive virus also showed a distict phenotype but was not temperature sensitive. Large plaque variants were occasionally found in one recombinant (R2.18) and these viruses were not temperature sensitive for growth in either serum. The isolation of a recombinant which showed a high frequency of reversion to wild type suggests that either ts C US4 contained a second ts lesion which has reverted to wild type and the reversion frequency observed in R2.18 represents the reversion frequency of the plaque morphology and sensitivity to gamma inhibitors phenotype. Alternatively the genetic composition of R2.18 may suppress a second ts lesion.By comparison of the mobility of polypeptides in SDS gels and neuraminidase typing it has been possible to show the genetic composition of R2.18 as

P1	P2	P3	HA	NP	NA	M	NS
US4	N	N	US4	N	US4	?	N

Sequence analysis using the same primer as in figure 2 showed that R2.18 and a non-temperature sensitive virus isolated from it had both reverted to Valine at residue 94.The role of the mutation of Val to Phe therefore appears to only partially determine the ts phenotype and the complete description of temperature sensitive growth in horse serum requires further analysis of mutant,wild type,and more recombinant and revertant viruses.

ACKNOWLEDGEMENTS

I thank Mrs.S.Broomfield for first class technical assistance,Dr.J.J.Skehel for neuraminidase typing of R2.18,Charles Hodgeman for writing and accessing his macro-spitbol sequence handling programs and Dr.B.W.J.Mahy for provision of all facilities and equipment.

REFERENCES

1 Wilson,I.A.,Wiley,D.C.& Skehel,J.J.(1981) Nature 289,366-373.
2 Rogers,G.N.,Paulson,J.C.,Daniels,R.S.,Skehel,J.J., Wilson,I.A.,

& Wiley,D.C. (1983) Nature 304, 76-78.

3 Almond,J.W. (1978) Ph.D. Thesis, University of Cambridge.

4 Almond,J.W.,McGeoch,D. & Barry,R.D. (1979) Virology 92, 407-415.

5 Gait,M.J.,Matthes,H.W.D.,Singh,M.,Sproat,B.S. & Titmus,R. (1982) Cold Spring Harbor Symp. Quant.Biol.XLVII,393-401.

6 Winter,G.,Fields,S. & Brownlee,G.G. (1981) Nature 292, 72-75.

7 Gething,M.-J.,Bye,J.,Skehel,J.J. & Waterfield,M.D. (1980) Nature 287, 301-306.

8 Verhoyen,M.,Fang,R.,Min Jou,W.,Devos,R.,Huylebroeck,D., Saman,E. & Fiers,W. Nature 296, 771-776.

9 Porter,A.G.,Barber,C.,Carey,N.H.,Hallewell,R.A., Threlfall,G., & Emtage,S. (1979) Nature 282, 471-477.

10 Air,G.M. (1981) Proc.Natl.Acad.Sci.USA 78, 7639-7643.

THE CARBOHYDRATES OF THE HEMAGGLUTININ OF INFLUENZA VIRUS[1]

Walter Keil[2], Rudolf Geyer[3], Heiner Niemann[2], Janusz Dabrowski[4], and Hans-Dieter Klenk[2]

Institut für Virologie[2] and Biochemisches Institut[3], Justus-Liebig-Universität, 63 Giessen and Max-Planck-Institut f. Medizinische Forschung[4], 69 Heidelberg, Germany

The carbohydrate side chains of the influenza virus hemagglutinin are attached by N-glycosidic linkages to the polypeptide (1). Complex (type I), mannose-rich (type II), and hybrid side chains are present (2,3), and each of these types exhibits a considerable amount of heterogeneity (1, 4). The biosynthesis of the hemagglutinin side chains involves assembly of the oligosaccharide (glucose)3-(mannose)9-(N-acetylglucosamine)2 by the dolichol pathway of glycosylation, its transfer to the nascent polypeptide, followed by the sequential removal of the glucose and most of the mannose residues and by the attachment of N-acetylglucosamine, galactose, and fucose residues (2, 5). The 3 oligosaccharide types found on the mature hemagglutinin differ therefore from each other in the completeness of processing. In various aspects the hemagglutinin carbohydrates have been found to be of biological significance. Evidence has been obtained that they protect the polypeptide from proteolytic degradation (6). Some, but not all of the side chains are antigenic determinants that cross-react with host specific carbohydrates (7). It is also reasonable to assume, although it has not been shown directly, that in some hemagglutinins oligosaccharides may prevent immune recognition of antigenic sites that are freely accessible in other hemagglutinins. The topography of the oligosaccharides on the hemagglutinin and the factors that

[1]This work was supported by the Deutsche Forschungsgemeinschaft (SFB 47, Virologie).

Copyright © 1984 by Academic Press, Inc.
All rights of reproduction in any form reserved.
ISBN 0-12-183501-4

determine their structure are therefore of considerable interest. We report here on studies on the structure of the oligosaccharides attached to the individual glycosylation sites of the hemagglutinin of fowl plague virus (FPV), an avian influenza A virus.

THE STRUCTURE OF THE OLIGOSACCHARIDES OF THE FPV HEMAGGLUTININ

The hemagglutinin of the FPV strain A/FPV/Rostock/34 (H7N1) has 7 potential attachment sites (8). The side chains were localized by a procedure involving fragmentation of the hemagglutinin by treatment with cyanogen bromide, V8 protease from *Staphylococcus aureus*, and trypsin. The positions of the fragments were determined by radioactive labeling of the sugars and of specific amino acids. Side chains of the complex type are attached to asparagine residues 12, 28, 123, 149, and 478. A mannose-rich oligosaccharide is linked to Asn(406). Asn(231) is not glycosylated (9). To obtain further information on the structure of the oligosaccharides attached to each glycosylation site glycopeptides derived by trypsin or Pronase degradation from

TABLE 1
LECTIN AFFINITY OF THE OLIGOSACCHARIDE SIDE CHAINS OF THE HEMAGGLUTININ OF FPV (ROSTOCK STRAIN)

	Asn(12)	Asn(28)	Asn(123)	Asn(149)	Asn(406)	Asn(478)
A^-L^+	30 %	55 %	17 %	38 %	--	25 %
A^-L^-	24 %	12 %	54 %	5 %	15 %	19 %
A^+L^+	16 %	22 %	3 %	--	--	22 %
A^+L^-	30 %	11 %	26 %	57 %	--	34 %
$A^{++}L^-$	--	--	--	--	85 %	--

A^-: no binding; A^+ (A^{++}): weak (tight) binding to Concanavalin A, glycopeptides are eluted with 10 mM (200 mM) α-methylmannoside.
L^-: no binding; L^+: binding to *Lens culinaris* agglutinin, glycopeptides are eluted with 500 mM α-methylmannoside.

the glycosylated cyanogen bromide fragments were subjected to consecutive affinity chromatography on Sepharose-bound Concanavalin A and *Lens culinaris* agglutinin. As shown in Table 1 the complex oligosaccharides attached to the individual glycosylation sites can be subdivided into 4 fractions by their differential affinity to these lectins.

The structures of the oligosaccharides liberated from the HA_1 subunit by alkaline hydrolysis in the presence of sodium borohydride were analyzed in detail by ^1H-NMR spectroscopy (data not shown) and by combined gas chromatography and mass spectrometry (GC/MS) (Table 2). The results of these studies combined with the data on lectin affinity allow the assignment of the structures shown in Fig. 1 to the major oligosaccharides present on HA_1.

TABLE 2
METHYLATION ANALYSIS OF COMPLEX OLIGOSACCHARIDES OF THE HA_1 OF FPV (STRAIN DUTCH) BY GC/MS

Alditolacetates	type Ia	type Ib
2,3,4-FucOH	0.37	0.32
2,3,4,6-ManOH	0.29	0.32
3,4,6-ManOH	1.72	1.86
2,4,6-ManOH	0.20	trace
3,6-ManOH	trace	trace
3,4-ManOH	0.28	0.14
2,4-ManOH	0.70	0.50
2-ManOH	0.23	0.30
2,3,4,6-GalOH	2.56	2.02
2(3)4,6-GalOH	trace	trace
2,3,4-GalOH	trace	trace
3,4,6-GlcN(Me)AcOH	1.09	0.85
3,6-GlcN(Me)AcOH	5,01	4,03
6-GlcN(Me)AcOH	0.14	--
1,3,5,6-GlcN(Me)AcOH	trace	0.27
1,3,5-GlcN(Me)AcOH	--	0.38

[a]Sulfated and neutral glycopeptides derived from HA_1 by Pronase digestion were separated by ion exchange chromatography (Keil et al., 1984). After methylation of the oligosaccharides, the alditolacetates of their constituent sugars were analyzed by GC/MS. The molar ratios, based on the sum of 3,4,6-ManOH and 3,4-ManOH = 2, are indicated.
Type Ia: sulfated glycopeptides. Type Ib: neutral glycopeptides.

These structures comprise bi- and triantennary oligosaccharides with or without bisecting N-acetylglucosamine and with or without fucose at the N-acetylglucosamine residue linked to asparagine. It should be pointed out that, after permethylation, only trace amounts of 3,6-dimethylmannose can be detected (Table 2). This indicates that tetraantennary oligosaccharides are not present on HA_1. This concept is supported by the NMR spectra which, in addition, exclude the presence of triantennary oligosaccharides containing bisecting N-acetylglucosamine.

The structures of the oligosaccharides attached to Asn(406) and Asn(478) have been analyzed also in some detail. The mannose-rich side chain Asn(406) exhibits considerable heterogeneity as demonstrated by the observation that 4 different oligosaccharides (HM1-4) can be discriminated after digestion with endo-ß-N-acetylglucosaminidase H (Fig. 2). Mass spectrometrical analysis of the permethylated oligosaccharides indicated for HM2 and HM3 the structures shown in Fig. 2. The glycopeptides derived from Asn(478) by Pronase digestion also show heterogeneity. One of these glycopeptides has a sugar composition (mannose:N-acetylglucosamine:galactose 3:4:2) compatible with a biantennary complex oligosaccharide. Permethylation analysis of the total carbohydrate attached to Asn(478) indicated that the biantennary structure can be modified by a bisecting N-acetylglucosamine, a fucose residue , and by additional antennae at both α-mannoses. At present, it cannot be decided yet, whether a tetraantennary oligosaccharide is linked to Asn(478) or whether triantennary oligosaccharides with the third antennae bound either to the α-1,6-linked mannose or to the α-1,3-linked mannose are present. It should be emphysized, however, that Asn(478) is the only attachment site of the FPV hemagglutinin that contains oligosaccha-

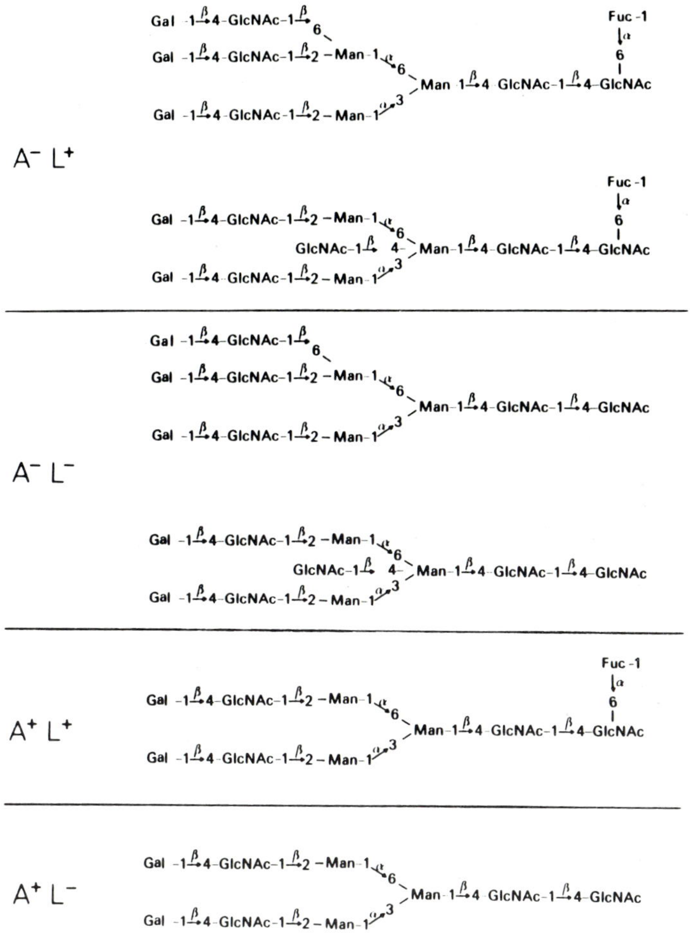

FIGURE 1. Structures of neutral oligosaccharides attached to HA_1. Their lectin affinities (see Table 1) are also indicated.

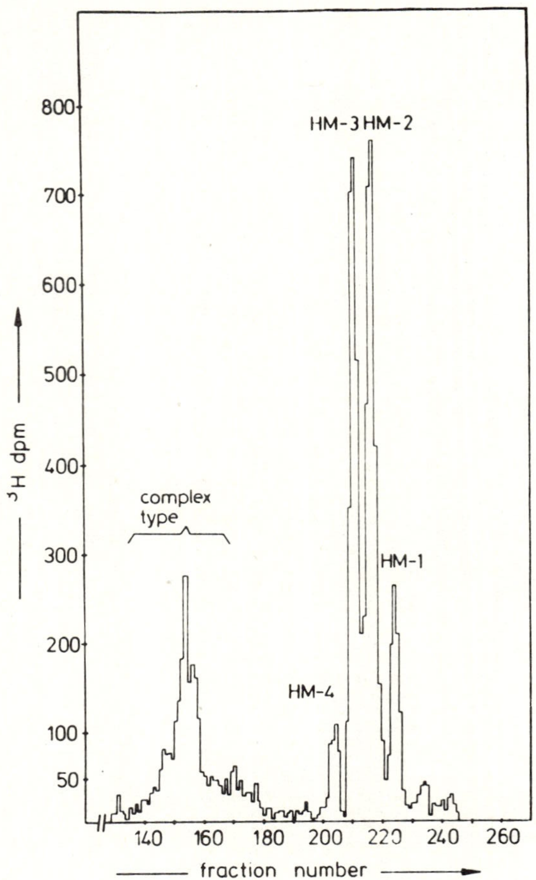

Structural Heterogeneity of the Oligosaccharide Attached to Asparagine 406 of the FPV Hemagglutinin

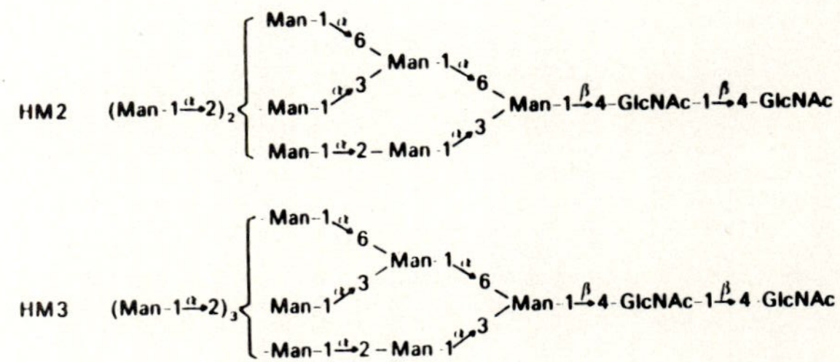

Structural Heterogeneity of the Oligosaccharide Attached to Asparagine 478 of the FPV Hemagglutinin

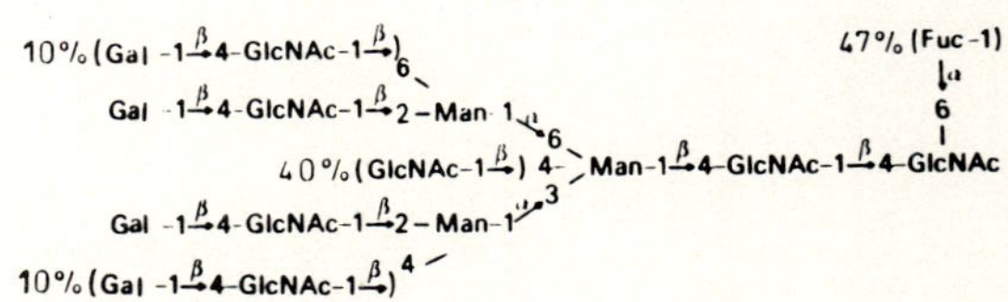

rides with the α-1,3-linked mannose substituted in position C-4. Taken together, these observations indicate that a specific set of heterogeneous oligosaccharides is attached to each glycosylation site.

FACTORS DETERMINING THE GLYCOSYLATION PATTERN OF THE HEMAGGLUTININ

It has been known for some time that the structure of the carbohydrate side chains of the hemagglutinin is influenced to some extent by the host cell (2, 10). Obviously, the carbohydrate pattern depends also on the presence of appropriate glycosylation sites Asn-X-Thr(Ser), and it has been shown that there are strain-specific variations in glycosylation that are determined by the primary structure of the polypeptide (10, 11). The data presented here provide some information on the factors that determine whether carbohydrate is attached to a potential glycosylation site and, if so, which type of oligosaccharide is attached. The lack of carbohydrate at Asn(231) of the FPV hemagglutinin is in keeping with the observation that attachment sites are usually not glycosylated if they contain, as is the case in this position, aspartic acid as the variable amino acid (12). Another amino acid that has been reported to interfere with glycosylation if occurring in this position is tryptophan (13). It has therefore been suggested that the FPV hemagglutinin is probably not glycosylated at Asn(406) (14). However, the data shown here demonstrate clearly that carbohydrate is present in this position. This indicates that merely from the primary structure of a polypeptide it cannot be predicted with certainty whether or not a potential attachment site is glycosylated.

In their crystallographic study on the hemagglutinin structure Wilson and coworkers (15) stress the fact that all carbohydrate attachment sites are located at the surface of the trimer. In

---

FIGURE 2. Structures of oligosaccharides attached to HA_2. Pronase digests of HA_2 were subjected to endo-β-N-acetylglucosaminidase H treatment. Complex glycopeptides (not susceptible to β-N-acetylglucosaminidase H) were separated from mannose-rich oligosaccharides (HM1-4) by Biogel P4 chromatography.

the light of this observation alone it is not easy to understand why the processing enzymes converting mannose-rich into complex oligosaccharides should have access to some but not to all side chains.

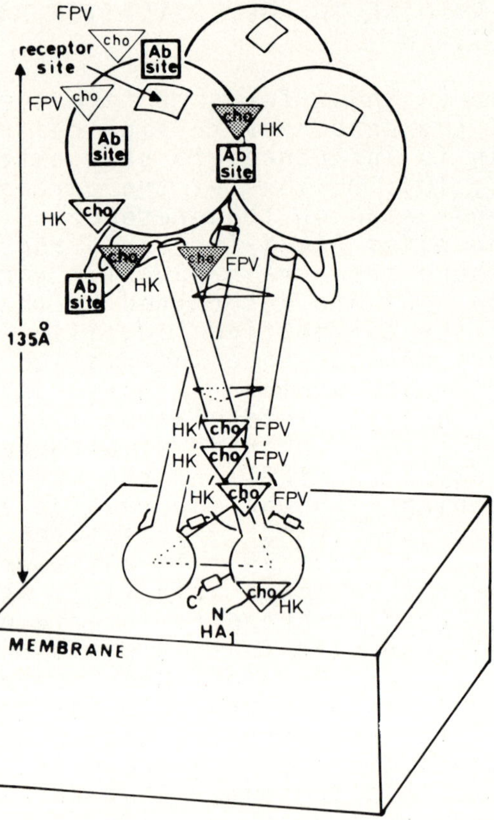

FIGURE 3. The topography of the carbohydrate side chains on FPV and Hong Kong (HK) hemagglutinins. The model of Wilson and coworkers (15) of the trimeric spike with the fibrous and globular domains is shown. Complex side chains are indicated by clear triangles, mannose-rich side chains by shaded triangles.

However, a comparison of the glycosylation patterns of the hemagglutinin of FPV and of several Hong Kong (H3) strains (16), reveals that all attachment sites for mannose-rich oligosaccharides are located in surface niches that may impose constraints on oligosaccharide processing (Fig. 3). These niches are formed by the interfaces between different monomers or between the two domains of a monomer. Thus, the mannose-rich side chain attached to Asn(285) of the H3 hemagglutinin is located at the junction between the fibrous and the globular regions of the monomers. The type II oligosaccharide attached to Asn(165) of the H3 hemagglutinin is located at an interface between different monomers. It is believed to form substantial contacts between the globular regions of the subunits (15). The mannose-rich oligosaccharide side chain of the FPV hemagglutinin is located at another interface between monomers. It is attached to the top half of the long α-helix of HA_2 (amino acids 401 - 424). Wilson and coworkers (15) have shown that this region is stabilized predominantly by extensive contacts with the other subunits of the hemagglutinin trimer. They have also pointed out that the polypeptide chain folding in this region of the monomer may only be completed on assembly of monomers to form trimers. It therefore appears that on the nacent unfolded polypeptide chain carbohydrate can be attached to Asn(406). However, once the conformational change resulting in trimer formation has occurred the carbohydrate side chain may no longer be accessible for further modifications. Furthermore, since the presence of oligosaccharides with 7, 8, and 9 mannose residues at Asn(406)(17) indicates that carbohydrate processing has been blocked after the initiation and well before the completion of the trimming process, and since trimming is believed to take place in the Golgi complex, it is reasonable to assume that this organelle is also the site of trimer assembly.

REFERENCES

1. Keil, W., Klenk, H.-D., and Schwarz, R. T. (1979). J. Virol. 31, 253.
2. Schwarz, R. T., Schmidt, M.F.G., Anwer, U., and Klenk, H.-D. (1977). J. Virol. 23, 217.
3. Nakamura, K., Brown, A. S., and Compans, R. W. (1980). Virology 107, 208.
4. Matsumoto, A., Yoshima, H., and Kobata, A. (1983). Biochemistry 22, 188.
5. Nakamura, K., and Compans, R. W. (1979). Virology 93, 31.
6. Schwarz, R. T., and Klenk, H.-D. (1974). J. Virol. 14, 1023.

7. Brown, L. E., Ward, C. W., White, D. O., and Jackson, D. C. (1981). In "Genetic Variation among Influenza Viruses" (D. P. Nayak, ed.), p. 233. Academic Press, New York.
8. Porter, A. G., Barber, C., Carey, N. H., Hallewell, R. A., Threlfall, G., and Emtage, J. S. (1979). Nature 282, 471.
9. Keil, W. (1982). Doctorial thesis, Justus-Liebig-Universität, Giessen.
10. Nakamura, K., and Compans, R. W. (1979). Virology 95, 8.
11. Schwarz, R. T., and Klenk, H.-D. (1982). Virology 113, 584.
12. Struck, D. K., and Lennarz, W. J. (1979). In "The Biochemistry of Glycoproteins and Proteoglycans" (W. J. Lennarz, ed.). p. 35, Plenum-Press, New York.
13. Clamp, J. R. (1975). In "The Plasma Proteins" (F. W. Putnam, ed.) Vol. II, p. 163. Academic Press, New York.
14. Waterfield, M., Gething, M. J., Scrace, G., and Skehel, J. J. (1980).In "Structure and Variation in Influenza Virus" (W. G. Laver, and G. Air, eds.). p. 11. Elsevier/North Holland, New York.
15. Wilson, I. A., Skehel, J. J., and Wiley, D. C. (1981). Nature 289, 366.
16. Ward, C., and Dopheide, T. A. (1981). In "Genetic Variation among Influenza Viruses" (D. P. Nayak, ed.). p. 323. Acacemic Press, New York.
17. Klenk, H.-D., Garten, W., Keil, W., Niemann, H., Bosch, F. X., Schwarz, R. T., Scholtissek, C., and Rott, R. (1981). In "Genetic Variation among Influenza Viruses" (D. P. Nayak, ed.). p. 193. Academic Press, New York.

HETEROGENEITY OF THE MEMBRANE (M1) PROTEIN OF INFLUENZA VIRUS[1]

Anastasia Gregoriades

The Public Health Research Institute of the City of New York
New York, N.Y. 10016

THE MEMBRANE (M) COMPONENT AND VIRUS ASSEMBLY

The mechanism by which the lipid containing viruses such as the rhabdo, myxo and paramyxoviruses assemble their components to produce infectious virus is not clear. The proteins of these viruses assemble at the cell plasma membrane where their close interaction with one another leads to the exclusion of host cell proteins. The viral glycoproteins are synthesized on rough endoplasmic reticulum (ER), pass to the smooth ER and through the cell plasma membrane, where they remain anchored (see 1,2). The internal viral proteins move through the cytoplasm by some unidentified mechanism and then onto the plasma membrane site, where the glycoproteins have assembled. Virus is produced by budding from the cell.

The major viral protein in the three groups of viruses is the membrane (M) component. This internal protein must play a pivotal role in assembly of the virus, since it must surround the ribonucleoprotein (RNP) at the same time that it recognizes the site on the plasma membrane where the glycoproteins have been assembled. The exact association of this protein with the lipid bilayer of the virus has only recently been described for the myxo (3) and rhabdoviruses (4).

In influenza virus, the membrane (M1) protein is very lipophilic, since it readily incorporates into lipid vesicles (3,5,6). The ease with which the protein is disaggregated by preformed lipid vesicles (6) suggests that the protein may insert into lipid in the infected cell, immediately after its synthesis. The availability of the sequence of the M1 protein (7,8) has allowed us to locate the sites on the protein which partition into lipid in vitro (3). The protein appears to be embedded in the lipid bilayer in the intact particle as well (3). The protein is characterized by many stretches of

---

[1]This work was supported by grant AI12567 from the NIH.

Copyright © 1984 by Academic Press, Inc.
All rights of reproduction in any form reserved.
ISBN 0-12-183501-4

neutral and hydrophobic amino acids (7,8) which are undoubtedly responsible for the solubility of M1 in organic solvent (9) and for its unusual electrophoretic properties on gels (10). We have studied this component further because it is such an important protein in the virus.

TWO FORMS OF M1 IN INFLUENZA VIRUS

We have observed that the M1 protein often splits into two bands when the viral proteins are analyzed on gels. It was possible that the second band was the hemagglutinin (HA2) molecule, the non-structural (NS1) which is found in the cell, the M1 or some unidentified protein. Therefore, we investigated this observation. S-35 methionine labeled WSN virus was grown on chick embryo fibroblasts and purified (6). The virus was solubilized with sodium dodecyl sulfate (SDS) and mercaptoethanol and analyzed on 15% polyacrylamide gels containing 1% cross-linker. The acrylamide was polymerized rapidly by prewarming the solution to 37°C. Under these conditions, the RNP separated into two bands, as described previously (11, 12). Two bands also appeared in the M1 area (Fig 1, lane 1). The second band appearing in the area was not the

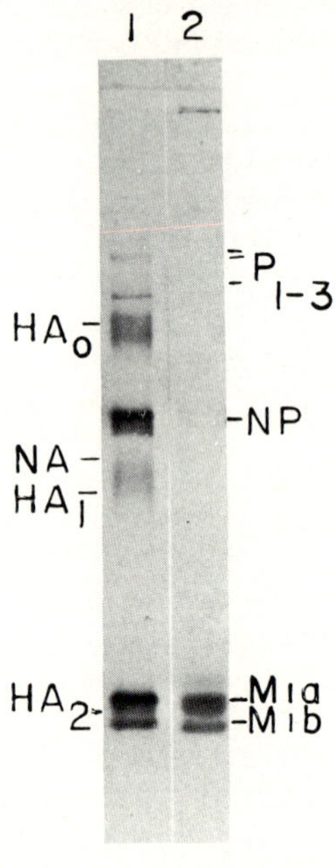

Figure 1. Separation of S-35 methionine labeled WSN virus proteins on 15% polyacrylamide gels. 1) WSN virus, 2) chloroform-methanol extract of WSN virus.

HA2, since chymotrypsin removed the glycoproteins but did not alter the appearance of the two proteins. The two proteins were extractable into acidic chloroform-methanol (Fig 1, lane 2), which is a property of the M1 component (9). The two proteins were designated M1a and M1b.

The possibility that M1a and M1b were a result of some aging process of the virus was investigated. Virus was harvested at 8 hours after infection, at 8-24 hours and after 40 hours of growth. Virus was purified and the glycoproteins were removed by incubating the virus with chymotrypsin overnight at room temperature. The viral proteins were then analyzed on 15% gels. The two bands in the M1 area were found in early as well as late virus and the ratio between the two proteins was the same in all preparations (Fig 2).

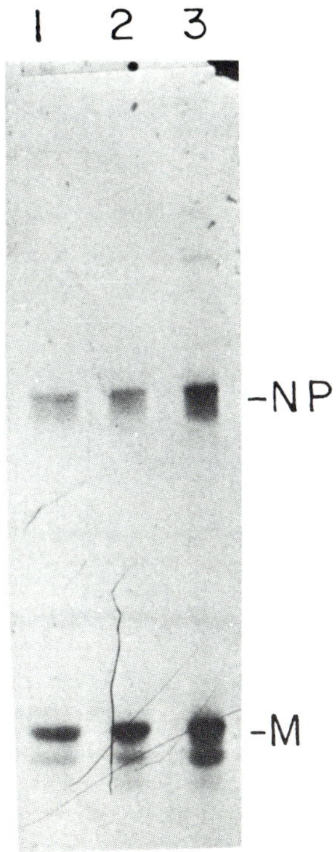

Figure 2. Polyacrylamide gel analysis of chymotrypsin treated WSN virus, labeled with S-35 methionine, which was harvested at different times after infection. 1) 8 hr virus, 2) 8-24 hr virus, 3) 40 hr virus.

The identity of the two proteins in the M1 area was determined by labeling the WSN virus with C-14 amino acids and analyzing the tryptic digests of M1a and M1b on silica gel thin layer plates. The two proteins were first extracted from the virus into chloroform-methanol. They were then separated on gels and the protein bands were detected by autoradiography. The bands were cut from the gel and the gel slices were fixed in methanol:water (1:1). Trypsin in 0.05 M ammonium bicarbonate was added and the slices were incubated for 2 hours at 37°C followed by one hour at room temperature. The supernatant containing the tryptic fragments was removed, lyophilized, resuspended in buffer and analyzed in two dimensions as described previously (6). The tryptic pattern of M1a and M1b was similar and a clear difference between the two patterns could not be discerned reproducibly (Fig 3). The proteins were clearly different from the NS1 component which was isolated from the nuclei of infected cells. Despite the fact that the tryptic pattern appeared identical, the results did not eliminate the possibility that a slight difference could exist between the two proteins which was due to a proteolytic cleavage that had occured in the cell or in the virus.

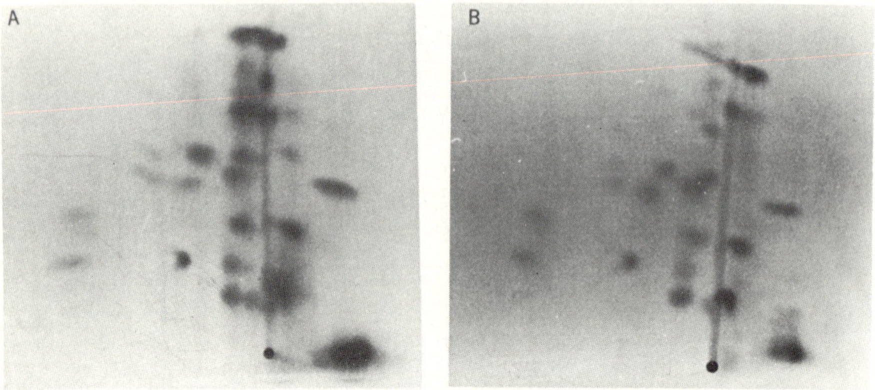

Figure 3. Two dimensional analysis of C-14 amino acid labeled tryptic digests of M1a and M1b on silica gel plates. A) M1a, B) M1b.

M1 IS A PHOSPHOPROTEIN

The non-structural NS1 and the RNP of influenza virus are phosphoproteins (12-15). Both proteins are found in more than one form in the virus or infected cell (11-15). The possibility that M1 was a modified protein and that the separation into more than one component was due to this modification was investigated. Chick embryo fibroblasts were starved of phosphate 17 hours, infected and incubated with medium lacking phosphate but containing P-32 phosphate. Virus was purified (6), the lipid was removed by extraction into chloroform-methanol (2:1). The viral proteins were solubilized with SDS, made 0.3 M with respect to NaCl and the RNA was hydrolyzed by treating the sample with ribonuclease for 17 hours at room temperature. The sample was then reduced with mercaptoethanol and analyzed on 15% gels. Phosphate was associated with both forms of M1 protein (Fig 4, lane 2).

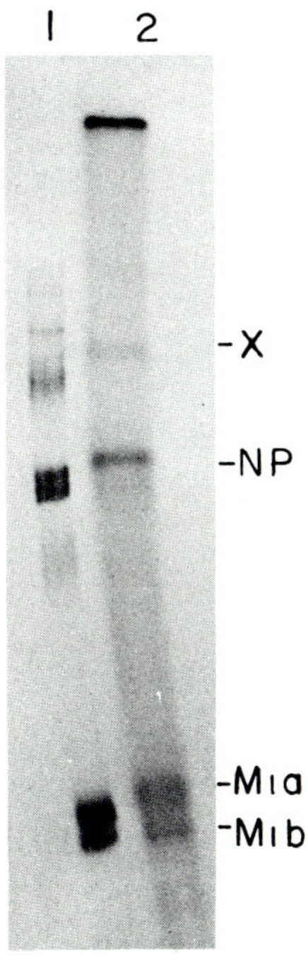

Figure 4. Polyacrylamide gel analysis of S-35 methionine and P-32 phosphate labeled WSN virus. 1) S-35 methionine labeled virus, 2) P-32 phosphate labeled virus.

Figure 4 also shows that the larger NP polypeptide is predominantly phosphorylated, while the smaller one contains less phosphate, as previously described (12). The component marked X in Fig 4, lane 2, could not be identified. It was, however, external on the virus, since it could be removed by treating whole virus with chymotrypsin before analyzing the proteins on polyacrylamide gels.

The phosphorylated amino acids in the M1 component were identified. The P-32 labeled M1a and M1b components were cut from gels, the slices were fixed in methanol-water (1:1) to remove the SDS and trypsin in 0.05M ammonium bicarbonate, pH 8.6 was added to the slices. These were incubated 2 hours at 37°C and then overnight at room temperature. The tryptic fragments in the fluid above the gel were removed, lyophilized and hydrolyzed for 3 hours at 110°C in 6 N HCl. The material was analyzed on Avicel plates (Analtech, N.J.), after addition of cold phosphoserine and phosphothreonine as markers. The samples were separated by electrophoresis in two dimensions, the first in acetic acid:pyridine:water (50:5:945), pH 3.5 and the second in formic acid:acetic acid:water (25:87:888) pH 1.9. Electrophoresis was at 400 v. P-32 labeled spots were detected by autoradiography, while the phosphoserine and phosphothreonine were detected by staining the same plate with 0.2% ninhydrin in acetone. The phosphorylated amino acid was serine in both M1a and M1b. A trace amount of phosphothreonine was also detected in M1b (Fig 5). The NS1 from infected cells contained only phosphoserine (Fig 5C), as was previously described (15). We do not know whether the separation of M1 into at least two components is due to phosphorylation of the protein. However, since the phosphoproteins NP and NS1 in influenza, as well as the phosphoproteins of other viruses (16,17) separate into multiple spots, based on the content of phosphate, it is possible that the heterogeneity of the M1 protein is based, in part, on its phosphorylation.

The M proteins of the rhabdo (16,18), paramyxo (19) and myxoviruses (this work) have been shown to be phosphoproteins. The manner by which phosphorylation of these proteins may influence the process of assembly of these viruses is not known. In other systems, phosphorylation of integral membrane proteins may affect the fluidity of the lipid bilayer which then leads to rearrangement of the membrane components (20). It is possible that the phosphate groups affect the interactions of the M proteins with the other viral components in the assembling structure. Alternatively, the phosphorylation itself may be a non-specific phenomenon which occurs because a parti-

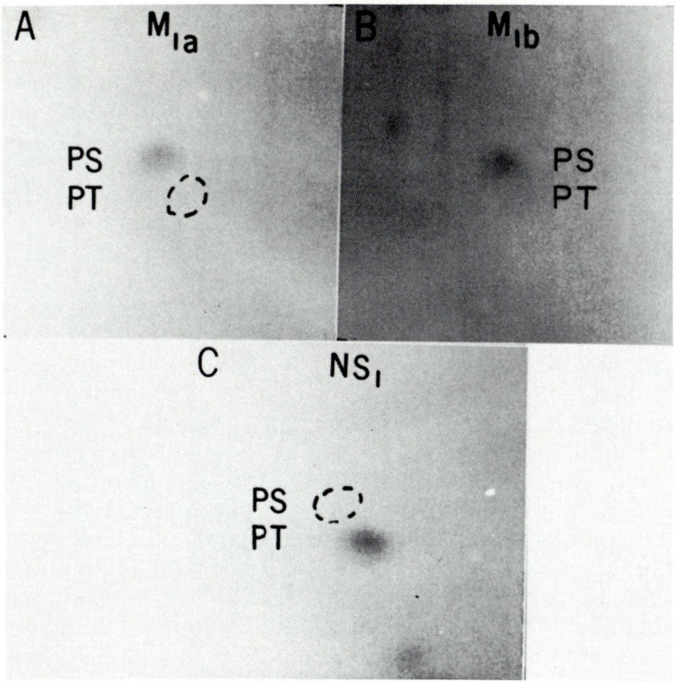

Figure 5. Two dimensional analysis on cellulose thin layer plates of P-32 labeled amino acids present in the M1 and NS1 proteins. A) M1a, B) M1b, C) NS1.

cular kinase recognizes a specific primary sequence in the protein. We are presently investigating these possibilities.

REFERENCES

1. Wickner, W. (1979). Ann. Rev. Biochem. 48, 23.
2. Sabatini, D.D., Kreibich, G., Morimoto, R. and Adesnik, M. (1982). J. Cell Biol. 92, 1.
3. Gregoriades, A. and Frangione, B. (1981). J. Virol. 40, 323.
4. Zakowski, J.J., Petri, W.A. and Wagner, R.R. (1981). Biochem. 20, 3902.
5. Bucher, D.J., Kharitonenkov, I.G., Zakomiridin, J.A., Grigoriev, V.B., Klimenko, S.M. and Davis, J.F. (1980). J. Virol. 36, 586.
6. Gregoriades, A. (1980). J. Virol. 36, 470.

7. Allen, H., McCauley, J., Waterfield, M. and Gething, M-J. (1980). Virology 107, 548.
8. Winter, G. and Fields, S. (1980). Nuc Acids Res. 8, 1965.
9. Gregoriades, A. (1973). Virology 54, 369.
10. Gregoriades, A. and Markarian, K. (1983). Arch. Virol. 76, 263.
11. Zhirnov, O.P. and Bukrinskaya, A.G. (1981). Virology 109, 174.
12. Petri, T. and Dimmock, N.J. (1981). J. Gen. Virol. 57, 185.
13. Privalsky, M.L. and Penhoet, E.E. (1977). J. Virol. 24, 401.
14. Privalsky, M.L. and Penhoet, E.E. (1978). Proc. Nat'l. Acad. Sci. 75, 3625.
15. Privalsky, M.L. and Penhoet, E.E. (1981). J. Biol. Chem. 256, 5368.
16. Clinton, G.M. and Huang, A.S. (1981). Virology 108, 510.
17. Hsu, C-H. and Kingsbury, D.W. (1982). Virology 120, 225.
18. Clinton, G.M., Burge, B.W. and Huang, A.S. (1978). J. Virol. 27, 340.
19. Lamb, R.A. and Choppin, P.W. (1977). Virology 81, 382.
20. Gazitt, Y., Ohad, I. and Loyter, A. (1976). Biochem. Biophys. Acta 436, 1.

Antigenic Characterization of Influenza A Matrix Protein with Monoclonal Antibodies

Kathleen L. van Wyke,[1] Jonathan W. Yewdell,[2], Suzanne M. Michalek[3], Jerry R. McGhee[3], and Brian R. Murphy[1]

[1]Laboratory of Infectious Diseases
National Institutes of Allergy and Infectious Diseases
National Institutes of Health
Bethesda, Md 20205

[2]Imperial College of Science and Technology, London, England

[3]University of Alabama, Birmingham, Alabama

PRODUCTION OF MONOCLONAL ANTIBODIES

The internal matrix protein of influenza viruses contains a type-specific antigen which allows serologic discrimination of types A, B, and C influenza viruses. The matrix proteins of influenza A viruses appear to be very similar. For example, the lengths of the genes coding for matrix proteins are identical between strains isolated 38 years apart (1), and several studies have found greater than 90% homology among matrix genes (1, 2). Peptide mapping has shown that although regions of the matrix protein are highly conserved, there are a small number of clear-cut differences among strains (3, 4). However, none of these differences has been detected in serologic assays using heterogeneous monospecific antiserum to the matrix protein (5).

Since antigenic variation has been readily demonstrated in the influenza A virus hemagglutinin (6), neuraminidase (7), and nucleo-protein (8) using monoclonal antibodies, we have produced monoclonal antibodies to the matrix protein.

Balb/c mice were hyperimmunized to the A/WSN/33 (H1N1) strain of influenza virus by two intraperitoneal injections of 200 µg of purified virus or 50 µg of purified matrix protein (9) in complete Freund's adjuvant followed by an intravenous boost 3 days before fusion. In addition, germ-free Balb/c mice received an oral immunization of formalin-inactivated X-73 vaccine virus followed by an intravenous injection one week later. One week later their spleen cells were fused with MOPC-21 myeloma cells which do not secrete immunoglobulins (NSI/1), using polyethylene glycol (10). The culture fluids were tested for antibody production by hemagglutination-inhibition and neuraminidase-inhibition, and enzyme-linked immunosorbent

Copyright © 1984 by Academic Press, Inc.
All rights of reproduction in any form reserved.
ISBN 0-12-183501-4

assays (ELISA). The specificity of hybridomas negative in the hemagglutination-inhibition and neuraminidase-inhibition tests but positive in ELISA was determined by immunoprecipitation assays. Hybridomas producing antibodies which specifically precipitated the matrix protein were cloned in soft agar. Oral immunization yielded only monoclonal antibodies directed to the internal nucleoprotein and matrix protein whereas intraperitoneal immunization yielded predominately antibodies to surface antigens but fewer to internal proteins. Three of nine influenza-specific monoclones produced after oral administration of antigen were of the IgA isotype, whereas after parenteral immunization no IgA monoclones were detected. The greater efficiency of oral immunization for the production of IgA monoclonals has been previously described (12).

ANTIGENIC VARIATION IN THREE MATRIX PROTEIN DETERMINANTS

In contrast to the hemagglutinin and neuraminidase surface antigens, the internal viral proteins were originally characterized as being antigenically invariant (5). Antigenic stability of the internal proteins seems reasonable since they would not be subject to immunologic selection pressure which is responsible for the variety of hemagglutinin and neuraminidase antigenic subtypes among influenza A viruses. Comparison of the number of nucleotide substitutions in genes coding for the hemagglutinin, neuraminidase, and matrix proteins suggests that these genes have similar mutation rates in nature, but these substitutions result in fewer amino acid changes in the matrix protein than in the hemagglutinin or neuraminidase (13).

To study antigenic variation in the matrix protein, monoclonal anti-matrix antibodies were tested in ELISA against a panel of influenza A viruses (Table 1). Three antibodies exhibited unique reactivity patterns indicating that each recognized a different epitope on the M protein molecule. These experiments clearly established that the A/WSN/33 matrix protein possesses a minimum of three antigenic sites, two of which independently undergo antigenic variation in other strains. The third epitope detected by our monoclonal antibodies was present on each of the 26 strains examined, and represents an influenza A type-specific matrix protein determinant. Monoclonal antibodies produced by oral immunization recognized the type-specific determinant.

To examine the extent of antigenic variation the three detectable matrix protein epitopes undergo, the monoclonal antibodies were tested in ELISA against 26 influenza A viruses isolated from animals and man over a 49-year period and representing most of the major antigenic subtypes.

TABLE 1
UNIQUE REACTIVITY PATTERNS OF THREE ANTI-MATRIX MONOCLONAL ANTIBODIES WITH INFLUENZA A VIRUSES, AS DETERMINED BY ELISA

	Reactivity of Monoclonal Antibody:		
Influenza virus strain	289/4	m2-1C6	904/6
A/WSN/33 (H1N1)	+	+	+
A/Duck/UKr/1/63 (H3N8)	-	+	+
A/Aichi/2/68 (H3N2)	+	+	-
A/USSR/90/77 (H1N1)	+	+	-
B/HongKong/5/72	-	-	-

Antigenic variation was detected in only three of these strains, namely A/Aichi/2/68 (H3N2), A/Duck/Ukraine/1/63 (H3N8), and A/USSR/90/77 (H1N1) (Table 1). The two 1968 H3N2 strains tested (Hong Kong and Aichi) possessed clearly distinguishable matrix proteins. This corroborates previous studies using monoclonal antibodies which demonstrated that the hemagglutinins and nucleoproteins of these strains are antigenically different (14, 11), and suggest that there were two distinct strains co-circulating during the first year of the H3N2 era. As a result of sequential, cumulative mutations in their hemagglutinins, each of these 1968 strains appear to have given rise to distinct antigenic families of H3N2 viruses (15). In contrast, the pattern of antigenic variation detected in the matrix protein is not likely to be caused by progressive accumulation of point mutations in the matrix gene. Rather, it is more likely that the matrix protein variation described here is an example of genetic dimorphism in influenza virus populations, which has been previously described for the influenza A hemagglutinin (16, 17, 18) and neuraminidase (19). An unexpected result of these experiments was that the M protein of the A/USSR/90/77 virus differed antigenically from that of the Fort Warren/1/50 strain since these two strains have not been distinguished previously using monoclonal antibodies to the hemagglutinin and nucleoprotein (20, 11). The fact that antigenic variation was detected in only 3 of 26 strains stands in sharp contrast not only to similar studies using monoclonal antibodies to the hemagglutinin and neuraminidase (20), but also to the other major structural protein of influenza viruses, namely the nucleoprotein (11). Although the number of monoclonal antibodies used in the present study is small, the results suggest that matrix protein undergoes a lesser degree of antigenic variation than the nucleoprotein. The reason for this is not clear, since both are internal proteins and neither are subject to antibody selection pressure in the population. It is possible that viruses with unaltered matrix proteins have a selective growth advantage since the matrix protein appears to play a role in virus assembly and maturation (21).

NUMBER OF NON-OVERLAPPING ANTIGENIC DOMAINS ON MATRIX PROTEIN

Competitive-binding ELISA assays were performed to determine the distribution of the three distinct antigenic determinants on the influenza A matrix protein. These experiments were designed to test the capability of each of three monoclonal antibodies to sterically inhibit the binding of antibodies to other epitopes on the matrix protein molecule. The results of these assays are illustrated by competitive-binding curves (Fig. 2). Each of two monoclonal antibody preparations (M2-1C6 and 289/4) was inhibited only by homologous antibodies, indicating that each of these monoclonal antibodies recognizes an epitope which occupies a physically distinct domain on the molecule. In contrast, the binding of antibody preparation 904/6 was also partially inhibited by high concentrations of antibody M2-1C6. This inhibition was not due to non-immunoglobulin contaminants in the preparations, since high concentrations of antibody 289/4 and monoclonal antibodies to the influenza A nucleoprotein or hemagglutinin produced only slight inhibition of 904/6 antibody binding (data not shown).

There are several possible explanations for non-reciprocal binding of monoclonal antibodies. For example, binding of one antibody might induce allosteric effects on the matrix protein, thus partially inhibiting the binding of a second antibody. Alternatively, antibodies of different binding affinities might react non-reciprocally in a competitive-binding assay. A third possibility is that the competing antibody, in addition to binding to its corresponding epitope, also binds with reduced affinity to a second cross-reactive determinant which is located in close proximity to the determinant recognized by the conjugated antibody. This situation could produce non-reciprocal competitive inhibition at high competing antibody concentrations, and has been suggested to explain similar data obtained with monoclonal antibodies to vesicular stomatitis virus (22). To discriminate among these possibilities, the antibodies were individually evaluated for their abilities to bind to fragments of the matrix protein, since this assay does not depend on competitive binding and would detect a cross-reactive determinant. Intact and proteolytically digested matrix protein was

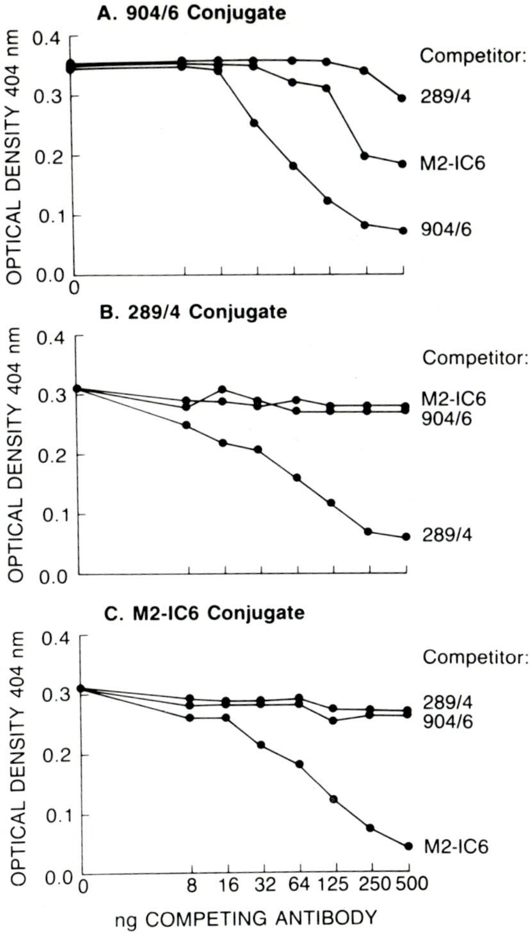

Figure 1. Determination of the number of nonoverlapping antigenic matrix protein domains by competitive-binding ELISA. Fixed concentrations of conjugated antibodies were mixed with increasing concentrations of purified unconjugated antibody competitors. These mixtures were tested in ELISA with A/WSN/33 (H1N1) virus.

transferred to nitrocellulose by the Western blot technique and immunoreactive polypeptides were identified using the three monoclonal antibody preparations (Fig. 2). Lanes a, b, and c (undigested matrix) demonstrate that each of the three antigenic determinants on the matrix protein remain intact following the process of Western blotting. Lane g illustrates that a relatively large peptide fragment (approximately 14 Kd) generated by V8 protease digestion of matrix protein possesses the epitope recognized by monoclonal antibody 904/6. In contrast, binding of the other two monoclonal antibodies to this or any other matrix protein fragment was not detected (lanes e and f). Although the reason for this negative result is not known, there are several possible explanations. For example, the proteolytic digestion could have produced very small immunoreactive peptides which were lost during electrophoresis. Indeed the predicted amino acid sequence of the matrix protein indicates that many di- and tri-peptides would be generated by v8 protease digestion (1). Alternatively, it is conceivable that the proteolytic degradation of the matrix protein altered the geometry of the binding sites recognized by these antibodies. The fact that only antibody 904/6 bound to the one matrix protein fragment suggests that the epitope recognized by antibody 904/6 may be topographically distinct from the epitope recognized by antibody M2-1C6 and that the non-reciprocal competitive binding in ELISA tests may be due to allosteric effects induced by antibody binding. The exact mechanism responsible for non-reciprocal binding of monoclonal antibodies to the matrix protein may be elucidated when the topologic location of the determinants involved is determined more precisely.

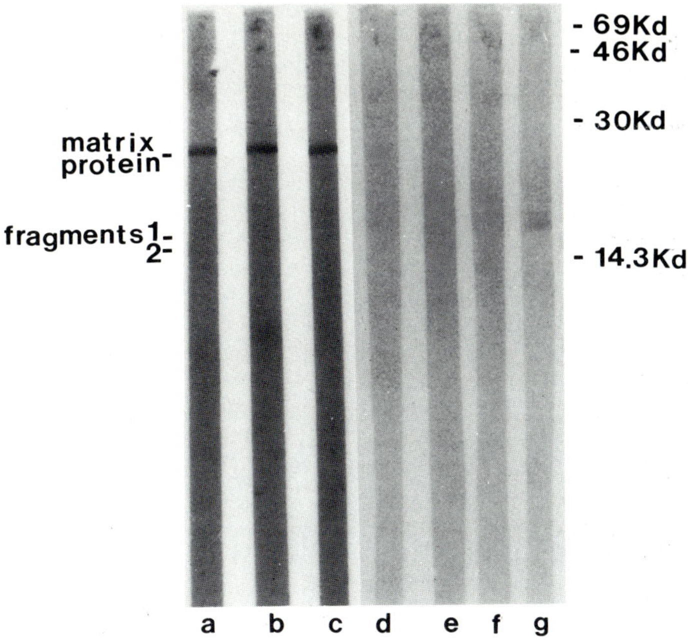

Figure 2. Western blot analysis of native and proteolytically digested matrix protein with three anti-matrix monoclonal antibodies. Purified matrix protein (undigested and V8 protease digested) were electrophoresed in 15% polyacrylamide SDS gels and then electrophoretically transferred to nitrocellulose sheets. Coomassie blue staining of the gel was used to locate positions of matrix protein and the 2 peptide fragments produced by proteolytic digestion. Nitrocellulose sheets were cut into strips and soaked separately in anti-matrix monoclonal antibodies. After washing the strips were soaked in ^{125}I-protein A for 30 min. After washing strips were air dried and exposed to Kodak XAR-5 film for 48 hours. Molecular weights were determined by co-electrophoresis of molecular weight standards. Lanes a-c, undigested matrix protein; lanes d-g V8 protease-digested matrix protein. Lane a, antibody m2-1C6; b, antibody 289/4; c, antibody 904/6; d, no antibody; e, antibody m2-1C6; f, antibody 289/4; g, antibody 904/6.

Acknowledgements

We acknowledge the capable technical assistance of Ms. E. Wu, and thank Mrs. L. Jordan for preparation of the manuscript and Dr. R. M. Chanock for helpful discussion.

REFERENCES

1. Lamb, R. A. and Lai, C.-J. (1981). Virol. 112, 746.
2. Scholtissek, C., Rohde, W., von Hoyningen, V., and Rott, R. (1978). Virol. 87, 13.
3. Brand, C. M., Stealy, V. M., and Rowe, J. (1977). J. Gen. Virol. 36, 385.
4. Dimmock, N. J., Carver, A. S., and Webster, R. G. (1980). Virol. 103, 350.
5. Schild, G. C. (1972). J. Gen. Virol. 15, 99.
6. Gerhard, W. (1976). J. Exp. Med. 144, 985.
7. Webster, R. G., Hinshaw, V. S., and Laver, W. G. (1982). Virol. 117, 93.
8. van Wyke, K. L., Bean, W. J., Jr., and Webster, R. G. (1981). J. Virol. 39, 313.
9. Yewdell, J. W., Frank, E., and Gerhard, W. (1981). J. Immunol. 126, 1814.
10. Kohler, G., and Milstein, C. (1976). Eur. J. Immunol. 6, 511.
11. van Wyke, K. L., Hinshaw, V. S., Bean, W. J. Jr., and Webster, R. G. (1980). J. Virol. 35, 24.
12. Colwell, D. E., Gollahon, K. A., McGhee, J. R., and Michalek, S. M. (1982). J. Immun. Meth. 54, 259.
13. Air, G. M. and Hall, R. M. (1981). In "Genetic variation among influenza viruses" (D. P. Nayak, ed.), p. 41. Academic Press, New York.
14. Webster, R. G. and Laver, W. G. (1980). Virol. 104, 139.
15. Webster, R. G., Laver, W. G., Air, G. M., Ward, C., van Wyke, K. L., and Gerhard, W. (1980). Annals of New York Acad. Sci. 354, 142.
16. Choppin, P. W., and Tamm, I. (1960). J. Exp. Med. 112, 921.
17. Kilbourne, E. D. (1978). Proc. Nat. Acad. Sci. USA 75, 6258.
18. Kilbourne, E. D., McGregor, S., and Easterday, B. C. (1979). Infect. Immun. 26, 197.
19. Mowshowitz, S., and Kilbourne, E. D. (1975). In "Negative strand Viruses" (R. D. Barry and B. W. J. Mahy, eds.), p. 765, Vol. 2. Academic Press, New York.
20. Webster, R. G., Kendal, A. P. and Gerhard, W. (1979). Virol. 96, 258.
21. Lenard, J. and Compans, R. W. (1974). Biochim. Biophys. Acta 344, 51.
22 LeFrancois, L., and Lyles, D. S. (1982). Virology 121, 157.

EVOLUTIONARY CONSERVATION OF INFLUENZA NUCLEOPROTEIN GENES IN HOST SPECIES[1]

William J. Bean
Division of Virology and Molecular Biology
St. Jude Children's Research Hospital
Memphis, TN 38101 USA

COMPARISON OF INFLUENZA GENES BETWEEN HOST SPECIES

Influenza viruses commonly infect swine, horses and numerous species of birds, as well as humans. The surface antigens of these viruses have been studied extensively and currently are classified into 13 hemagglutinin subtypes and 9 neuraminidase subtypes. Representatives of each hemagglutinin and neuraminidase subtypes have been found among the avian influenza strains. Three subtypes are associated with human influenza, and two are found in swine and horses (1,2). Far less is known about the genetic diversity of the non-surface proteins and how these proteins influence host specificity and other biological characteristics of these viruses. Among the non-surface proteins of influenza viruses, the nucleoprotein (NP) has been the most extensively characterized. Several investigations have compared the NPs in different influenza virus strains by various methods. Schild et al. (3) compared the antigenic cross-reactivity of different strains by immunodiffusion and demonstrated differences among human H1N1 and H3N2 influenza NPs. van Wyke et al. (4) used a panel of monoclonal antibodies and found six different reactivity patterns among the human influenza virus strains and two among the non-human strains. Two studies have used oligopeptide mapping methods to compare NP proteins from different strains (5,6). In these studies, the human and non-human viral NPs were classified into two and four groups, respectively. In neither study was a clear distinction seen between the human and non human strains. Scholtissek et al. (7), however, has

[1]This work was supported by Public Health Service Grant AI 16841 from the National Institute of Allergy and Infectious Diseases, Public Health Service Childhood Center Support (CORE) Grant CA 21765 from the National Cancer Institute, and by ALSAC.

Copyright © 1984 by Academic Press, Inc.
All rights of reproduction in any form reserved.
ISBN 0-12-183501-4

shown by hybridization that the RNA segment coding for the NPs of the three human influenza virus subtypes were closely related to each other and readily distinguishable from some avian and equine influenza virus strains.

The discrepancies among the studies described above in which the NP polypeptides were analyzed, probably occurred because the various methods may have examined different parts of a highly conserved protein that may have a limited number of functional configurations (6). Analysis of the RNAs coding for the NP polypeptide has the advantage of examining the entire molecule and will include RNA changes that do not affect the polypeptide sequence. Although some of these changes may not have any functional significance, they will reflect the evolutionary relationships among the various virus isolates. This study compared the NP genes of a large panel of influenza isolates from human and non-human hosts by RNA-RNA hybridization to determined the extent of genetic diversity of this protein and if related nucleoproteins are consistently found in viruses from certain hosts.

HYBRIDIZATION ANALYSIS OF THE NP GENES OF INFLUENZA STRAINS FROM DIFFERENT HOSTS.

The panel of influenza virus strains chosen for this study included avian virus isolates from a variety of wild and domestic species and representatives of all hemagglutinin and neuraminidase subtypes. The mammalian virus isolates analyzed included equine viruses of the H3 and H7 serotypes, swine viruses of the H1 and H3 subtypes and representatives of the human H1, H2, and H3 subtypes. Genetic homologies were measured using a competitive RNA-RNA hybridization assay (9) in which individual RNA segments are isolated from selected reference strains and labeled with ^{125}I. The labeled probes are allowed to anneal with homologous complementary RNA in the presence of increasing amounts of competing viral RNA from other virus strains. The efficiency with which the added RNA competes with the annealing of the labeled RNA with its cRNA provides a measure of relative homology between the strains being compared.

Initially, all of the influenza virus strains were compared in hybridization assays with the RNA segment coding for the nucleoprotein of a recent avian virus A/Ty/Minn/833/79. The competition curves for nearly all of the 60 avian influenza strains formed a continuum between the homologous competition curve and that of Tern/So. Africa/61, which under these hybridization conditions left 35% of the reference RNA uncompeted. These avian viruses were desig nated as "group 1" with respect to their nucleoprotein. Other strains, e.g., Equine/Miami/1/63, competed less efficiently, leaving 70% or more of the reference RNA uncompeted.

The only mammalian virus strain that was included in group 1 was the previously described isolate from seals that was shown to have derived all of its genetic information from one or more avian influenza viruses (10). The only avian influenza virus strains not falling into this group were viruses from gulls of the H13 serotype (2). Hybridization analysis with one of these H13 strains, Gull/Md/704/77, showed that they share a nucleoprotein that is not closely related to that of any of the other influenza virus strains in the panel (group 2).

Similar analysis using the RNA coding for the NP of Equine/Miami/1/63 indicated that all of the equine virus strains of the panel, with the exception of Equine/Prague/1/56, share a closely related nucleoprotein (group 3). The NP of Equine/Prague/1/56 is not closely related to any of the other strains of the panel and was designated as group 4.

The remaining viruses, those of human and swine origin, all had closely related NPs and were designated as group 5. The results are summarized in Table 1.

TABLE 1
NUCLEOPROTEIN GROUPS

NP Group	Species	Associated Hemagglutinin Subtypes
1	Avian	H1-H12
2	Gull	H13
3	Equine	H3,H7
4	Equine	H7
5	Human, Swine	H1, H2, H3

The results show a clear correlation between NP grouping and host species, and suggests that the strains that normally infect certain species have been genetically isolated from strains infecting other species for a period of time that is probably much longer than the period over which these strains have been isolated.

Perhaps the most surprising finding is the apparent degree of conservation of the avian virus strains. With strains of 13 HA subtypes continuously co-circulating, one might expect that over a long period of time, their NP genes would have diverged into numerous genetically distinct forms. Only two were found and one has a very restricted distribution.

Particularly interesting is the association of the group 2 NP only with isolates of the H13 serotype in gulls. One possibility is that the RNAs coding for these proteins may have evolved together over a long period of time and have developed a functional dependency that inhibits recombination with other strains and may not be well suited for infecting other species. The maintenance of this and other host restrictions indicates that there may be significant functional differences between these NPs that favor replication in certain hosts. Although the mechanisms that would give a strain with a particular NP gene a selective advantage in a particular host are unknown, it is likely that interactions with other gene products are also involved.

The earliest of the equine influenza virus isolates, A/Equine/Prague/1/56 (H7N7), has a nucleoprotein that is not closely related to that of any other strain in the panel. Later equine influenza isolates of the H7N7 serotype share the group 3 NP with the H3N8 equine strains, suggesting that this gene may have been replaced by the group 3 nP. However, the number of equine viruses in the panel was limited, and the group 4 NP may still exist in nature.

The human and swine influenza virus strains have been extensively analyzed using the NP gene of each of the prototype human influenza strains as the labeled reference RNA and several human influenza virus strains of each antigenic type. The results, in all cases, are consistent with a continuous accumulation of mutations in the human epidemic strains. It was also seen in these experiments that the NP of the swine influenza viruses were more closely related to early human viruses than to later isolates. When the NP of Swine/Iowa/15/30 was used in competitive hybridization assays with other strains of this group. The early human influenza strains and A/NJ/8/76 (a human isolate very similar to recent swine viruses) (11,12), showed similar levels of homology while the later human isolates showed progressively less homology. This suggests that the NP genes of the human and swine strains share a common and relatively recent progenitor and is consistent with serological studies, indicating that the influenza strain of the 1918 pandemic was antigenically more similar to swine influenza than to later human strains (13).

The hybridization assays used in this study can provide only a measure of relative genetic relatedness among the strains being tested. The absolute number of nucleotide differences between strains of the different groups cannot be reliably determined by hybridization since the efficiency of hybridization is highly dependent upon the distribution of base substitutions within the RNAs and upon the annealing conditions (14,15,16). To date only two influenza virus NPs have been sequenced (17,18), A/PR/8/34 and A/NT/60/68, both human isolates (group 5). These differ in nucleotide sequence by about 8%. Sequencing selected members of the other groups will provide an estimate of the evolutionary distances between them, and may provide some help in elucidating the mechanisms involved in species specificity of influenza viruses.

ACKNOWLEDGMENTS

I wish to thank Janet Ulm for excellent technical assistance, Dr. J.J. Skehel for kindly providing purified fowl plague virus RNA, and Drs. R.G. Webster and V.S. Hinshaw for access to virus strains and many invaluable discussions.

REFERENCES

1. Hinshaw, V.S., Webster, R.G., and Rodriguez, R.J. (1982). Arch. Virol. 67:191-201.
2. Hinshaw, V.S., Air, G.M., Gibbs, A.J., Graves, L., Prescott, B., and Karunakaran, D. (1982). J. Virol. 42:865-872.
3. Schild, G.C., Oxford, J.S., and Newman, R.W. (1979). Virology 93:569-573.
4. van Wyke, K.L., Hinshaw, V.S., Bean, W.J. Jr., and Webster, R.G. (1980). J. Virol. 35:24-30.
5. Dimmock, N.J., Carver, A.S., and Webster, R.G. (1980). Virology 103:350-357.
6. Sokolov, B.P. and Zhdanov, V.M. (1982). J. Gen. Virol. 58:199-203.
7. Scholtissek, C., Rohde, W., Von Hoynigen, V., and Rott, R. (1978). Virology 87:13-20.
8. Rohde, W., Harms, E., and Scholtissek, C. (1977). Virology 15:393-404.
9. Bean, W.J., Jr., Cox, N.J., and Kendal, A.P. (1980). Nature 284:638-640.
10. Webster, R.G., Hinshaw, V.S., Bean, W.J., Jr., van Wyke, K.L., Geraci, J.R., St. Aubin, D.J., and Petursson, G. (1981). Virology 113:712-724.

11. Palese, P. and Schulman, J.L. (1976). Nature 263:528-530.
12. Hinshaw, V.S., Webster, R.G., Bean, W.J., Jr., Downie, J., and Senne, D.A. (1983). Science 220:206-208.
13. Davenport, F.M., Hennessy, A.V., and Francis, T. (1953). J. Exp. Med. 98:641-656.
14. McConaughty, B.L. and McCarthy, B.J. (1970). Biochemical Genetics 4:409-424.
15. Scholtissek, C. (1979). Virology 93:594-597.
16. Sriram, G., Bean, W.J., Jr., Hinshaw, V.S., and Webster, R.G. (1980). Virology 105:592-599.
17. Winter, G. and Fields, S. (1981). Virology 114:423-428.
18. Huddleston, J.A. and Brownlee, G.G. (1982). Nuc. Acid Res. 10:1029-1038.

VIRUS-SPECIFIC POLYPEPTIDES IN CELLS INFECTED WITH INFLUENZA C, DHORI, AND THOGOTO VIRUSES

John P.M. Clerx and Herbert Meier-Ewert

Institute for Medical Microbiology and Hygiene
Technical University, 8000 Munich 40, FRG

INTRODUCTION

Influenza C viruses have been isolated worldwide from humans with mild upper respiratory tract disease. Only recently, additional isolates from pigs were reported from China (1). The virus has several properties in common with influenza A and B viruses, but differs, among other things, in having 7 RNA segments and 1 surface glycoprotein which shares both hemagglutinating and receptor destroying activities (2,3,4). Dhori (DHO) and Thogoto (THO) viruses were originally considered bunyavirus-like viruses on the basis of morphology, but biochemical analyses have shown that these viruses share structural characteristics with established orthomyxoviruses (5). Several THO virus isolates have been made from mammals, but most isolates were recovered from ticks. The virus has been implicated in severe human disease in Africa (6,7). Although antibodies against DHO virus have been found in human, camel and other domestic animal sera (8,9,10), all virus isolations were made from ticks. Unlike influenza A and B viruses, but like influenza C virus, DHO and THO have a genome consisting of 7 unique, single-stranded RNA segments (5, unpublished observations). All three viruses are pleomorphic, bud from the plasma membrane of infected cells, and have a clearly defined fringe of peplomers (Fig. 1). Influenza C virus peplomers are arranged in a hexagonal pattern (Fig. 1), a feature not seen in DHO or THO viruses. Their replication is inhibited by actinomycin D and a-amanitin, and the in vitro transcription of influenza C virus is primer-dependent (3).

Copyright © 1984 by Academic Press, Inc.
All rights of reproduction in any form reserved.
ISBN 0-12-183501-4

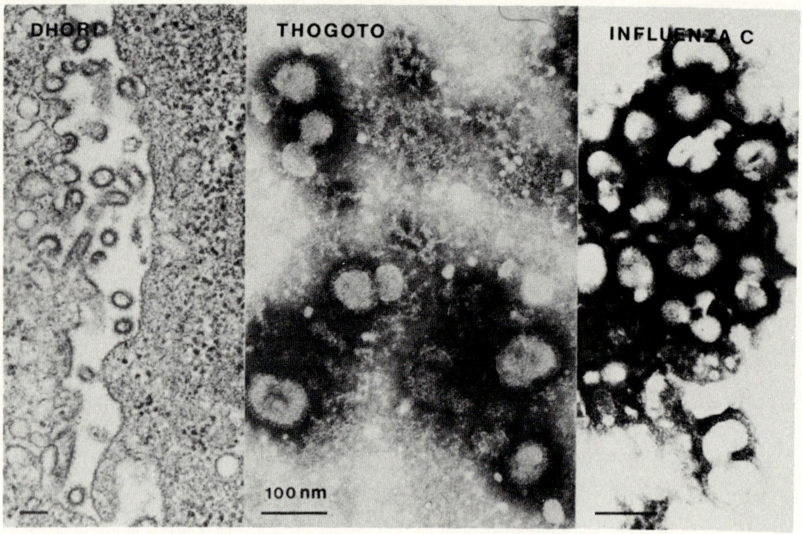

FIGURE 1. Electron micrographs of DHO, THO and Influenza C viruses. The left panel shows prototype DHO (I-611313) virus budding into the interstitial space of BHK cells. The central panel shows negatively stained THO (SiAr 126) virus recovered from the supernatant of infected Vero cells by pelleting. Influenza C virus (JHB/1/66) with its characteristic surface structure is shown in the right hand panel. Bars indicate 100 nm.

Since substantiation of a common replication strategy of influenza C, DHO and THO viruses with influenza A and B viruses could open up new insights into the evolution, infection potential and adaptation of orthomyxoviruses, we have performed experiments aimed at the polypeptide composition of THO virus and the detection of virus-induced polypeptides in infected cells.

INFLUENZA C VIRUS

Viral Polypeptides

Several papers from our laboratory and others have been published dealing with the polypeptide composition of influenza C virus particles (3,4,11,12). Virus particles are composed of 3 major polypeptides, designated NP (the major nucleocapsid protein), M (inner membrane bound), and the external HA protein, which, depending upon the host system, is proteolytically cleaved into disulfide bridge-bonded gp65 and gp30. The small subunit anchors the spike in the membrane. Recent experiments have determined that the influenza C peplomer is a trimer (A. Nagele and H. Meier-Ewert, submitted for publication). Purified preparations of virions contain 3 minor polypeptides of MW 80-90,000 which are presumed to be the equivalent of the polymerase proteins of influenza A and B viruses (3).

Intracellular polypeptides.

In addition to the three major viral polypeptides NP, M and gp88 (the uncleaved precursor of gp65 and gp30, Fig. 2), two additional bands of molecular weight 24,000 and 14,000 can be detected in infected cell extracts (13), which were absent in uninfected cells and not immune precipitated by antiserum against purified virus. These polypeptides are found early in infection but not at 15 hrs post-infection (Fig. 2 and 4). Isoelectric focusing of labeled cell extracts indicated that the presumptive NS1 polypeptide (13) migrates just to the acidic side of the M protein in Fig. 4 and migrates slightly faster than the M protein. Identification of the presumptive NS2 polypeptide has not yet been possible although three polypeptides (which are absent in infected cells) can be detected in infected cell extracts (see arrows in upper right-hand panel of Fig. 4). Additional experiments are needed to determine if these polypeptides are viral encoded.

DHORI VIRUS

Viral Polypeptides.

Purified virus preparations were found to contain, in decreasing order of abundance (as revealed by staining or autoradiography), vp 28, the internal, phosphorylated membrane protein; gp 65, the outer glycoprotein; vp 54-56, the major nucleocapsid protein, and 2 or 3 minor polypeptides of 85-90,000 MW (Fig. 3A). Upon fractionation of purified virus with Triton X-100 in 0.25M NaCl, viral nucleocapsids could be separated from membrane associated proteins (Fig. 3B). All of the glycoprotein and over 95% of vp 28 remained at the top of the gradient, whereas all of vp54-56 and most of the minor 85-90,000 MW proteins were recovered in the nucleocapsid fraction, indicating a similarity to the architecture of influenza viruses. Two dimensional (2D) gel electrophoresis of viral proteins (14) showed the three major viral polypeptides to focus over a broad pH range: gp 65 between pH 7.5 and 6, vp 54-56 streaking around pH 7.3, and vp 28 from the cathodic and of the gel to pH 6.5. Occasionally vp 28 appeared as "beads on a string" in which each bead towards the anodic end of the gel could represent additional phosphate molecules. The major nucleocapsid doublet proteins vp 54-56 had similar isoelectric points and our data suggest that the smaller polypeptide could be a

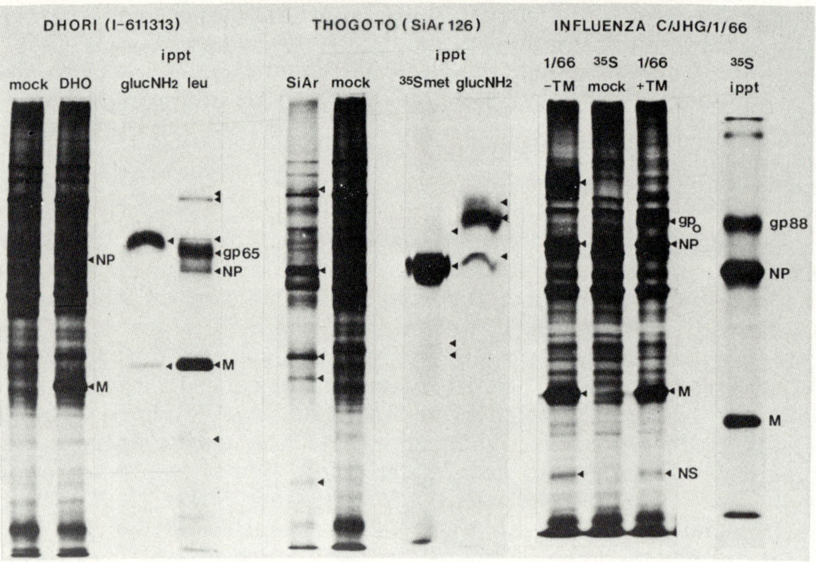

FIGURE 2. SDS-polyacrylamide gel electrophoresis of ^{3}H-leucine labeled cell extracts from cells infected with DHO virus (lane 2), THO virus (lane 5), and influenza C virus (lanes 9 and 11) alongside uninfected cells (lanes 1, 6, and 10). Arrowheads point to bands not present in uninfected cells. Notice the different mobility of gp 88 of influenza C virus in the absence (lane 9) or presence of tunicamycin (lane 11). Lanes 3 and 4 show the polypeptides immune precipitated from glucosamine (lane 3) or leucine labeled extracts (lane 4) from DHO infected vero cells. Lanes 7 and 8 show methionine and glucosamine labeled polypeptides immune precipitated from THO virus infected vero cells. Lane 12 is an immune precipitate of influenza C virus infected chicken fibroblasts.

breakdown product of the larger one since it increased relative to the other upon storage of virus preparations. 2D gel analyses have not yet given an unequivocal answer on the question if there are 2 or 3 minor viral polypeptides 85-90,000 in MW, due to insufficient dissociation from other proteins during IEF. Data thus far suggest that vp 89 and vp 85 are neutral in character (pI 7), whereas a polypeptide of 85,000 MW is more acidic (pI 5). However, the latter polypeptide was present in varying amounts in virus preparations and sometimes absent in immune precipitates, raising the question if it is virus specific.

Intracellular Polypeptides.

One dimensional gel electrophoresis of leucine or methionine labeled cell extracts showed vp 28 to be the most abundant viral polypeptide in cells, with lesser amounts of gp 65 and vp 54 (Fig. 3). Glucosamine or mannose labeled cell extracts revealed two glycosylated polypeptides, gp 65 and a 27-29,000 MW polypeptide, both being recognized by specific-antibodies (Fig. 3). The smaller polypeptide comigrates with the major viral polypeptide vp 28 and therefore is undetectable when using amino acid labeled cell extracts on 1 dimensional gels (Fig. 3). Two dimensional gel electrophoresis of leucine or methionine labeled cell extracts (Fig. 4) revealed next to vp 54-56 and vp 28, several lower MW polypeptides. Previous studies (5) showed the presence of two polypeptide bands at MW 24,000 and 14,000 in immune precipitates from infected cells. Since infected mouse brain was used to prepare the antiserum, antibodies against non-structural proteins may be present and account for this observation. 2D gel analyses indicated that the 24,000 MW band is composed of three polypeptides of basic mature (pI 8.5-9), whereas the 14,000 MW band is a single polypeptide spot at pH 6.5. We do not know yet if the three basic polypeptides (Fig. 4, arrowheads) are related to each other or are breakdown products. Peptide mapping studies are required to resolve this question.

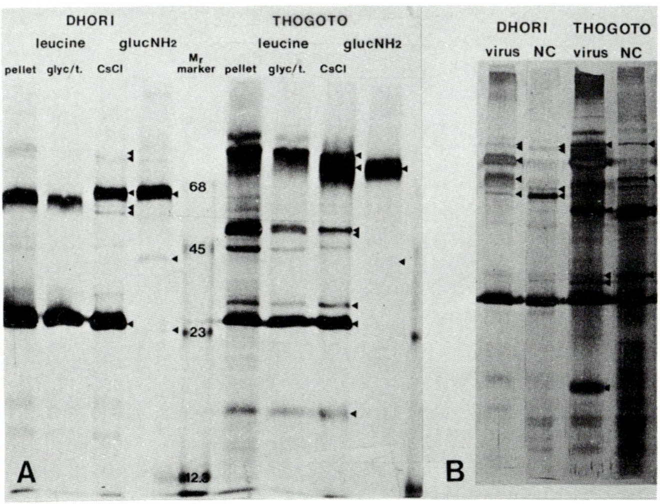

FIGURE 3. A fluorograph of a 10% SDS-polyacrylamide gel showing the ^{3}H leucine-labeled polypeptides in DHO and THO virus preparations at different purification stages. Viruses were pelleted

from clarified BHK cell supernatants (lanes 1 and 6), followed by one cycle each of glycerol-tartrate (lanes 2 and 7) and CsCl gradient centrifugation (lanes 3 and 8). Lanes 4 and 9 show the ^{3}H glucosamine labeled polypeptides of similarly purified preparations. Fig. B shows silver stained polypeptides in preparations of DHO and THO virus (lanes 1 and 3) and nucleocapsids obtained after detergent treatment of DHO virus (lane 2) or from spontaneously disrupted THO virus (lanes 4).

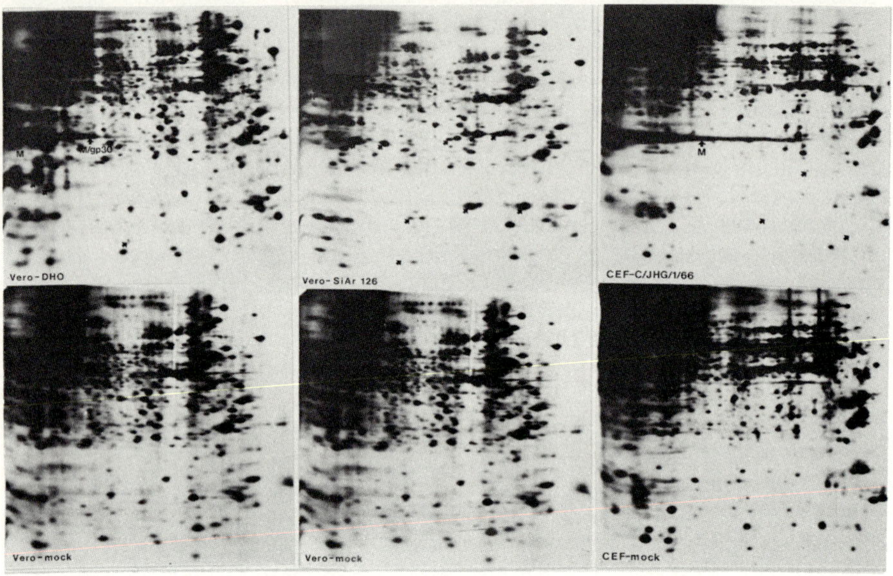

FIGURE 4. Two-dimensional gel electrophoresis of extracts of infected cells (top row) or uninfected cells (bottom). Vero cells were infected with DHO (left) or THO virus (center) at a m.o.i. of 1 p.f.u./cell and labeled form 8-16 hrs. p.i. at 37°C with ^{35}S methionine. Chicken embryo fibroblasts were similarly infected with influenza C virus and labeled from 15-16 hrs p.i. at 33°C (right). Isoelectric focusing was from left to right in 4% polyacrylamide gels as described by Bravo et al. (14). Second dimensions were run in 12% Laemmli type gels (top to bottom). Arrowheads point to polypeptide spots in the infected cells which are absent from uninfected cells.

Analysis of glucosamine labeled purified virus and glucosamine labeled cell extracts on 2D gels indicates that the 27-29,000 MW glycoprotein found in infected cells and immune precipitates from cells, is absent in purified virus preparations and therefore could be a glycosylated non-structural proteins.

THOGOTO VIRUS

Viral Polypeptides.

Studies on the polypeptide composition of THO virus were hampered by the lability of virus particles during purification. Following the purification steps by running samples of a) crude virus pellets, b) virus band from a glycerol-tartrate gradient, and c) virus band from a 10-40% CsCl gradient on a 10% SDS-polyacrylamide gel, it became obvious that large amounts of vp 50 were lost in the purification steps (Fig. 3A). Depending on storage and method of pellet resuspension (sonication versus vortexing), a sharp band at high density was sometimes found in glycerol-tartrate and CsCl gradients (1.29 g/ml). A sample of such material alongside complete virus is shown in Fig. 3B. The main polypeptide in such fractions was vp 50. Additional polypeptides in these preparations were 90,000, 48,000 and 34,000 daltons as well as a small proportion of vp 29. The 62,000 polypeptide in this preparation is of unknown origin: it was not found in purified virus. Clearly, the polypeptides lost from the presumptive nucleocapsids were gp 85, vp 29 and vp 18. Immune precipitation revealed that the ascitic fluid prepared against SiAr 126, the THO isolate used in this study, was predominantly directed against vp 50, but from glucosamine labeled cell extracts, additional polypeptides were precipitated (Fig. 3). The polypeptides migrating ahead of gp85 in this sample were not observed in purified virus, and probably represent precursors to gp 85 (compare Figs. 2 and 3A).

In summary, purified THO virus particles contained 3 major proteins gp 85, vp 48-50 and vp 29 and 4 less abundant polypeptides vp 90, a nucleocapsid-associated vp 34, vp 32 and vp 18.

Intracellular Polypeptides.

SiAr 126 infected Vero cells contained relatively large amounts of virus-induced polypeptides but did not produce more than $5x10^6$ p.f.u./ml. In contrast, infected BHK cells contained less virus-induced polypeptides, but produced up to $2x10^8$ p.f.u./ml supernatant. Two-dimensional gel electrophoresis extracts made from infected Vero cells showed that the major viral nucleocapsid polypeptide vp 50 forms an alkaline streak (cathode is up in 1st dimension). In the 29-35,000 MW range, 5 polypeptides are found which are not present in uninfected cells (Fig. 4). Four of these are also present in purified viruses: the putative M protein vp 29 is found as two discrete spots at approximately pH 8 and pH 7.5, vp 32 at pH 6 and vp 34 at pH 5.5. The polypeptide migrating with MW of 34,000 at pH 5 is not only labeled with leucine or methionine but also with glucosamine and mannose. Since we have never observed a glycosylated polypeptide smaller than 42,000 daltons in several independent glucosamine-labeled virus preparations, (Fig. 3A) the possibility exists that it could be non-structural. In addition to the aforementioned polypeptides, 4 smaller polypeptides are found in infected cells: two have molecular weights of about 18,000, one of 16,000 and one of approximately 12,000 daltons. Either one or both of the 18,000 MW polypeptides could be the same as p18 found in virus preparations, but peptide mapping studies have to resolve this question.

ACKNOWLEDGEMENTS

This work was supported by a research stipend from the Alexander von Humboldt Foundation, Bonn, FRG (J.C.) and The Deutsche Forschungsgemeinschaft.

REFERENCES

1. Kuo, Y.C., Fengen, J., Wang, M., Wang, P., and Chu, C.M., 1982, Kexue Tongbao 27,1118.
2. Palese, P., Racaniello, V.R., Desselberger, U., Young, J., and Baes, M., 1980, Phil. Trans. Soc. Lond. B288, 299.
3. Meier-Ewert, H., Nagele, A., Herrler, G., Basak, S., and Compans, R.W., 1981, In "The Replication of Negative Strand Viruses" (D.H.L. Bishop and R.W. Compans, eds.), p. 173. Elsevier North-Holland, New York.
4. Herrler, G., Compans, R.W. and Meier-Ewert, H., 1979, Virology, 99, 49.
5. Clerx, J.P.M., Fuller, F., and Bishop, D.H.L., 1983, Virology, 127, 205.
6. Causey, O.R., Kemp, G.E., Madbouley, M.H., and Lee, V.H., 1969, Bull. Soc. Path. Exot. 62, 249.
7. Moore, D.L., Causey, O.R., Carey, D.E., Reddy, S., Cooke, A.R., Akinkugbe, F.M., David-West, T.S., and Kemp, G.E., 1975, Ann. Trop. Med. Parasit. 69, 49.
8. Berge, T.O., ed. (1975). "International Catalogue of Arboviruses Including Certain Other Viruses of Vertebrates". DHEW Publication no. CDC 75-8301, Washington, D.C.
9. Sokhey, H., Dandawate, C.N., Gogate, S.S., Ghosh, S.N., and Gupta, N.P., 1977, Indian J. Med. Res. 66, 726.
10. Anderson, C.R., and Casals, J., 1973, Indian J. Med. Res. 61, 1416.
11. Cox, N.J., and Kendal, A.P., 1976, Virology, 74, 239.

12. Sugawara, K., Nakamura, K., and Homma, M., 1983, J. Gen. Virol. 64, 579.
13. Petri, T., Herrler, G., Compans, R.W., and Meier-Ewert, H., 1980, "FEMS Microbiol. Lett.", 9, 43.
14. Bravo, R., Small, J.V., Fey, S.J., Larsen, P.M., and Celis, J.E., 1982, J. Mol. Biol. 154, 121.

Biology

MECHANISM OF PERSISTENT LCMV INFECTION IN MICE: ROLE OF LCMV VARIANTS IN SUPPRESSION OF LCMV SPECIFIC CYTOTOXIC T CELL RESPONSE AND MAINTENANCE OF THE CARRIER STATE[1]

Rafi Ahmed, Aimo Salmi,[2] and M.B.A. Oldstone

Department of Immunology
Scripps Clinic and Research Foundation
La Jolla, California 92037

Mice infected with lymphocytic choriomeningitis virus (LCMV) as neonates or *in utero* become carriers with life-long viremia. This is the natural infection. Such carrier mice make an anti-viral antibody response, but contain no detectable LCMV specific cytotoxic T cells (CTL) (1,2,3,4). Since the classic studies of Traub, almost 50 years ago, the mechanism of suppression of CTL responses and its possible role in the persistence of LCMV, have remained a central question about the biology of LCMV infection (5). In this report, we describe our studies on the mechanism of LCMV persistence and suppression of CTL responses in Balb/c wehi mice persistently infected with LCMV Armstrong CA1371. We have found that: (1) Spleen cells from carrier mice actively suppress the induction of LCMV specific CTL response of normal adult mice. (2) The suppression is specific for the CTL response and has no effect on anti-LCMV antibody response. (3) Suppression of LCMV specific CTL response is associated with the establishment of persistent infection. (4) Suppression of LCMV specific CTL response is not mediated by a suppressor cell but is due to the presence of genetic variants of LCMV in spleens of carrier mice. We propose that these LCMV variants play a crucial role in the maintenance of the persistent infection in mice.

Spleen Cells from Carrier Mice Actively Suppress the Induction of LCMV Specific CTL Response of Normal Adult Mice but have No Effect on Anti-LCMV Antibody Response. To examine the effect of spleen cells from carrier mice on

---

[1]This work was supported by USPHS Grant NS-12428 and NIH Grant AI-09484.

[2]Present address: Department of Virology, University of Turku, Turku, Finland

Copyright © 1984 by Academic Press, Inc.
All rights of reproduction in any form reserved.
ISBN 0-12-183501-4

the generation of LCMV specific CTL and antibody responses, spleen cells from 6-12 week old Balb/c wehi carriers were mixed with spleen cells from normal 6-12 week old Balb/c mice and transferred intravenously (I.V.) into irradiated normal Balb/c mice. The mice were challenged with LCMV Armstrong at the same time and at various times post infection the mice were checked for LCMV specific CTL activity, LCMV specific antibody, and presence of infectious LCMV in spleen (see Table 1). Irradiated mice reconstituted with normal spleen cells made LCMV specific antibody and CTL responses, and cleared the infection by day 10. In contrast, mice reconstituted with carrier spleen cells made no detectable CTL response, a minimal amount of antibody and were unable to clear the infection. Since the irradiated normal recipient mice provide the antigen presenting cells, this result shows that the unresponsiveness of carrier spleen cells to LCMV is not simply due to a defect in antigen presenting cells. Mice reconstituted with a mixture of carrier spleen cells and normal spleen cells made a vigorous anti-LCMV antibody response but a poor LCMV specific CTL response and did not clear the virus infection. Thus, our results show that spleen cells from carrier mice actively suppress the induction of LCMV specific CTL response of normal mice. The suppression is specific for the CTL response and has no effect on anti-LCMV antibody response. This effect is also virus specific since carrier spleen cells do not suppress the induction of vaccinia virus specific CTL response (data not shown).

Suppression of LCMV Specific CTL Response is Associated with the Establishment of Persistent LCMV Infection. Irradiated mice reconstituted with mixtures of carrier and normal spleen cells were examined for a period of 3 months to see if they cleared the infection or not. Infectious LCMV was checked in serum, spleen, lymph nodes and thymus. LCMV was present in high titers in the organs (10^5-10^6 PFU) and in the serum (10^4-10^5 PFU) of all mice that were tested. A summary of the transfer experiments is shown in Table 2. There is a striking correlation between suppression of CTL responses and inability to clear the virus infection. There have been other reports showing that carrier spleen cells suppress the induction of LCMV specific CTL responses of normal mice (6, 7, 8). However, this is the first study demonstrating the important biological consequences of this specific suppression - namely that this leads to viral persistence. When irradiated mice were reconstituted with carrier spleen cells plus LCMV specific CTL (day 8 spleen cells from acutely infected normal Balbc/wehi mice) the infection was

TABLE 1

EFFECT OF CARRIER SPLEEN CELLS ON THE GENERATION OF LCMV SPECIFIC CTL AND ANTIBODY RESPONSES: SUPPRESSION OF CTL RESPONSE IS ASSOCIATED WITH INABILITY TO CLEAR VIRUS INFECTION

Source of spleen cells transferred[a]	LCMV specific CTL in spleen[b] % specific ^{51}Cr release from infected targets (H-2^d)				LCMV specific antibody in serum[c] ELISA titer (log2)				LCMV titer in spleen[d] PFU/spleen (log10)			
	day3	day6	day8	day10	day3	day6	day8	day10	day3	day6	day8	day10
5 x 10^7 cells	0	18	51	23	<4.7	5.7	14.0	15.6	5.7	4.7	2.3	<1.6
Normal Balb	0	13	65	20	<4.7	5.3	14.8	15.9	5.6	3.5	<1.6	<1.6
	2	12	64	19	<4.7	6.8	16.0	16.2	5.4	3.9	<1.6	<1.6
5 x 10^7 cells	0	1	0	0	<4.7	<4.7	<4.7	<4.7	6.0	5.9	5.2	4.5
Carrier Balb	0	0	0	2	<4.7	<4.7	<4.7	5.6	6.2	6.1	5.8	5.8
	0	2	1	0	<4.7	<4.7	7.5	<4.7	6.1	5.8	5.7	5.7
5 x 10^7 cells	0	3	4	0	<4.7	9.7	14.3	13.7	6.0	5.5	5.0	5.7
Normal Balb	2	0	11	1	<4.7	7.9	15.3	13.8	6.2	5.7	5.7	4.9
+	0	4	7	0	<4.7	7.8	14.7	14.1	6.0	6.1	5.4	5.2
5 x 10^7 cells												
Carrier Balb												

[a]Spleen cells were transferred I.V. into irradiated (600R) normal Balb/c wehi mice. All recipient mice were infected I.V. with 2 x 10^5 PFU of LCMV Armstrong CA1371 at the same time.

[b]CTL activity was checked on LCMV infected and uninfected Balb Cl7 (H-2^d) fibroblasts in a 6-8 hr ^{51}Cr release assay. The data shown is % specific killing of LCMV (Armstrong) infected targets at an effector:target ratio of 50:1. Samples assayed in triplicate.

[c]Determined by ELISA test using purified LCMV (Armstrong) as antigen.

[d]Spleens were homogenized and titrated on vero cell monolayers.

TABLE 2
SUMMARY OF TRANSFER EXPERIMENTS

Source of spleen cells transferred	LCMV specific immune response		Outcome of infection[a]
	CTL	Antibody	
Normal Balb (N)	++++	++++	Clearance
Carrier Balb (C)	nil	±	Persistence
N + C	±	++++	Persistence

[a]Mice were checked for presence of infectious LCMV in serum, spleen, lymph nodes and thymus for a period of 3 months post infection.

cleared within a week (data not shown). Using a different protocol Zinkernagel and Welsh have also shown the important role of LCMV specific CTL in anti-viral protection against acute LCMV infections in mice (9).

Active Suppression of LCMV Specific CTL Response by Carrier Spleen Cells is not Mediated by a Suppressor Cell. Table 3 shows that carrier spleen cells following anti-theta + complement treatment, or 3000R X-irradiation, or sonication, retain the ability to suppress LCMV specific CTL response. Thus, the effect is not due to a suppressor cell. As in the earlier experiments, anti-LCMV antibody response was not effected, and associated with the CTL suppression was inability to clear the virus infection.

Suppression of LCMV Specific CTL Response is Due to the Presence of Genetic Variants of LCMV in Spleens of Carrier Mice. We next considered the possibility that suppression of LCMV specific CTL response may be due to LCMV present in carrier spleen cells. Approximately 5×10^5 LCMV infectious centers (IC) are transferred with 5×10^7 carrier spleen cells since about 1% of spleen cells score as LCMV IC. In our transfer experiments this has ranged from 1.4×10^5 to 1.2×10^6 IC. In experiments where carrier spleen cells were sonicated, 4-8 $\times 10^5$ PFU of LCMV were transferred. Injection of 5×10^6 PFU of LCMV Armstrong CA 1371 induced a strong CTL response and the infection was cleared within 8-10 days. This suggested that if suppression was due to the carried over virus, the LCMV present in spleens of carrier mice may be a genetic variant or in a form that induces suppression. To check this possibility, clones of LCMV were isolated from the spleen of a 2 month old Balb carrier mouse,

TABLE 3
ACTIVE SUPPRESSION OF LCMV SPECIFIC CTL RESPONSE BY CARRIER SPLEEN CELLS IS NOT MEDIATED BY A SUPPRESSOR CELL

Source of spleen cells transferred[a]	Treatment of carrier spleen cells[b]	LCMV specific CTL in spleen[c] % specific ^{51}Cr release from infected targets (H-2[d]) (E:T 50:1)	LCMV specific antibody in serum[c] ELISA titer (log 2)	LCMV titer in spleen[c] PFU/ spleen (log 10)
Normal Balb (N)	None	60	14.8	<1.6
		57	16.0	<1.6
		51	13.3	<1.6
Carrier Balb (C)	None	0	<4.9	4.3
		0	<4.9	4.5
		0	<4.9	5.3
N + C	None	9	16.3	5.5
		8	15.8	5.3
		5	14.7	5.5
N + C	α - Theta + C'	16	15.9	5.6
		8	16.4	5.2
		13	16.0	5.1
N + C	X-irradiation 3000 R	0	13.7	4.1
		0	14.0	3.7
N + C	Sonication	9	15.5	5.2
		9	16.4	5.3
		8	14.7	4.7

[a]See footnote for Table 1.

[b]5×10^7 carrier spleen cells were treated as indicated and then mixed with 5×10^7 normal spleen cells and transferred I.V. into irradiated normal recipients.

[c]LCMV specific CTL, antibody and virus titer were checked 8 days post infection. Assays were performed as described in footnotes for Table 1.

by two cycles of plaque to plaque purification on vero cells, and then stocks were grown up in BHK cells. Four of these cloned LCMV spleen isolates were tested for their ability to induce LCMV specific CTL responses. All four isolates induced poor CTL responses, normal antibody responses, and were present in high titers in spleens 8 days post infection.

TABLE 4

ACTIVE SUPPRESSION OF LCMV SPECIFIC CTL RESPONSE BY CARRIER SPLEEN CELLS IS DUE TO THE PRESENCE OF GENETIC VARIANT OF LCMV IN CARRIER MICE.

Source of spleen cells transferred[a]	Challenge virus	Dose (PFU)	LCMV specific CTL in spleen[b] % specific ^{51}Cr release from infected targets (H-2^{d}) (E:T 50:1)		LCMV specific antibody in serum[b]	LCMV titer in spleen[b]
			LCMV Armstrong	LCMV clone 13	ELISA titer (log2)	PFU/spleen (log10)
5×10^7 cells Normal Balb	LCMV Armstrong CA 1371	2×10^5	62	52	12.5	<1.6
		"	41	38	12.9	2.3
		2×10^6	53	40	14.0	2.7
		"	59	47	13.6	<1.6
5×10^7 cells Normal Balb	LCMV Clone 13	2×10^5	7	2	11.6	5.2
		"	6	0	12.4	5.1
		2×10^6	14	18	12.8	5.0
			19	22	11.9	5.1

[a]Spleen cells were transferred I.V. into irradiated normal Balb mice. At the same time recipient mice were infected I.V. with either LCMV Armstrong or Clone 13 at the indicated dose. LCMV Clone 13 was isolated from the spleen of a 2 month old Balb carrier mouse infected at birth with triple plaque purified LCMV Armstrong CA 1371.

[b]LCMV specific CTL, antibody and virus titer were checked 8 days post infection. Assays were performed as described in footnotes for Table 1.

Data obtained with one spleen isolate - LCMV clone 13 - is shown in Table 4. When mice were challenged with both LCMV Armstrong CA 1371 and Clone 13 there was a strong suppression of the CTL response (data not shown). Our results show that suppression of CTL responses is due to the presence of genetic variants of LCMV in carrier mice. The LCMV spleen isolates used in these experiments were derived from a 2 month old Balb carrier mouse infected at birth with 3x plaque purified LCMV Armstrong CA 1371. Thus, it appears likely that the LCMV variants found in the spleen emerged during the persistent infection.

In conclusion, we have shown that LCMV undergoes mutation during persistent infection in its natural host, and there is selection of LCMV variants in spleens of carrier mice. These LCMV variants play a crucial role in the suppression of LCMV specific CTL responses and in the maintenance of the persistent infection.

ACKNOWLEDGEMENTS

This is Publication No. 3199-IMM from the Department of Immunology, Scripps Clinic and Research Foundation, La Jolla, California 92037. We thank Rita Concepcion for her excellent technical assistance.

REFERENCES

1. Oldstone, M. B. A., and Dixon, F. J. (1967). Science 158, 1193.
2. Benson, L., and Hotchin, J. (1969). Nature 222, 1045.
3. Marker, O., and Volkert, M. (1973). J. Exp. Med. 137, 1511.
4. Cole, G. A., Prendergast, R. A., and Henney, C. S. (1973). In "Lymphocytic Choriomeningitis Virus and other Arenaviruses" (F. Lehmann-Grube, ed.), p. 61. Springer-Verlag, Berlin.
5. Traub, E. (1936). J. Exp. Med. 63, 847.
6. Zinkernagel, R. M., and Doherty, P. C. (1976). In "Immunological Tolerance. Mechanisms and Potential Therapeutic Applications" (D. H. Katz and B. Benacerraf, eds.), p. 403. Academic Press, New York.
7. Dunlop, M. B. C., and Blanden, R. V. (1977). J. Exp. Med. 145, 1131.
8. Cihak, J., and Lehmann-Grube, F. (1978). Immunology 34, 265
9. Zinkernagel, R. M., and Welsh, R. M. (1976). J. Immunol. 117, 1495.

THE AFRICAN ARENAVIRUSES LASSA AND MOPEIA: BIOLOGICAL AND IMMUNOCHEMICAL COMPARISONS

J.C.S. Clegg and G. Lloyd

Special Pathogens Reference Laboratory
Public Health Laboratory Service
Centre for Applied Microbiology and Research
Porton Down, Salisbury SP4 OJG, United Kingdom.

ORIGIN OF VIRUS STRAINS

Lassa virus, the etiological agent of Lassa fever, has been isolated from patients from several countries in West Africa, and also from the rodent Mastomys natalensis, in which it is enzootic in these areas. The closely related virus Mopeia (also known as Mozambique virus), which is apparently not associated with human disease, and which is able to protect experimental animals against subsequent challenge with Lassa virus (1), has been isolated from the same rodent species in Mozambique and Zimbabwe (2, 3). We have compared the characteristics of several strains of these viruses (see Table 1) with respect to the pathology induced on experimental infection of guinea pigs and primates and the immunological and molecular properties of their structural proteins.

EXPERIMENTAL INFECTIONS WITH LASSA AND MOPEIA VIRUSES

All the strains of Lassa virus so far tested induce a characteristic disease course in Dunkin-Hartley guinea pigs and in Rhesus monkeys, with pyrexia, viremia and high concentrations of virus in liver, kidney and spleen. Damage to liver and kidney and impairment of their function are indicated by changes in several serum enzyme and metabolite levels, e.g. glutamate oxaloacetate transaminase and urea. The disease is uniformly fatal in monkeys and causes 60-70% mortality in the guinea pig. No significant differences have been noted between the different virus strains used, including those isolated direct from Mastomys rather than from human cases of Lassa fever. In contrast, none of the strains of Mopeia virus tested induce clinically detectable disease, with the exception of a small transient temperature rise seen in some guinea pigs infected with strain 152. No

Copyright © 1984 by Academic Press, Inc.
All rights of reproduction in any form reserved.
ISBN 0-12-183501-4

TABLE 1
ORIGIN OF VIRUS STRAINS

Strains	Area of Origin	Species of Origin
GA 391, PB1, LP, SA 549, 134, 331	W. Africa	Man
523, 535	W. Africa	*Mastomys*
148, 150, 152, 20410, 478	S.E. Africa	*Mastomys*

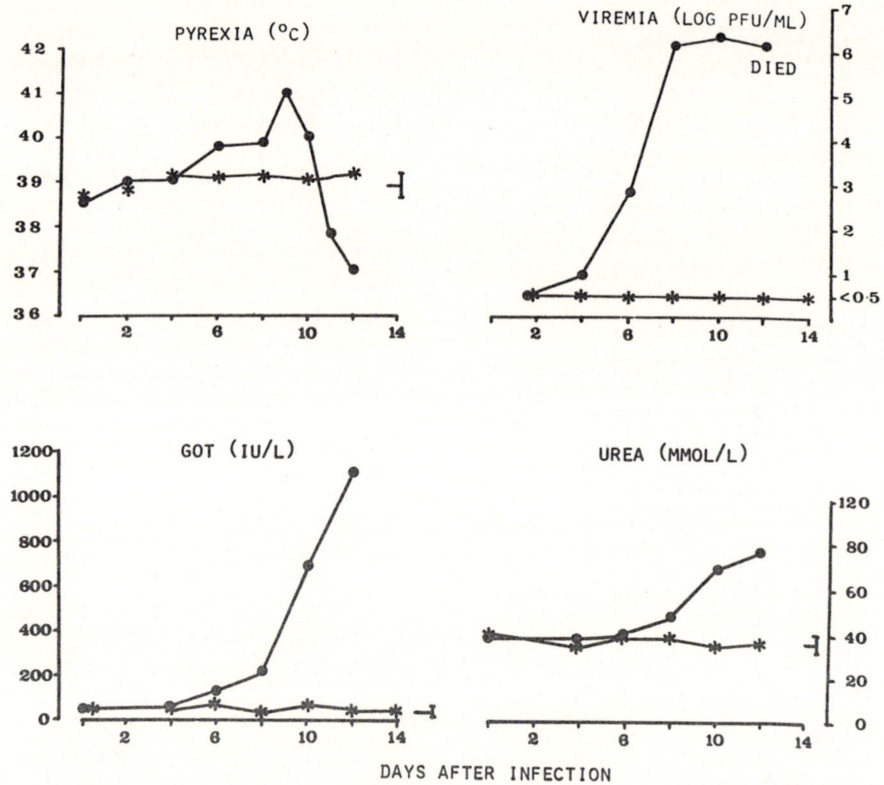

FIGURE 1. Experimental infections of primates with Lassa or Mopeia viruses. Rhesus monkeys weighing 4-5 Kg were injected subcutaneously with 10^3 pfu of Lassa virus strain GA 391 (●——●) or Mopeia virus strain M150 (*——*). Vertical bars indicate normal ranges.

viremia is observed in monkeys, and only occasional, low level, short-lived viremia in guinea pigs. Specimen data typical of infection of monkeys by Lassa and Mopeia viruses are shown in Fig. 1. Infection with either virus results in production of similar levels of antibody detectable by immunofluorescence by day 3-4 after infection and reaching a maximum by day 8-9. These biological properties indicate that the virus strains isolated in W. Africa and S.E. Africa fall into two different and relatively homogeneous groups.

IMMUNOCHEMICAL DIFFERENTIATION OF LASSA AND MOPEIA VIRUSES

Cross-reactions between many different strains of virus can be conveniently analysed at the level of individual polypeptide species by separation of lysates of infected cells on SDS-polyacrylamide gels followed by transfer to nitrocellulose blots and incubation with appropriate antisera. Previous work has shown that the structural proteins N and G2, and the glycoprotein precursor GPC are detected in Lassa virus-infected cells by homologous antisera (4). When blots containing proteins from cells infected with a set of W. African and S.E. African virus strains are incubated with antisera raised in guinea pigs against any of a number of strains of Lassa virus, antibody binds to the N and G2 proteins of every virus strain tested (Fig. 2a). Thus there are antigenic determinants common to W. and S.E. African arenaviruses on the internal nucleocapsid protein and on at least one of the envelope glycoproteins. No conclusions concerning G1 can be drawn since it is antigenically inactive in this system (4). When antisera raised against S.E.African arenaviruses are used a completely different pattern of reactivity is found (Fig. 2b). The N protein of the S.E. African viruses reacts strongly with antibody, but W. African N proteins reacted weakly or not at all, and no antibody binds to the glycoprotein even of the homologous virus. Thus the viruses again fall into two groups; those from W. Africa, which cause disease in experimental animals and give rise to antisera with broad reactivity with the N and G2 proteins of all the African arenaviruses tested, and those from S.E.Africa, which are non-pathogenic and give rise to antisera with reactivity almost completely restricted to the N protein of the S.E. African arenaviruses. The anti-glycoprotein antibody level in sera raised against Mopeia virus strains is currently being investigated in other systems involving less drastic denaturation of the antigens.

Some monoclonal antibodies raised against Lassa virus are also able to differentiate between these two groups of

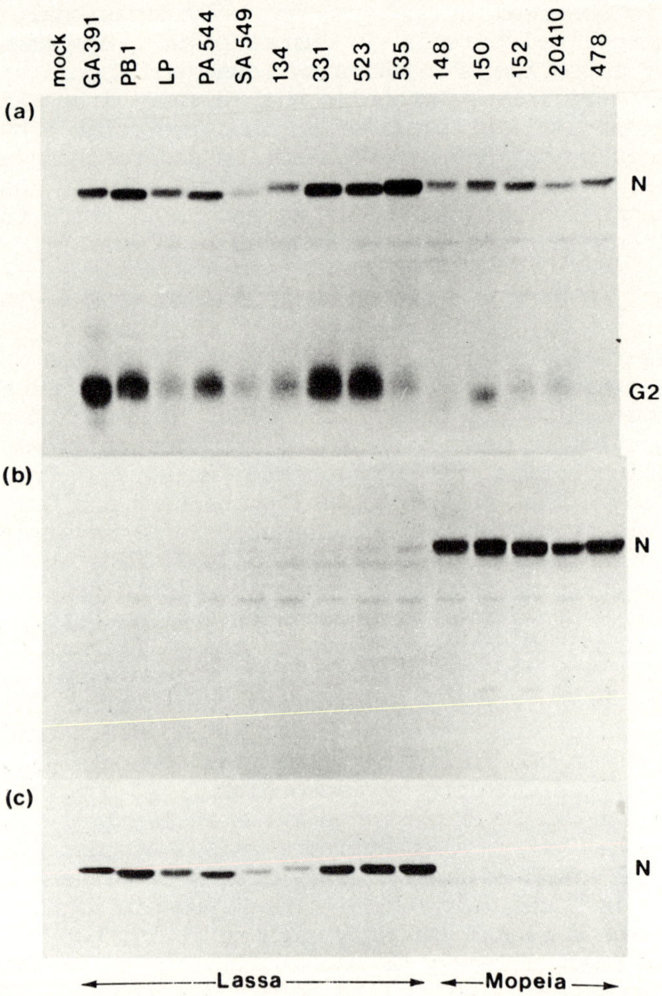

FIGURE 2. Immunoreactivity of Western blots of proteins of cells infected with African arenaviruses. Monolayers of CV-1 cells were infected at a multiplicity of about 0.1. Three days after infection the cells were lysed and the proteins separated by electrophoresis on 10% acrylamide SDS gels. The proteins were then electrophoretically transferred to nitrocellulose membranes which were probed with (a) guinea pig antiserum to Lassa virus strain GA 391, (b) guinea pig antiserum to Mopeia virus strain 148, or (c) monoclonal antibody 1.52, raised against Lassa virus strain GA 391. Bound antibody was located by subsequent incubation with peroxidase-conjugated second antibody and the peroxidase substrate 4-chloronaphthol.

virus strains, reacting strongly with W. African virus N protein, but only weakly, or not at all, with N protein of S.E. African origin (Fig. 2c). Other monoclonal antibodies active against the Lassa virus N protein react with all the virus strains tested, as might be expected from the broad reactivity of conventional Lassa virus antisera.

SIZE HETEROGENEITY OF STRUCTURAL PROTEINS

During the course of these investigations, it became apparent that there were small differences in the mobilities of the N and G2 proteins between the different strains of viruses (see Fig. 2a and 2c). When extracts of cells infected with two different strains of virus were mixed (e.g. PA 544 and 134) the two sizes of N protein could be clearly resolved after electrophoresis (not shown), indicating that the mobility differences do reflect differences in size. Lassa virus N proteins ranged from 58,000 to 62,000 in molecular weight, while the Mopeia virus strains all had N protein molecular weights of 62,000.

The variations in G2 mobility are most clearly shown up after fractionation on gels with a higher concentration of acrylamide (Fig. 3). Under these conditions G2 is in some cases resolved into a doublet set in a broader band of

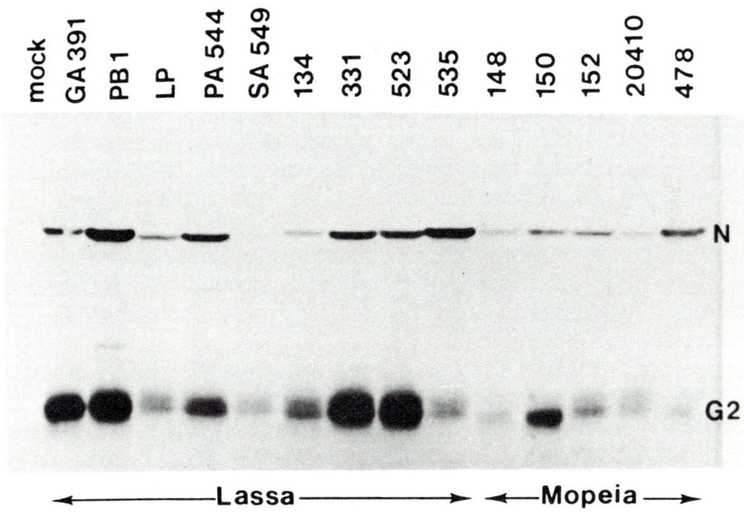

FIGURE 3. Apparent size variation in G2 between different strains of African arenaviruses. The blot was made as in Figure 2 except that a 13% acrylamide gel was used. It was probed with guinea pig antiserum to Lassa virus strain PB1.

immunoreactive material possibly as a result of different degrees of glycosylation of the protein in the intracellular environment. Some of the inter-strain variation in G2 mobility may be attributed to differences in the proportion of different forms of G2, but in some cases, e.g. Mopeia virus strains 148, 150 and 478, the mobility shift seems more extreme. Further work is required to determine the significance of these differences. The mobility of glycoprotein G1, as determined by staining blots with concanavalin A (5), appears to be relatively constant (not shown).

We can conclude that there are significant molecular differences between different strains of Lassa virus in terms of the size of the nucleocapsid protein, and possibly also that both Lassa and Mopeia virus strains are heterogeneous with respect to the apparent size of the envelope protein G2.

CONCLUSIONS

The data presented here support the division of the African arenaviruses into two groups. One group consists of virus strains isolated in W. Africa from Lassa fever patients or from Mastomys and which cause serious haemorrhagic disease in experimental animals. Although this group appears to be fairly homogeneous in its pathogenicity and antigenicity using the sera and monoclonal antibodies tested so far, there are inter-strain variations in the sizes of two of the viral structural proteins which may result in subtle antigenic variations not yet detected. The other group consists of virus strains isolated from Mastomys in S.E. Africa, which are homogeneous with regard to their lack of acute pathogenicity and have their own apparently uniform pattern of antigenicity which is different from that of Lassa virus. However the possibility of antigenic variation is again raised by the observation of apparent size differences in the envelope protein G2. It is proposed that this group of viruses be denoted Mopeia, rather than Mozambique (B. McIntosh and D.I.H. Simpson, personal communication).

Lassa and Mopeia viruses have previously been separable by the use of a monoclonal antibody specific for G2 of lymphocytic choriomeningitis virus which reacted with Mopeia virus but not Lassa virus by immunofluorescence (6). The present results with both conventional antisera and N-specific monoclonal antibodies indicate that antigenic differences between the two groups are also carried on the N protein. However, Western blots using Lassa virus antisera clearly also demonstrate the existence of common antigenic determinants on both the N and

G2 proteins, thus confirming and extending the close relationship between the viruses previously observed by immunofluorescence (7).

ACKNOWLEDGMENTS

Thanks are due to Nicky Jones, John Chamberlain, Gill Stagg and Ron Raymond for their expert assistance, and to Nancy Ellwood for preparing the manuscript.

REFERENCES

1. Kiley, M.P., Lange, J.V. and Johnson, K.M. (1979). Lancet ii, 738.
2. Wulff, H., McIntosh, B.M., Hamner, D.B. and Johnson, K.M. (1977). Bull. WHO 55, 441.
3. Johnson, K.M., Taylor, P., Elliott, L.H. and Tomori, O. (1981). Amer. J. trop. Med. Hyg. 30, 1291.
4. Clegg, J.C.S. and Lloyd, G. (1983). J. gen. Virol. 64, 1127.
5. Clegg, J.C.S. (1982). Analyt. Biochem. 127, 389.
6. Buchmeier, M.J., Lewicki, H.A., Tomori, O. and Johnson, K.M. (1980). Nature 288, 486.
7. Wulff, H., Lange, J.V. and Webb, P.A. (1978). Intervirol. 9, 344.

REPLICATION OF BUNYAVIRUSES IN A XENOPUS LAEVIS CELL LINE

Gillian E. Watret, Craig R. Pringle
and Richard M. Elliott

MRC Virology Unit,
Institute of Virology,
Church Street,
Glasgow G11 5JR, Scotland

The family Bunyaviridae comprises over 200 arthropod borne viruses which are classified into 4 genera: Bunyavirus, Nairovirus, Uukuvirus and Phlebovirus (1). We have been analysing the genetic diversity of bunyaviruses throughout the family and to facilitate comparisons of viruses from different genera we sought a cell line that was permissive for all viruses. A number of mammalian cells, including BHK, Vero, BSC-1, and chick embryo fibroblasts (CEF) were permissive for some but not all of the viruses studied. Prompted by the observation of Leake *et al*., (2) that a continuous cell line, XTC-2, derived from *Xenopus laevis* (3) was permissive for a wide range of arboviruses, we investigated whether XTC-2 cells would support the replication of viruses from all four genera of the Bunyaviridae. This study also afforded the opportunity to characterize two Scottish bunyaviruses- St Abb's Head (SAH, Uukuvirus genus; 4) and Clo Mor (CM, Nairovirus genus; 5).

GROWTH OF BUNYAVIRUSES IN XTC-2 CELLS.

Initially, the patterns of growth of viruses representing three of the bunyavirus genera were compared in XTC-2 cells. Cells were infected with 1-5 pfu/cell of virus and cultures harvested at various times after infection for plaque titration as described by de Madrid and Porterfield (6). The growth curves obtained for Bunyamwera virus (BUN, Bunyavirus genus), St Abb's Head virus and Clo Mor virus are displayed in Figure 1, and show that XTC-2 cells will support the productive replication of all

Copyright © 1984 by Academic Press, Inc.
All rights of reproduction in any form reserved.
ISBN 0-12-183501-4

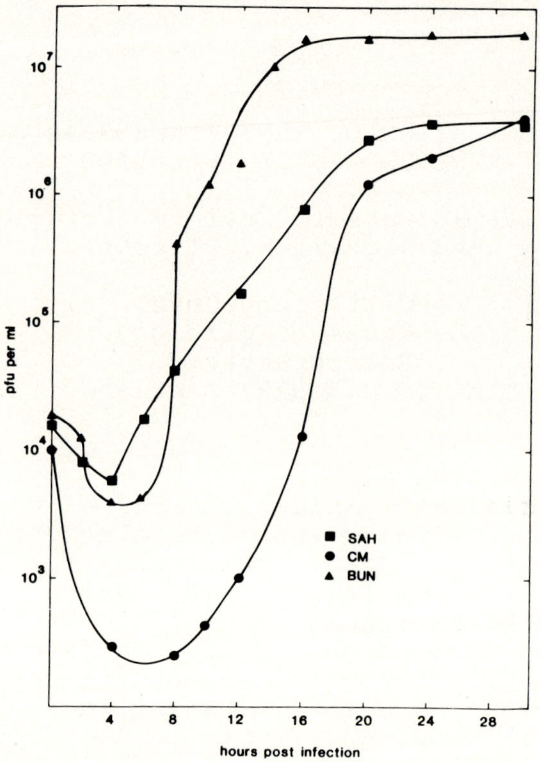

FIGURE 1. Growth curves of Bunyamwera virus, St Abb's Head virus and Clo Mor virus in XTC-2 cells.

three viruses. The kinetics of growth of the viruses in XTC-2 are similar to those in other cells - Bunyamwera virus in BHK, St Abb's Head virus in CEF and Clo Mor virus BSC-1 (data not shown). The final yields of virus per cell were also similar between the different virus/cell combinations.

The efficiency of plaquing of these three viruses in XTC-2 cells and another appropriate cell line was compared, and similar titres were obtained. Since XTC-2 cells appeared suitable for plaque assay of the three representative viruses, a number of other bunyaviruses were tested; these results are sumarized in Table 1. To date only the two viruses representing the Phlebovirus genus failed to replicate in XTC-2 cells; other phleboviruses will have to be examined to see if this is a general characteristic of this genus.

TABLE 1
BUNYAVIRUS REPLICATION IN XTC-2 CELLS

Genus	Virus	Plaque production	Protein synthesis
	Bunyamwera	+	+
	Maguari	+	ND[1]
	Batai	+	+
	Main Drain	+	+
	Northway	+	ND
Bunyavirus	Germiston	+	+
	Kairi	+	ND
	Guaroa	+	+
	Lumbo	+	+
	La Crosse	+	+
	Sathuperi	+	+
Uukuvirus	St Abb's Head	+	+
	Potepli	+	+
Nairovirus	Clo Mor	+	+
	Dugbe	+	+
Phlebovirus	Naples sandfly fever	-	-
	Sicilian sandfly fever	-	-

[1]ND = not determined

PROTEIN SYNTHESIS IN BUNYAVIRUS-INFECTED CELLS.

The proteins induced by Bunyamwera, St Abb's Head or Clo Mor viruses in XTC-2 cells are shown in Figure 2, and their molecular weights are compared in Table 2. Identification of Bunyamwera proteins is facilitated because of the marked inhibition of host cell protein synthesis, in contrast to the relative lack of inhibition by St Abb's Head or Clo Mor viruses. The use of different radioactive precursors or immunoprecipitation using specific antisera identified the St Abb's Head virus and Clo Mor virus proteins. Similar labelling experiments

TABLE 2
COMPARISON OF THE MOLECULAR WEIGHTS OF THE PROTEINS INDUCED BY BUNYAVIRUSES IN XTC-2 CELLS.

Protein	BUN	CM	SAH
L	180[1]	ND[2]	ND
glycoprotein	115 (30)	90	62-75
N	24	50	25
non-structural protein	60 50 16 13	45	58 30

[1] molecular weight x 10^{-3}
[2] ND= not detected

with these viruses in other cells revealed the same spectra of induced proteins, further confirming the suitability of XTC-2 cells for comparative studies.

Bunyamwera virus induced 7 detectable polypeptides, including 4 non-structural (NS) proteins. The identification of these proteins is generally in agreement with the recent results of Short et al., (7); however it should be noted that in our hands the G2 protein of Bunyamwera virus is expressed more variably and less reproducibly than by other members of the genus, though G2 can be detected in purified virus preparations (not shown). The temporal synthesis of proteins in Bunyamwera infected XTC-2 cells is in agreement with observations of infected BHK cells (7, 8). The synthesis of NSp13 follows that of the nucleocapsid protein N, and the synthesis of NSp16 follows that of G1; this conforms with our unpublished observations that NSp13 is encoded by the S RNA segment and NSp16 is encoded by the M RNA segment.

In addition to identifying the G1/G2 doublet and N protein in St Abb's Head virus infected cells, two non-structural proteins have also been detected: NSp58 and NSp30. The coding assignments of these proteins has yet to be determined.

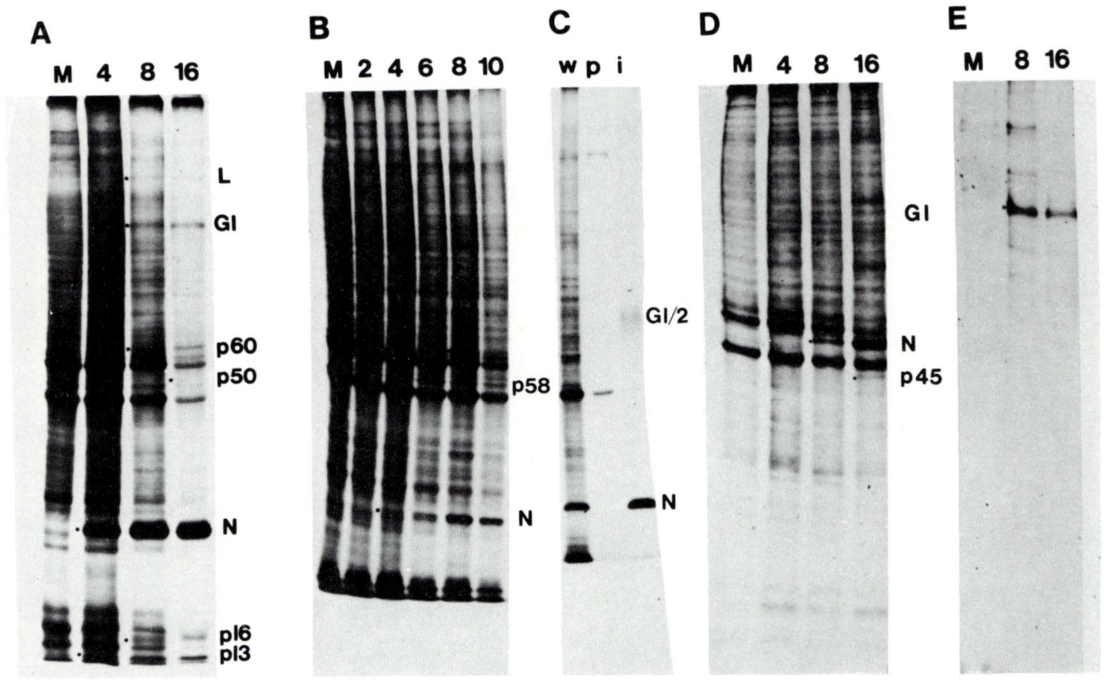

FIGURE 2. Proteins synthesized in bunyavirus-infected XTC-2 cells. 1 hour pulses with ^{35}S-methionine (A-D) or ^{3}H-mannose (E) at indicated times post-infection. M = mock infected. A = Bunyamwera virus. B and C = St. Abb's Head virus; in C, w = whole cell extract, p = preimmune control and i = immuneprecipitation. D and E = Clo Mor virus.

Clo Mor virus is distinguished by having a nucleocapsid protein about twice as large (50,000 mol. wt.) as that of Bunyamwera or St Abb's Head virus (25,000). One glycoprotein 90,000 in molecular weight and one NS protein (45,000) were also detected. These results provide physio-chemical evidence supporting the inclusion of Clo Mor virus in the Nairovirus genus.

As shown in Table 1, a number of other bunyaviruses have been shown to induce the synthesis of viral polypeptides in infected XTC-2 cells, the only exceptions being the two phleboviruses. Our results indicate therefore that the XTC-2 cell line represents a suitable host to compare the macromolecular synthesis of bunyaviruses from three at least of the four genera.

ACKNOWLEDGEMENTS

We thank Drs J.S. Porterfield and P.A. Nuttall for providing some of the viruses used in this study. G.E.W. received a post-graduate studentship from the Medical Research Council.

REFERENCES

1. Bishop, D. H. L., Calisher, C. H., Casals, J., Chumakov, M. P., Gaidamovich, S. Ya., Hannoun, C., Lvov, D. K., Marshall, I. D., Oker-Blom, N. K., Pettersson, R. F., Porterfield, J. S., Russell, P. K., Shope, R. E., and Westaway, E. G. (1980). Intervirology 14, 125-143.
2. Leake, C. J., Varma, M. G. R., and Pudney, M. (1977). J. Gen. Virol. 35, 335-339.
3. Pudney, M., Varma, M. G. R., and Leake, C. J. (1973). Experientia 29, 466-467.
4. Nuttall, P. A., Carey, D., Reid, H. W., and Harrap, K. A. (1981). J. Gen. Virol. 57, 127-137.
5. Main, A. J., Downs, W. G., Shope, R. E., and Wallis, R. C. (1976). J. Med. Ent. 13, 309-315.
6. de Madrid, A.T., and Porterfield, J. S. (1969). Bull. W.H.O. 40, 113-121.
7. Short, N. J., Meek, A. D., and Dalgarno, L. (1982). J. Virol. 43, 840-843.
8. Pennington, T. H., Pringle, C. R., and McCrae, M. A. (1977). J. Virol. 24, 397-400.

INTERACTION OF NEUTRALIZED INFLUENZA VIRUS WITH AVIAN AND MAMMALIAN CELLS

N. J. Dimmock, H. P. Taylor and A. S. Carver

Department of Biological Sciences,
University of Warwick,
Coventry, CV4 7AL, U.K.

Although neutralization by antibody has long been a crucial test of the identity of a virus and is probably also the keystone of protective immunity, very little is known of the mechanism by which viruses lose infectivity. One popular view is that antibody prevents the early stages of virus-cell interaction. However we have shown that the kinetics of attachment of a neutralized avian influenza virus (A/FPV/Rostock/34; H7N1) to primary chick embryo fibroblast (CEF) cells were little affected by the presence of antibody. Similarly penetration, uncoating and transport of the virus genome to the nucleus was also unimpaired (1,2).

ASSOCIATION OF NEUTRALIZED FPV WITH AVIAN AND MAMMALIAN CELLS

One criticism of the use of primary CEF cells was the possibility that uptake of neutralized virus was mediated by cells bearing Fc receptors, such as macrophages which are the commonest cell of this type. This possibility can be investigated by culturing CEF cell monolayers through several passages as macrophages, which do not divide in vitro, will be diluted out. We calculated that three such passages resulted in a 27-fold dilution of macrophages, assuming that they all survived this 8 day period. Using cells held on ice and methods described before (1,2) Table 1 shows that there was little change in the ability of virus neutralized with rabbit antiserum to attach to passage 3 CEF cells and for its RNA to become associated with their nuclei. Thus we find no evidence that macrophages contribute to the observed interactions of CEF cells and neutralized influenza virus.

In further experiments we examined the ability of neutralizing antibody to interfere with the binding of FPV to mammalian cells. Figs. 1 and 3a show that neutralizing antibody did not significantly alter the rate of attachment of virus to mouse (L), human (HFF 4), hamster (BHK 21) or chick

[1]This work was supported by the Science and Engineering Research Council and the Medical Research Council.

Copyright © 1984 by Academic Press, Inc.
All rights of reproduction in any form reserved.
ISBN 0-12-183501-4

TABLE 1
CELL ASSOCIATION AND NUCLEAR ACCUMULATION OF ^{32}P-LABELLED FPV IN THIRD PASS CEF CELLS.

	^{32}P-FPV		
	+PBS	+ heterologous† antiserum	+ αH7*
Cell associated	1.0	0.6‡	0.7
Nuclear accumulation	1.0	0.8	0.8

† Hyperimmune α-concanavalin A.
* Neutralization was 99.97%.
‡ Relative to the PBS control.

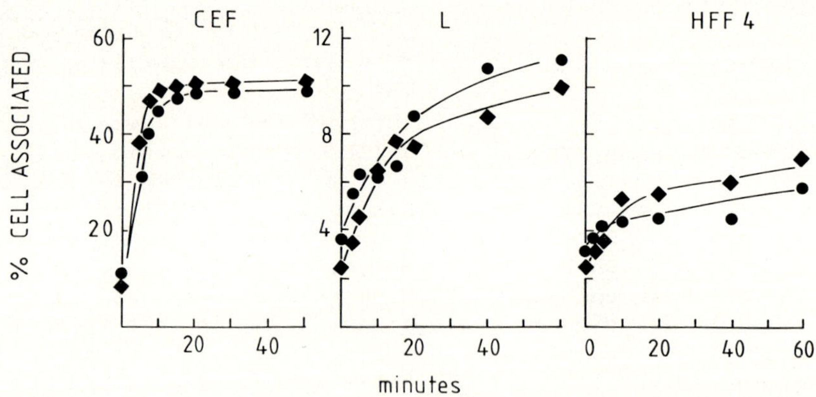

FIGURE 1. Association of purified ^{32}P-labelled A/FPV/-Rostock/34 (H7N1) with monolayers of CEF (~10^7), L (~10^6) and HFF4 (~10^7) cells. Approximately 2 x 10^7 pfu (containing 200 HAU and 1.7 x 10^5 cpm) were incubated with a 1:10 dilution of polyclonal αHA (●—●; rabbit α A/FPV/Dutch/27 (H7N7) or preimmunization serum (◆—◆) for 1h at 25°C. Mixtures were then cooled to about 4°C and inoculated to monolayers held on ice. At intervals monolayers were rinsed (x3) with cold PBS and the radioactivity associated with cells and in washes was determined by binding to Whatman DE 81 filter discs. This is virion RNA. Neutralization was > 99.9%. Preimmune serum did not neutralize.

cells. However there were differences both in the kinetics and overall efficiency of attachment between cell types such that CEF > BHK 21 > L >HFF 4.

ACCUMULATION OF VIRION RNA IN NUCLEI

Preliminary work indicated that in BHK, as with CEF cells (1,2,4), the majority of cell-associated virion RNA accumulates in the nucleus, and this is consistent with the view that at least the initial transcriptional events are nuclear (3). In these experiments it is convenient to separate the non-synthetic from the synthetic stages of infection and this is easily done by using a temperature of around 0-4°C (4). In CEF cells temperature has relatively little affect on the uptake into the cell of either influenza virus or chicken low-density lipoprotein (LDP) and a rise in temperature from 4 to 37°C gave an increase of 1.3 and 2.4-fold respectively (unpublished data).

Validation of the Nuclear Fractionation Methodology. Demonstration whether or not influenza virion RNA becomes nuclear depends on the isolation of nuclei uncontaminated by cytoplasmic constituents. One way in which spurious nuclear localization could occur would be if plasma or other membrane fragments collapsed around the nucleus and cosedimented with it. To test this possibility we inoculated BHK cells with Semliki Forest virus which multiplies excusively in the cytoplasm. Like influenza virus, SFV has an envelope and a genome of RNA. Fig. 2a shows the rate of attachment at 4, 25 and 37°C and Fig. 2b demonstrates that irrespective of temperature 95% of SFV RNA remains cytoplasmic. We assume that the 5% of virus RNA in the nucleus represents the level of contamination by cytoplasmic constituents. The 'nuclear monolayer' method for isolation of nuclei is quick and simple and is invaluable in this type of study as it allows frequent samples to be made. We find the nuclear monolayer method performs equally as well as Dounce homogenization coupled with sedimentation through sucrose, and the latter has the real disadvantage of taking over an hour to produce purified nuclei.

RNA of Neutralized FPV Accumulates in the Nuclei of BHK Cells. When we measured the accumulation of RNA from neutralized FPV into BHK nuclei we found this was identical to that of infectious virus (Fig. 3) although the kinetics of both were slower than with CEF cells. Thus we conclude that all stages of infection upto and including the presence of virion RNA at the site of transcription are unaffected by the process of

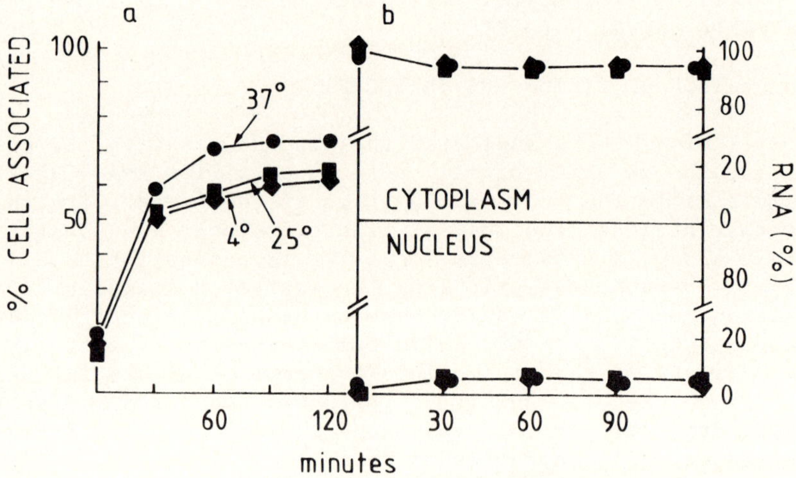

FIGURE 2. Validation of the nuclear monolayer fractionation method using BHK cells (~10^7/monolayer) inoculated with purified ^{32}P-labelled Semliki Forest virus (SFV). Monolayers were inoculated with 6 x 10^5 pfu containing 6 x 10^4 cpm with replicates on ice (4°C), at 25°C and 37°C. Radioactivity binding to Whatman DE 81 filter discs (which is virus RNA) is presented here. After incubation monolayers were washed extensively with PBS and all placed at 4°C. Nuclei were isolated at 4°C as described previously (2) except that 0.5% Nonidet P40 was found to be optimal for BHK cells. (a) Represents virus attached to cells and (b) the distribution of virion RNA between the cytoplasmic and nuclear fractions.

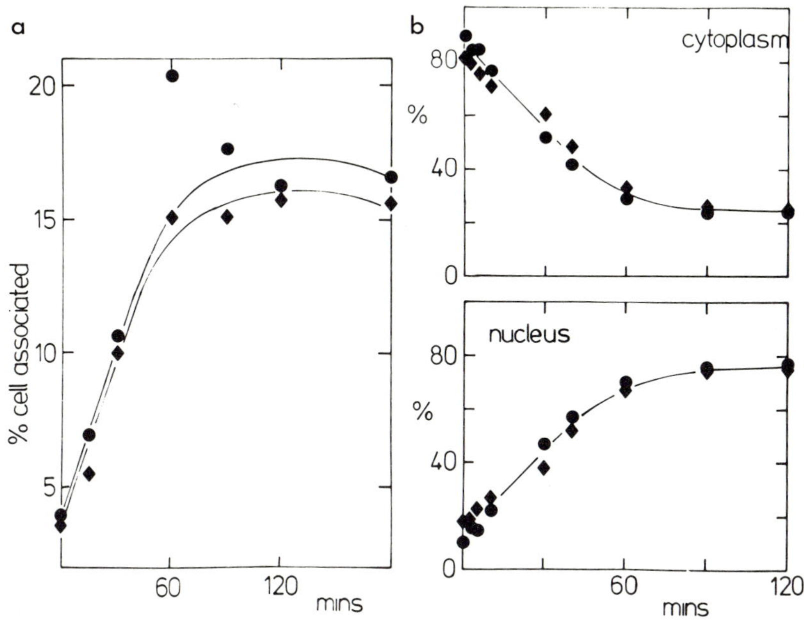

FIGURE 3(a). Association of non-neutralized and neutralized P-FPV with BHK cell monolayers (as described in Fig 1). (b) Fractionation of BHK cells into nucleus and cytoplasm as described in Fig 2. Radioactivity binding to DE 81 filters is presented here. Neutralized virus ●; non-neutralized virus ◆.

neutralization in both CEF and BHK cells. The problem of accounting for the loss of infectivity has not yet been solved.

REFERENCES

1. Possee, R. D., and Dimmock, N. J. (1981). In "Genetic Variation among Influenza Viruses". ICN-UCLA Symposium XXI (D. P. Nayak & C. F. Fox, eds.), p. 473. Academic Press, London.
2. Possee, R. D., Schild, G. C., and Dimmock, N. J. (1982). J. Gen. Virol. 58, 373.
3. McAuley, J. W., and Mahy, B. W J. (1983). Biochem. J. 211, 281.
4. Stephenson, J. R., and Dimmock, N. J. (1975). Virology 65, 77.

RESTRICTED MOBILITY OF INFLUENZA HEMAGGLUTININ ON HELA CELL PLASMA MEMBRANES[1]

Sukla Basak[a], Richard W. Compans[a], and Michael B. A. Oldstone[b]

[a]Department of Microbiology
University of Alabama in Birmingham
Birmingham, Alabama 35294

[b]Department of Immunology
Scripps Clinic and Research Foundation
La Jolla, California 92037

In the presence of specific antibody, many viral as well as cellular surface antigens are observed to redistribute in the plane of the membrane, forming a discrete cap. This phenomenon has been used to study the lateral mobility of antigens in the plasma membrane, and the interaction of antigens with one another as well as with the cytoskeleton (1-5). We have been interested in the interactions between surface antigens of influenza virus during virus assembly. In order to study such interactions, we have examined the lateral mobility of influenza viral antigens on surfaces of infected HeLa cells.

INFLUENZA HEMAGGLUTININ IS IMMOBILE ON HELA CELL SURFACES

The redistribution of measles viral antigens on the surfaces of infected HeLa cells in response to specific antibody has been well documented (1, 2). When cells are exposed to a mouse monoclonal antibody to the measles hemagglutinin, followed by fluorescein-conjugated anti-mouse immunoglobulin, the measles HA is observed to form a unipolar distribution (Fig. 1A). However, under similar conditions the HA glycoprotein of influenza virus (A/WSN strain) was not found to be capped; the antigen remained uniformly distributed over surfaces of infected cells (Fig. 1B).

To investigate the basis for the restricted mobility of influenza HA in HeLa cells, we have doubly infected HeLa cells with measles and influenza virus. At times when both measles and influenza antigens were expressed, measles HA was found to form caps on

[1]This work was supported by NIH grants AI 07007, AI 12680, CA 09202, and NS 12428.

Copyright © 1984 by Academic Press, Inc.
All rights of reproduction in any form reserved.
ISBN 0-12-183501-4

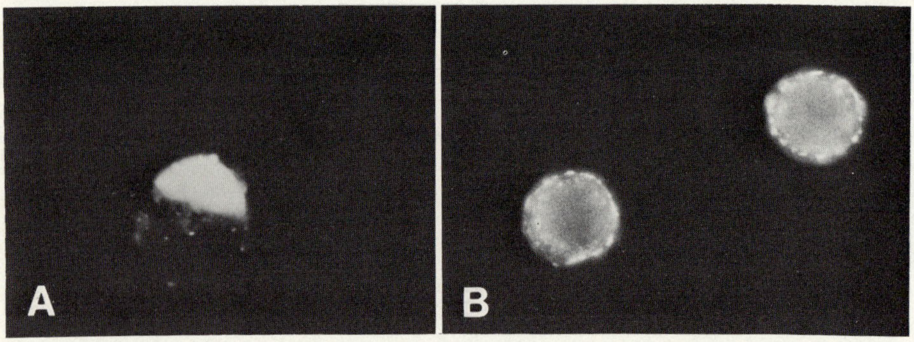

FIGURE 1. Distribution of viral antigens on the surfaces of infected HeLa cells. Cells infected with either measles virus (24 h.p.i.) or A/WSN strain influenza virus (5 h.p.i.) were treated on ice for 30 min with mouse monoclonal antibodies to the respective HA glycoproteins, followed by fluorescein conjugated goat anti-mouse antibody at 37° for 60 min. A. Capping of measles virus HA. B. Uniform surface distribution of influenza HA.

doubly-infected cells in response to specific antibody. However, no lateral mobility of the influenza HA was observed (Fig. 2A). Since the cells were doubly infected, the observed differences are not due to differences in the HeLa cell membrane resulting from virus infection. A lack of association of measles and influenza HA glycoproteins was also evident when cells exhibiting capped measles HA were doubly stained to localize influenza HA (Fig. 2B and C).

We have observed restriction of HA mobility in HeLa cells infected with several other influenza virus strains including A/Jap/305/57, A/USSR/90/77 and A/Mallard/Alb/279/77, using either monoclonal antibodies or polyvalent sera. In addition, no capping of influenza HA was observed in HeLa cells infected with the ts 51 mutant of A/WSN virus at non-permissive temperature. This is a mutant in the M protein, although the exact nature of the functional defect is not known (6).

ROLE OF CYTOSKELETAL COMPONENTS

We have examined the effects of compounds known to affect cytoskeletal structures on the lateral mobility of influenza viral antigens. The cytoskeleton has been reported to be important for the maturation of several enveloped viruses, and its alteration has been observed to either enhance or inhibit capping of viral or cellular antigens in various systems (4, 5). When influenza virus-infected

HeLa cells were treated with cytochalasin D or colchicine to disrupt either the microfilament or the microtubular network, the mobility of HA was still found to be completely restricted. These results indicate that the integrity of the cytoskeleton is not essential for the restriction of influenza HA mobility in HeLa cells.

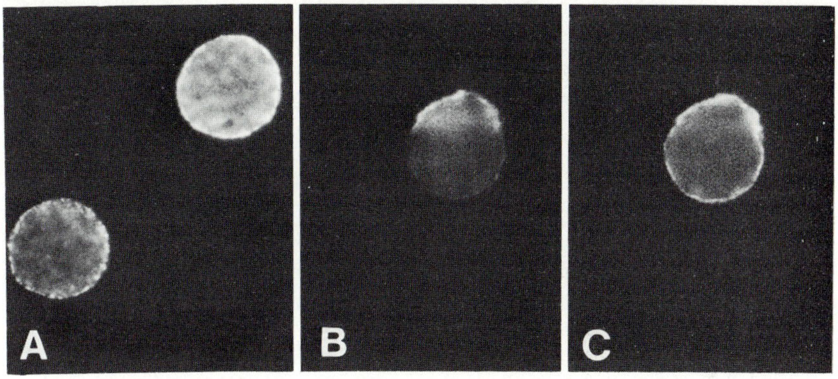

FIGURE 2. Surface immunofluorescence of measles HA or influenza HA on doubly-infected HeLa cells. Cells were infected with measles virus and superinfected with influenza virus after 16 hrs. At 24 hr, the cells were treated at 4° with monoclonal antibody to influenza HA (A) or measles HA (B) followed by fluorescein-tagged anti-mouse immunoglobulin (37°). To the cells shown in B, goat anti-mouse antibody was added to saturate any free sites, and a monoclonal antibody to influenza HA was then added and tagged with rhodamine-conjugated antibody at 4°. The same cell is shown in panels B and C; C shows the distribution of influenza HA.

MOBILITY OF HA INSERTED INTO HELA CELL SURFACES BY VESICLE FUSION

To further investigate the basis for the restriction of capping in HeLa cells, we have prepared lipid vesicles containing purified influenza virus glycoproteins . The vesicles were then fused to HeLa cell membranes at pH 5.5, and the lateral mobility of HA was examined. Fig. 3 shows that under such conditions influenza HA can be capped on HeLa cell membranes. No capping was observed, however, when purified virions were fused to the cell membrane under identical conditions. This observation suggests that the restricted mobility of influenza HA on HeLa cell surfaces may involve viral internal proteins.

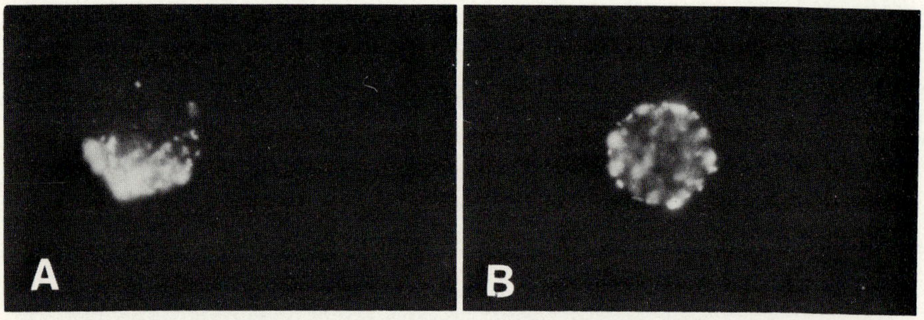

FIGURE 3. Antibody induced redistribution of influenza HA glycoprotein inserted into HeLa cell surfaces by fusion of glycoprotein vesicles. Cells were treated with antisera as described for Fig. 1B. A, Distribution of HA after treatment with second antibody at 37°; B, Same at 4°.

CONCLUSIONS

Our observations suggest that restriction of lateral mobility of influenza HA is probably due to interactions of influenza HA with other viral or cellular proteins. Since in doubly infected cells capping was observed with measles HA but not influenza HA, the difference between the two viruses is not due to an effect of measles virus on the HeLa cell membrane. The fact that influenza HA exhibits mobility on the HeLa cell membrane after insertion by fusion with vesicles suggests that the M protein may play a role in the restriction of lateral mobility. It will be of interest to determine whether the mobility of influenza HA on cell surfaces varies in different cell types. Influenza virus replication in HeLa cells is abortive, and the restricted mobility of HA may be related to an apparent defect in virus budding, which has previously been reported in this cell type (8).

REFERENCES

1. Joseph, B. S. and Oldstone, M. B.A. (1974). J. Immunol. 113, 1205.
2. Lampert, P. W., Joseph, B. S., and Oldstone, M. B. A. (1975). J. Virol. 15, 1248.
3. Dales, S. and Oldstone, M. B. A. (1982). J. Exp. Med. 156, 1435.
4. Sundqvist, K, G., Otteskog, P., and Ege, T. (1978). Nature 274, 915.
5. Mousa, G. Y., Trevithick, J. R., Bechberger, J., and Blair, D. G. (1978). Nature 274, 808.
6. Ritchey, M. B. and Palese, P. (1979). J. Virol. 21, 1196.
7. Herrler, G., Nagele, A., Meier-Ewert, H., Bhown, A. S., and Compans, R. W. (1981). Virology 113, 439.
8. Caliguiri, L. A. and Holmes, K. V. (1979). Virology 92, 15.

REASSEMBLY OF MEMBRANOUS PARTICLES WITH INFLUENZA VIRUS HEMAGGLUTININ AND RESTORATION OF BIOLOGICAL AND IMMUNOLOGICAL ACTIVITIES [1]

Yasuhiro Hosaka

Department of Preventive Medicine, Research Institute for Microbial Diseases, Osaka University, Suita, Osaka 565, Japan

Influenza virus hemolysis and cell fusion

Recently, influenza viruses were found to induce hemolysis and cell fusion (1,2,3). We presented two lines of evidences that influenza virus hemagglutinins (HA) are responsible for these phenomena: 1. liposomes of influenza virus HA reassembled from solubilized envelopes induced hemolysis and cell fusion (4), and 2. hemolysis and cell fusion were not affected if virus neuraminidase activity was heat-inactivated (5).

We designed to determine what sort of organization of HA is responsible for inducibility of hemolysis and cell fusion and further for humoral and cell-mediated immunogencities. For this purpose, we isolated HA, then reassembled membrane-associated HA with lipids, and compared biological and immunological properties between membrane-bound and -free HA.

Influenza viruses, PR8 (H1N1), WSN (N1N1), and Bangkok (abbreviated BKK) (H3N2), were grown in the chorioallantoic cavity of 10 or 11 days old chick embryos at 35 C for 3 days. Infected fluids were directly used for target cell infection. Virus purification was done by differential centrifugation, followed by velocity sedimentation in sucrose gradients (6).

Solubilization of virus envelopes and isolation of hemagglutinin (HA).

Purified PR8 virions were treated with 0.25% NP40, and the solubilized virions were centrifuged at 100,000 x g for 45 min. The supernatant was mixed with solid CsCl and centrifuged at 46,000 rpm for 40 h in Beckman SW50.1 rotor. Fig. 1 shows the isopycnic sedimentation pattern of solubilized envelopes: in Fig. 1B, all the peaks of distributions of HAU,

[1] This work was supported by Grant No. 58440032 (1983) from the Ministry of Education, Science, and Culture of Japan.

Copyright © 1984 by Academic Press, Inc.
All rights of reproduction in any form reserved.
ISBN 0-12-183501-4

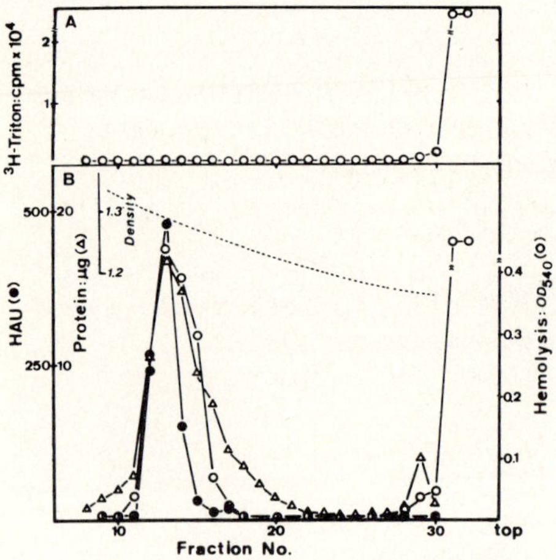

Fig. 1. Isopycnic sedimentation pattern of NP40-solubilized envelope. In a parallel exp., 2 μCi of ^{3}H-Triton-supplemented NP40 was used for solubilization (Fig. 1A). Fractionation was done dropwise from the tube bottom. Hemolysis was measured for 10 μl aliquots, and HAU and protein, as values for the total (0.15 ml) of each fraction (Fig. 1B).

hemolytic activity, and protein content coincide with each other at the buoyant density of about 1.28, and in Fig. 1A, ^{3}H-radioactivity is confined to the top fractions. This pattern of ^{3}H-Triton distribution was considered to represent the pattern of distribution of NP40, because NP40 is a non-ionic detergent similar to Triton X-100. Lipids (phospholipid-P and cholesterol) were also found to be localized to the top fractions. These findings indicated that the protein peak was free from lipids and detergent, and that the hemolytic activity observed was not due to contamination of detergent.

Properties of lipid-free hemagglutinins

Fig. 2 shows the polyacrylamide gel electrophoresis patterns of the protein peak, and top fractions thus separated (Fig. 1). The protein fraction contained predominantly HA (HA_1 + HA_2) polypeptides, while the top fractions contained no detectable protein. Thus, we used this protein peak fractions as lipid-free hemagglutinins (HA).

This HA fraction exhibited hemolytic activity as shown in Fig. 1B, but was found not to cause cell fusion of human erythrocytes at pH 5.1-5.3. The pH optimum for hemolysis by this HA fraction was found to be a little lower than those for PR8 virions and membrane-associated HA (HA virosomes).

Electron microscopy of the HA fraction revealed that it consisted of aggregates of spikes of 12-13 nm long, and the top one, membranous particles of various sizes without spikes.

The finding that lipid-free HA of influenza virus caused hemolysis is in contrast to the lipid requirement of Sendai

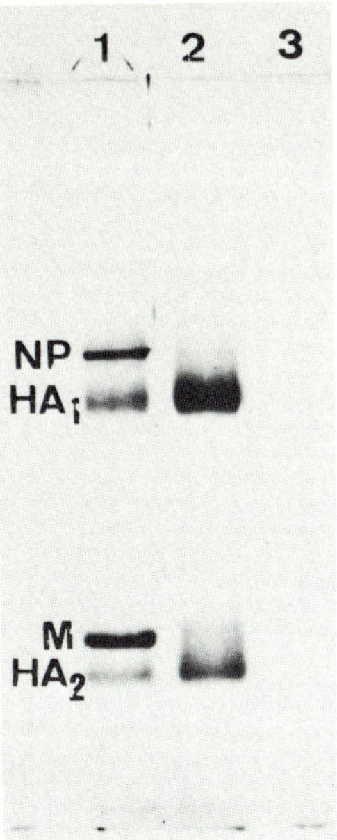

Fig. 2 Gel electrophoresis pattern of HA peak (lane 2) and top (lane 3) fractions of Fig. 1.

Lane 1: reference of PR8 virions. Virions (15 μg protein (6)), HA peak (8 μg protein), and top (20 nM phospholipid-P (7)) were applied. Electrophoresis was done in 10% gel by the method of Laemmli (8). Bands are designated according to Lamb and Choppin (9).

glycoproteins for hemolysis (10,11). This difference was explained in a submitting report (12). Briefly, for viral hemolysis, adsorption and lysis functions of virus glycoprotein(s) are required. Whereas the two functional determinants are associated with one glycoprotein, HA, of influenza virus (13), they are associated with different glycoproteins, HN and F, separately(14,15). Thus, Sendai virus hemolysis is only manifested when the two determinants are close together with help of membrane organization. For cell fusion, plural envelope fusion by a connecting viral component between two cells are required (16,17,18), and membrane-associated glycoproteins are needed for that.

Reassembly of virosomes of influenza virus hemagglutinins

Fig. 3 shows reassembly of virosomes from a constant dose of HA and various amounts of the top lipids by using 3 day dialysis method. HAU of assembled preparations increased with lipid doses used. However, hemolytic activity and in vitro secondary immunogenicity for cytotoxic T lymphocyte (CTL) generation were not so changed with lipid doses used. The both activities were increased with the minimum dose of lipids and then decreased. At present, we do not know why they altered in parallel. The reason why these two activities were not so much changed seems to be due to their dependency on the total numbers of HA spikes. The increase of HAU was explained by the finding of formation of larger sized particles with larger amounts of lipids. It seems likely because larger reassembled particles more firmly attached to erythrocytes, resulting in the increase of agglutinated cells.

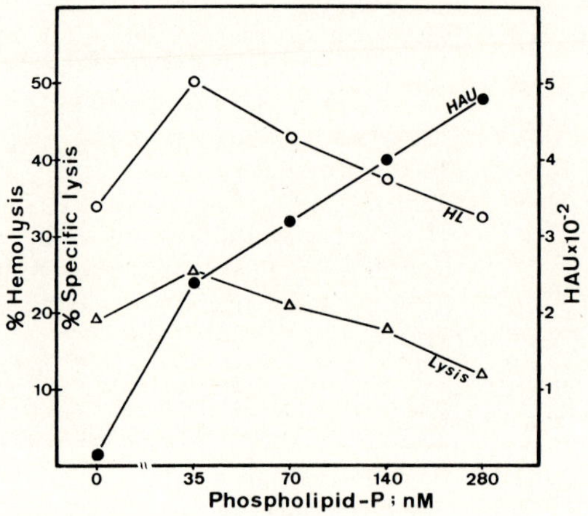

Fig. 3 Reassembly of virosomes from PR8·HA and various amounts of top lipids.

HA (300 μg protein) and indicated amounts of lipids were mixed and dialyzed to PBS containing 2 mM $MgCl_2$ in the presence of SM2 beads (12,19) for 3 days. The final volumes were adjusted to 1.5 ml of PBS, and hemolysis and HA were expressed as described in Fig. 1. In vitro cellular immunogenicity was represented by % Specific lysis of WSN-infected L929 targets (50 HAU/10^6 cells) by effectors secondarily generated in the presence of virosomes (60 μg protein) from WSN-primed CBA/j mouse spleen cell cultures. The culture medium was RPMI1640 with 10% fetal calf serum (FCS), $5x10^{-5}$M 2-mercaptoethanol, 10 mM Hepes, 2 mM L-glutamine, and 5 μg/ml of Gentamycin. ET ratio=50:1. % Specific lysis of normal L929 targets was about 10%, and the values presented in this figure had been subtracted by this normal lysis.

Target susceptibility of ultraviolet (UV)-inactivated influenza virus- and virosome-coated murine cells to CTL-mediated lysis and complement-dependent antibody-mediated lysis (CDAMC)

Since influenza viruses and virosomes were found to cause fusion from without of human erythrocytes, and cell fusion was considered to be mediated by envelope fusion, we expected that if murine targets were coated with UV-influenza virus or virosomes at low pH, they exhibited target susceptibility to CTL-mediated lysis, similarly as the case of Sendai virus and its virosomes (20-23). Thus, we employed the procedure to coat cells with virus at acidic medium as shown in Fig. 4.

Mixture (0.2-0.5 ml) of virus and murine cells in a glass tube

↓ kept in the cold for 20 min (neutral RPMI1640+2%FCS), centrifuged.

Virus+cell mixture: medium exchange for coating medium (RPMI1640, no bicarbonate) of various pH

↓ incubated at 37 C for 3 min, then addition of several vol. of neutral RPMI1640+10%FCS, further incubated at 37 C for 27 min.

Virus antigen-associated target cells

↓ washed by centrifugation distributed into Falcon wells (No. 3072).

Interaction of target cells with CTL or with anti-virus antibody+complement

Fig. 4 Procedure to prepare target cells coated with UV-influenza virus at various pH

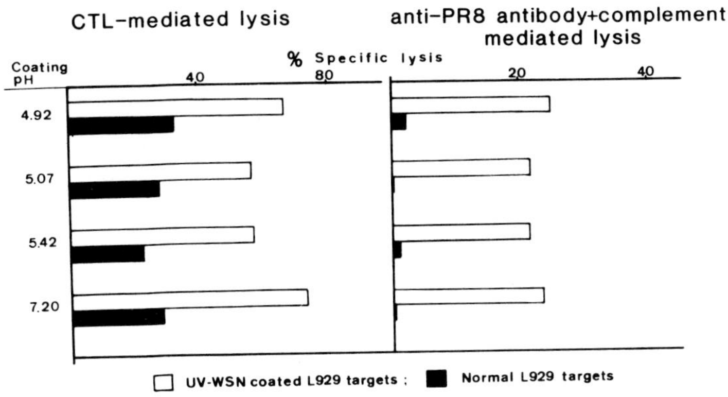

Fig. 5 Effect of coating pH on induction of target susceptibilities of UV-WSN-coated L929 cells to CTL-mediated lysis and complement-dependent antibody-mediated lysis.

Virus coating: UV-WSN (400 HAU)/10^6 cells

CTL: anti-WSN effectors in vitro secondarily generated during 5 day culture from PR8-primed CBA/j mouse spleen cells. 6 h ^{51}Cr-release assay was used at the ET ratio=28:1.

CDAMC: UV-WSN-coated targets were reacted with 1/20 dilution of anti-PR8 rabbit serum at room temperature for 1 h and incubated at 37 C for 1.5 h in the presence of 1/10 dilution of rabbit complement (serum)(Cedarlane, Canada). Cytotoxicity was calculated from ^{51}Cr released during this period.

Fig. 5 shows that L929 targets, which had been detached from walls with EDTA, coated with UV-WSN at acidic pH exhibit a good susceptibility to CTL-mediated lysis, but those at neutral pH the highest one, and that all the targets display good susceptibility to CDAMC, which confirms that the virus antigens anchored well on the target cell surfaces under the condition used. Although we do not have a good explanation for the susceptibility induced at neutral pH, it is not unreasonable to expect that during influenza virus entry the virus envelope meet at some stage of endocytosis fusion with cell membranes, which is unlikely to be the fusion at the initial stage of infection like Sendai virus. When P815 mastocytoma cells were coated with UV-WSN at various pH, they did not exhibit a definite susceptibility to CTL-mediated lysis, although they did that to CDAMC. However, when P815 cells were coated with UV-N influenza virus at neutral pH, they exhibited a high susceptibility to CTL-mediated lysis (24). These findings seem to indicate that depending on targets and virus strains used, the susceptibility of virus-coated targets to CTL-mediated lysis is induced or not. In the system of A/Jap (H2N2)-coated P815 targets, no susceptibility was reported to occur (25).

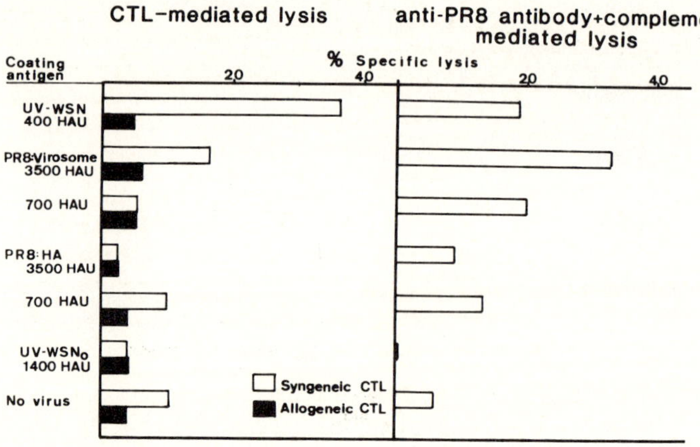

Fig. 6 Target susceptibilities of UV-WSN, PR8·virosomes, and PR8·HA-coated L929 cells to CTL-mediated lysis and CDAMC.
Assay methods are the same as described in Fig. 5 except use of secondary effectors of 4 day culture from PR8-primed (9 days earlier) CBA/j and BALB/c mouse spleen cells at the ET ratio=25:1. Antigen dose/10^6 cells.

Fig. 6 shows the following four findings: 1. the CTL-susceptibility of UV-WSN-coated L929 targets is of H-2 restricted nature (26). 2. PR8·virosome-coated L929 targets exhibit some but not significant CTL-susceptibility (WSN·virosome showed a similar result), 3. PR8·HA-coated ones exhibit no CTL-susceptibility, and 4. all the L929 targets coated with virus antigens used except PR8·HA (3500 HAU) exhibit CDAMC-susceptibility, which means that the virus antigens used anchored well on the target cell surfaces. Nowadays we do not know the exact reason why virosome-coated L929 targets did not exhibit good CTL-susceptibility, whereas virion-coated ones did.

Elicitation of memory CTL with virosomes

Although PR8 or WSN·virosome-coated L929 targets did not show a definite CTL-susceptibility, these virosomes were found to elicit memory CTL in mice: when they were intraperitoneally injected into mice, no detectable virus-specific cytotoxicity was observed 1 week later, but when primed mouse spleen cells were in vitro stimulated with virions but not with virosomes for 4-5 days, generated effectors exhibited H-2 restricted high cytotoxicity. These effectors had cross cytotoxicity in terms of HA subtype (12). The virosomes used were 60-300 ug proteins/mouse and CBA/j mouse showed better response.

The virosomes were also found to elicit stronger humoral response than the same protein dose of HA in terms of generation of hemagglutination-inhibition antibody (12).

Conclusion

1. Lipid-free hemagglutinins of influenza virus was isolated from NP40-solubilized envelopes using isopycnic centrifugation, and was found to induce hemolysis but not cell fusion.
2. When the isolated HA was assembled with lipids to make virosomes using dialysis method, they recovered cell fusion activity and also the capability to elicit in vivo memory CTL.
3. When L929 cells but not P815 cells were coated with UV-inactivated WSN, the target susceptibility to CTL-mediated lysis was induced even at neutral pH-coating. When L929 cells were coated with virosomes, a small but not significant CTL-susceptibility of the targets was induced. This reason was unknown

Acknowledgments

The author wish to thank Miss Fuyoko Sasao for her assistance.

References

1. Maeda,T., and Ohnishi,S. (1980). FEBS Letters 122, 238.
2. Lenard,J., and Miller,D.K. (1981). Virology 110, 479.
3. Huang,R.T.C., Rott,R., and Klenk,H.-D. (1981). Virology 110, 243.
4. Hosaka,Y., Yasuda,Y., and Fukai,K. (1983). J.Virol. 46,1014.
5. Hosaka,Y., Seriburi,O., Moran,M.G., Yasuda,Y., and Fukai,K. (1982). Biken J. 25, 51.
6. Lowry,O.H., Rosebrough,N.J., Farr,A.L., and Randall,R.J. (1951). J.Biol.Chem. 193, 265.
7. Bartlett,G.R. (1979). J.Biol.Chem. 234, 466.
8. Leammli,U.K. (1970). Nature 227, 680.
9. Lamb,R.A., and Choppin,P.W. (1976). Virology 74, 504.
10. Hosaka,Y., and Shimizu,Y.K. (1972). Virology 49, 627.
11. Fukami,Y., Hosaka,Y., and Yamamoto,K. (1980). FEBS Letters 114, 342.
12. Hosaka,Y. in submitting
13. Wilson,I.A., J.J.Skehel, and Wiley,D.C. (1981). Nature 289, 366.
14. Homma,M., and Ohuchi,M. (1973). J.Virol. 12, 1457.
15. Scheid,A., and Choppin,P.W. (1974). Virology 57, 475.
16. Apostolov,K., and Almeida,J.D. (1972). J.Gen.Virol. 15, 227.
17. Hosaka,Y., and Shimizu,K. (1977). In "Virus infection and cell surface" (G.Poste and D.Nicolson, eds.), p. 129. North-Holland Pub. Co., Amsterdam.
18. Knutton,S. (1980). J.Cell Sci. 43, 103.
19. Volsky,D.J., and Loyter,A. (1978). FEBS Letters 92, 190.
20. Schrader,J.W., and Edelman,G.M. (1977). J.Exp.Med. 145, 523.
21. Sugamura,K., Shimizu,K., and Bach,F.H. (1978). J.Exp.Med. 148, 276.
22. Koszinowski,U., Gething,M.J., and Waterfield,M. (1977). Nature 267, 160.
23. Fukami,Y., Hosaka,Y., Yasuda,Y., and Bonilla,J.A. (1979). Infect.Immun. 26, 815.
24. Kurrle,R., Wagner,H., Röllinhoff,M., and Rott,R. (1979). Eur.J.Immunol. 9, 107.
25. Braciale,T.J., and Yap,K.L. (1978). J.Exp.Med. 147, 1236.
26. Zinkernagel,R.M., and Doherty,P.C. (1979). Adv. in Immunol. 27, 51.

INCORPORATION OF M-PROTEIN OF INFLUENZA VIRUS INTO LIPOSOMES. FURTHER CHARACTERIZATION OF LIPOSOMAL STRUCTURES[1]

D.J. Bucher, R.R. Dourmashkin,[2] and J.A. Greenberg

Department of Microbiology
Mount Sinai School of Medicine
of The City University of New York
One Gustave L. Levy Place
New York, New York 10029

M-protein purified from influenza virus will readily incorporate into the lipid bilayer in artificial liposomal systems as shown from earlier work from our laboratory(1,2) as well as by A. Gregoriades.(3) In our earlier reports(1,2), we described liposomes of two discrete densities formed in the presence of M-protein as analyzed by gradient centrifugation analysis; a low density fraction (1.04-1.06g/cm^3) with little or no M-protein and a heavy density fraction (1.15-1.22g/cm^3) rich in M-protein. Electron microscopic analysis of the heavy density fraction following freeze fracture preparation showed the presence of 80-90Å particles decorating the surface. Only one membrane surface could be visualized, unlike the multilamellar structures seen for those liposomes which contained no M-protein. The lack of fracturing suggested the membrane had been altered by the incorporation of M-protein and could not be fractured along the hydrophobic plane of the lipid bilayer, as observed for control liposomes containing no M-protein. In the present report we describe additional electron microscopic characterization undertaken to provide further elucidation of the liposomal structures formed in the presence of M-protein.

FORMATION OF LIPOSOMES CONTAINING M-PROTEIN

Preparation of M-protein. M-protein was purified by SDS gel chromatography according to previously described techniques employing unused swine influenza virus vaccine which had been inactivated with β-propiolactone as the source of viral protein.(4,5) The virus used in this strain has been designated as the recombinant strain X-53a; the HA and NA are derived from the parent strain A/NJ/11/76 (H1N1) and its internal proteins including the M-protein are derived from A/PR8/34 (H1N1)

---

[1]This work was supported in part by grant AI-16073 from the National Institute of Allergy and Infectious Diseases.

[2]Present address: Laboratory of Paediatric Gastroenterology, The Medical College of St. Bartholomew's Hospital, London, England.

Copyright © 1984 by Academic Press, Inc.
All rights of reproduction in any form reserved.
ISBN 0-12-183501-4

virus.[6,7] SDS was removed from the M-protein preparation by extensive dialysis versus distilled water at 4°.

Preparation of Liposomes. Liposomes were prepared as described earlier with solubilization of the lipids by octyl glucoside (1-O-N-octyl-β-D-glucopyranoside) and subsequent dialysis to incorporate M-protein into liposomes.[1,2] Liposomes were separated into low density and heavy density fractions following sucrose gradient centrifugation as previously described.[1]

Electron Microscopic Analysis. Liposomal fractions from the sucrose gradient were fixed for one hour in 0.1M cacodylate buffer, pH 7.4, containing 1% tannic acid and 3% glutaraldehyde. Following fixation, liposomes were pelleted by centrifugation at 100,000g for one hour.

Pellets were embedded in Sparr resin for thin sectioning. Sections were stained with uranyl acetate and Reynold's lead citrate solution and examined with a Philips 301 electron microscope.

ELECTRON MICROSCOPIC ANALYSIS OF M-PROTEIN LIPOSOMES.

Under the conditions of fixation and high centrifugal force, control liposomes containing no protein lost all vesicular structure and formed sheet-like structures as viewed in cross section (shown in Figure 1A). Heavy density liposomes containing M-protein were found to be multilamellar structures as seen in Figure 1B; earlier freeze fracture preparation and electron microscopic analysis permitted visualization of only the surface features of the heavy density liposomes.[1] Liposomes from the low density fraction did not resemble control vesicles but were visualized as hollow vesicular structures surrounded by one or a few lamellae as shown in Figure 2. Evidently a sufficient quantity of M-protein was present in the light fraction to produce extensive cross-linking. This cross-linking prevented collapse of these vesicles at 100,000gs as was observed for control vesicles.

Our earlier observation of low M-protein content in the low density liposomes was based on protein content as related to turbidity. The multilamellar structures which are solid protein-lipid structures will evince the same level of turbidity (or light scattering) as 'empty' vesicles seen in the low density fraction. The concentration of M-protein in the heavy density liposomal fraction relative to turbidity measured as A_{260nm} was 0.25mg protein per 1.0 unit A_{260nm}; the concentration of M-protein in the low density liposomal fraction relative to turbidity was found to be 0.02mg protein per 1.0 unit

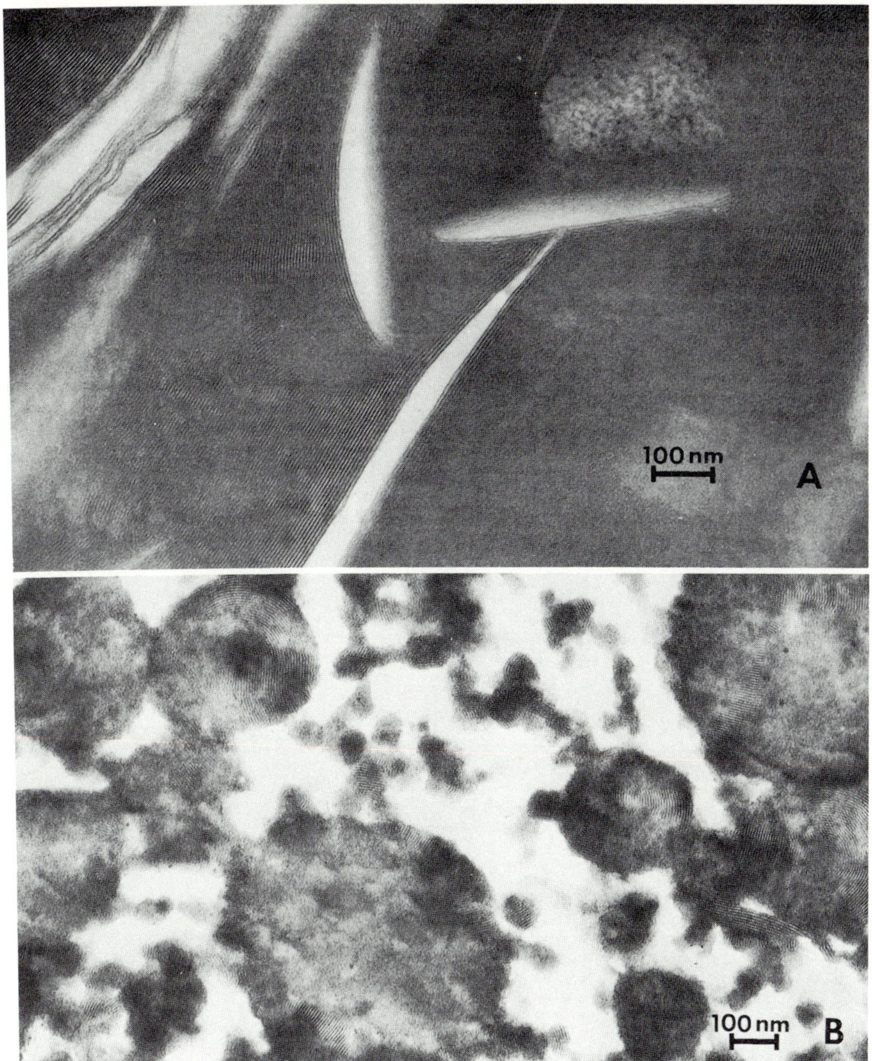

Figure 1. Electron micrographs of (A) control liposomes and (B) heavy density liposomes formed in the presence of M-protein. Specimens were prepared by thin section of pelleted liposomes following embedding (see text). (A) Upper panel; instrument magnification is 36,000 with final magnification of 81,000. (B) Lower panel; instrument magnification is 25,000 with final magnification of 56,000.

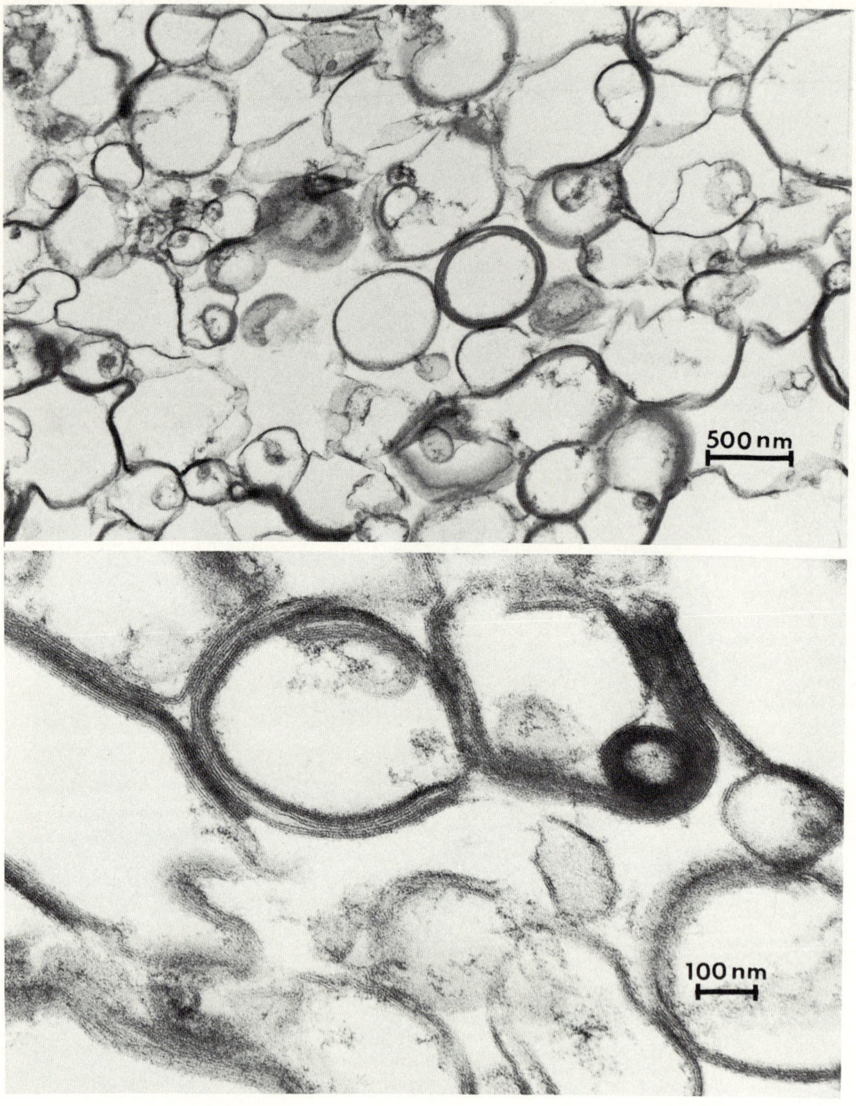

Figure 2. Electron micrographs of low density liposomes formed in the presence of M-protein with specimens prepared by thin section of pellet following embedding (see text). Upper panel, instrument magnification is 9800 with final magnification of 22,000. Lower panel, instrument magnification is 36,000 with final magnification of 81,000.

IN VITRO AND IN VIVO PROPERTIES OF AN INFLUENZA A HOST RANGE VIRUS[1]

Dan C. DeBorde, Armen M. Donabedian,
Susan M. Peters, and H. F. Maassab

Department of Epidemiology
School of Public Health
University of Michigan
Ann Arbor, Michigan 48109

HOST RANGE PHENOTYPE AND GENOTYPE

CR43-3 virus was made during a reassortment experiment using the cold-adapted (ca) and temperature sensitive (ts) A/Ann Arbor/6/60 H2N2 virus and the wild type (wt) A/Alaska/6/77 H3N2 virus (1). CR43-3 is an influenza A cold-adapted reassortant virus that has a limited host range when compared to either parent virus or to its congenic reassortant, CR31-10. It grows well in primary chick kidney (PCK) cells, but very little, if at all, in Madin-Darby canine kidney (MDCK) cells or in ferrets after intranasal inoculation (2). CR43-3 infected MDCK cells showed high levels of CPE at 48 hr pi (post-infection), suggesting that they had been infected, and that some virus replication functions were active resulting in CPE and cell death (2). Media removed from these MDCK cells infected with 0.1 to 1.0 PFU/ml of CR43-3 virus, contained 10^3 PFU/ml when plaqued in PCK cells (2). This low level of virus could represent a small amount of new virus growth or residual virus remaining from the inoculum. It was also not known whether noninfectious virus particles were being released or not.

This host range virus has an aberrant NS gene which migrates differently from the NS gene of either parent on polyacrylamide and agarose gels used for genotyping the reassortants (2,3). Reassortants made with experiments using A/Cal/10/78 H1N1 virus and CR43-3 virus showed that the host range property and the aberrant NS gene cosegregated with one another (2). Sequence studies on this NS gene (performed in collaboration with Dr. Peter Palese, Mt. Sinae Hospital NY)

---

[1]This work was supported by Contract No. 1-AI-72521, National Institute of Allergy and Infectious Diseases, Development and Applications Branch, Bethesda, MD 20205

Copyright © 1984 by Academic Press, Inc.
All rights of reproduction in any form reserved.
ISBN 0-12-183501-4

TABLE 1
GENOTYPE

Virus	Pb2	Pb1	Pa	HA	NP	NA	M	NS	Gene
A/AA/6/60	A[a]	A	A	A	A	A	A	A	
A/Ala/6/77	W[b]	W	W	W	W	W	W	W	
CR43-3	A	A	A	W	A	W	A	Wd[c]	
CR31-10	A	A	A	W	A	W	A	W	

[a] represents gene derived from A/AA/6/60 virus.
[b] represents gene derived from A/Ala/6/77 virus.
[c] Wd represents the NS deletion mutant of A/Ala/6/77 virus.

have demonstrated that the CR43-3 NS gene is a deletion mutant of the NS gene of the A/Alaska/6/77 wt parent (3). The deletion is 36 bases long and occurs from nucleotides 222-257 of the A/Alaska/6/77 NS gene in the region coding for NS1 protein (3).

Three types of experiments were performed to further characterize the restricted host range phenotype of CR43-3 virus: 1) a growth curve analysis using HA activity and EID_{50} assays to monitor the infections; 2) the analysis of 48 hr pi media from cells infected with virus in the presence of either ^{35}S-methionine or ^{32}P-phosphate; and 3) an analysis of the time course of protein production in virus infected cells. Positive controls consisted of CR43-3 virus grown in PCK cells, and its congenic virus, CR31-10, whose genotype varies only by the NS1 deletion (as determined by electrophoretic migration on polyacrylamide and agarose gels) grown in both MDCK and PCK cells. Table 1 presents the genotypes of CR43-3 and CR31-10 and their parent viruses.

VIRUSES AND VIRUS CULTURE

The viruses used were CR43-3 and CR31-10. Both viruses were made by reassortment in this laboratory using A/Ann Arbor/6/60 and A/Alaska/6/77 viruses in two different experiments. Procedures for the reassortment experiments have been previously described (1,4). Both reassortant viruses have been plaque-to-plaque purified three times in PCK tissue culture. Virus pools used in these experiments were made by high dilution passages in embryonated eggs. MDCK cells, obtained from Flow Laboratories, Rockville, MD, and PCK cells were grown and maintained as described earlier (5,6). Confluent monolayers of either cell type of approximately 4 x 10^6 cells each were infected with 10-20 plaque-forming units (PFU) per cell of virus for 1 hr at room temperature and the inoculum removed prior to growth curve or radioisotope labeling experiments.

VIRUS GROWTH CURVE

Both PCK and MDCK cells were separately infected with CR43-3 or CR31-10 virus as above, the cells washed 3 times with Hank's balanced salt solution (HBSS), and 5 ml of 2x Eagle's media (2) added to the infected cells. The infected cells were incubated at 33C for 48 hr, and 0.2 ml was removed at specific times pi. An equal volume (0.2 ml) of fresh 2x Eagle's media was added to the cells immediately after each sample was taken. Each sample was processed for HA activity, and for virus titer using Egg Infectious Dose/50 (EID_{50}). Four replicate eggs were used at each dilution, and the EID_{50} determined by the Reed-Muench method (7).

Figure 1a and 1b show the HA activity for each virus and cell combination during infection. While CR43-3 virus showed a normal pattern of growth in either PCK or MDCK cells, CR43-3 virus produced almost no HA activity in MDCK cells, finally showing one $\log_2$ unit of activity at 36 and 48 hr pi. CR31-10 virus showed a later pattern of growth in MDCK cells than in PCK cells, but it reached the same final HA titer in either cell line.

Figure 1C and 1D show the EID_{50} titer for the same infections as above. The infected cells had been washed three times with fortified Eagle's media after the original inoculum was removed, but there was still a substantial residual virus background. CR43-3 virus-infected MDCK cells produced less than one $\log_{10}$ of new infectious virus, while it produced over two $\log_{10}$ of virus in PCK cells, and CR31-10 virus produced two to two and one-half $\log_{10}$ of virus in either cell type.

RELEASE OF VIRUS PARTICLES FROM INFECTED CELLS

Both cell types were infected with each virus line in duplicate, one of which was radioactively labeled with ^{32}P-phosphate, while the other was labeled with ^{35}S-methionine. At 1 hr pi, infected cells were covered with methionine-free 2x Eagle's media containing 150 µCi of ^{35}S-methionine at 100 µCi/ml, and incubated at 33C for 4 hr. At this time 3.5 ml of 2x Eagle's media was added to each flask of infected cells and the incubation continued until 48 hr pi. Viruses were labeled with ^{32}P-phosphate in the same manner except Eagle's MEM lacking inorganic phosphate was used during the initial 4 hr labeling period. The media was removed and clarified by centrifugation at 1000xg for 10 min at 4C, and

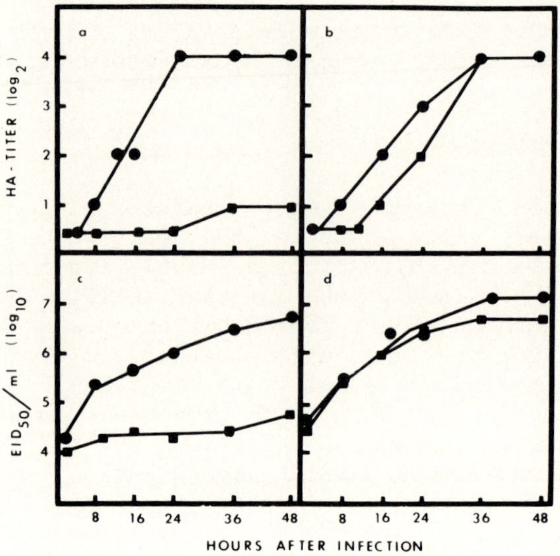

Figure 1. Virus growth curve. Closed circles represent PCK cells; closed squares represent MDCK cells. Virus used: a and c = CR43-3; b and d = CR31-10

subsequently at 10,000xg for 30 min at 4C. The virus was then pelleted by centrifugation at 22,500 RPM at 4C in a type 30 Beckman rotor for 2 hr. The virus-containing pellet was resuspended in TNE (0.01M Tris-HCl, pH 7.4, 0.1M NaCl, and 0.001M EDTA) and analyzed on 18 ml, 30-60% (w/v) sucrose equilibrium gradients in TNE, centrifuged at 17,000 RPM for 16 hr in a Beckman SW27.1 rotor at 4C. Figure 2 shows the gradient profiles for CR43-3 virus grown in both cell types and labeled with both radioisotopes. Even in the permissive PCK cells, CR43-3 virus makes a lot of material of lower densities than infectious virus which bands at 1.18 g/cc. CR31-10 virus gave results essentially identical to those seen for CR43-3 virus infected PCK cells, whether it was grown in MDCK or in PCK cells (results not shown). Unfractionated pelleted material and material from all three pools of each gradient were assayed in the appropriate gel system. The three pools for every gradient were: pool I - fractions 3-6; pool II - fractions 7-11; and pool III fractions 12-16. ^{32}P-Phosphate labeled RNA was analyzed in 1.5% agarose gels in TBE buffer (8) containing 0.1% SDS as described earlier (2). ^{35}S-methionine labeled proteins were analyzed on 7-14% polyacrylamide gradient gels using standard Laemmli buffers and acrylamide-bis stock solutions (9). Figure 3 shows the protein gel analysis of the unfractionated sample and three gradient pools for ^{35}S-methionine labeled CR43-3 virus

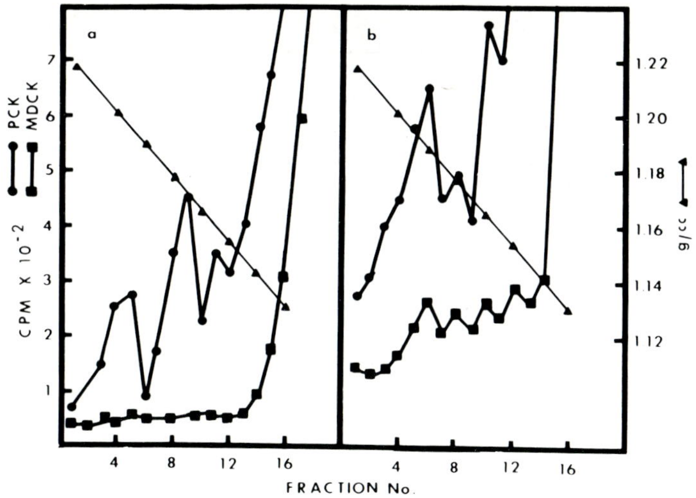

Figure 2. Equilibrium sucrose gradient analysis of CR43-3 virus-infected cells, 48 hr pi, media. Radioactive tracer used: a) ^{32}P-phosphate; b) ^{35}S-methionine

infected cell, 48 hr pi media, together with the unfractionated samples of ^{35}S-methionine labeled CR31-10 virus infected cell, 48 hr pi media. The unfractionated material contained a large amount of both virus- and cell-specific proteins. The virus peak, pool I, contained proportionately less cellular-specific proteins. The virus proteins from CR43-3 virus infected MDCK cells looked similar to the virus protein from infected PCK cells except that there was almost no M protein present. The CR43-3 congenic virus control, CR31-10, gave results in both PCK and MDCK cells similar to those of CR43-3 virus in PCK cells. Pools #II and III for all viruses still contained predominantly viral proteins, but with increasing amounts of the cellular proteins in the unfractionated aliquots. ^{32}P-phosphate labeled CR43-3 grown virus in PCK cells and CR31-10 virus grown in MDCK or PCK cells each had all eight vRNA segments in all three pools and in the unfractionated sample. CR43-3 virus grown in MDCK cells did not produce enough ^{32}P-phosphate labeled vRNA for analysis as expected from the gradient profile in Figure 2.

INTRACELLULAR VIRAL PROTEIN PRODUCTION

The time course of viral protein production was analyzed from 2 to 8 hr pi for both viruses in both cell types. Infected cells were incubated at 33C until 15 min prior to the designated labeling time. Then the maintenance media was

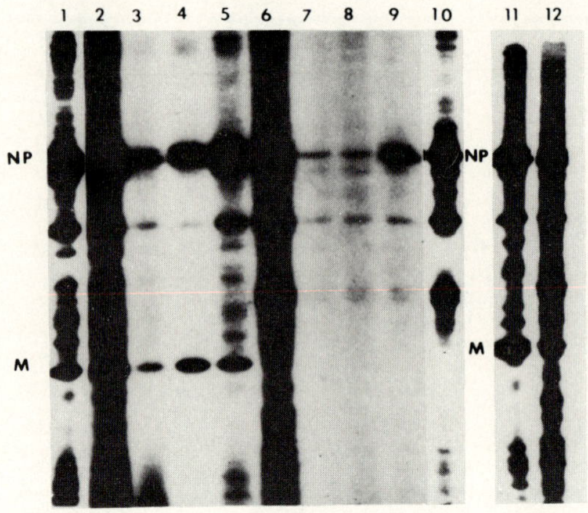

Figure 3. Gradient protein gel analysis of ^{35}S-methionine labeled, 48 hr pi media from virus infected cells. Virus-cell combinations: CR43-3 in PCK cells = lanes 1-5; CR43-3 in MDCK cells = lanes 6-10; CR31-10 in PCK cells = lane 11; and CR31-10 in MDCK cells = lane 12. Lanes 1, 2, 6, 10, 11, and 12 are unfractionated samples prior to gradient analysis, while lanes 3 and 7, 4 and 8, and 5 and 9 are pools I, II and III respectively. Lanes 1 and 10 are lighter exposures of lanes 2 and 6 respectively.

removed and 2.0 ml of 0.15M NaCl was added to each flask. This solution was removed after 15 min and 1.5 ml of methionine-free 2x Eagle's media containing 100 µCi/ml of ^{35}S-methionine was added to the cells and incubated at 33C for 30 min. The radioactive media was removed, cells washed with phosphate buffered saline, and lysed into Laemmli electrophoresis sample buffer (9). Samples were boiled for 5 min prior to electrophoresis. Figure 4a shows the time course of CR43-3 viral protein production in both PCK and MDCK cells. There is considerably less M protein produced intracellularly in MDCK cells than in PCK cells. Figure 4b shows the proteins produced by CR31-10 in both PCK and MDCK cells. Spectrophotometric tracings of the 8 hr pi gel lanes revealed that the ratio of M to NS1 in MDCK cells was only 34% compared to that ratio in PCK cells, and compared to the ratio of M to NS1 protein in CR31-10 infected MDCK cells it was 38%. The M protein deficiency is the only obvious protein anomaly apparent between the permissive and non-permissive host cell systems. NS2 protein is made in larger quantities in CR43-3 virus infected PCK and MDCK cell lines than it is in either of the CR31-10 virus infected cell lines, especially in the MDCK cell lines, but how or whether

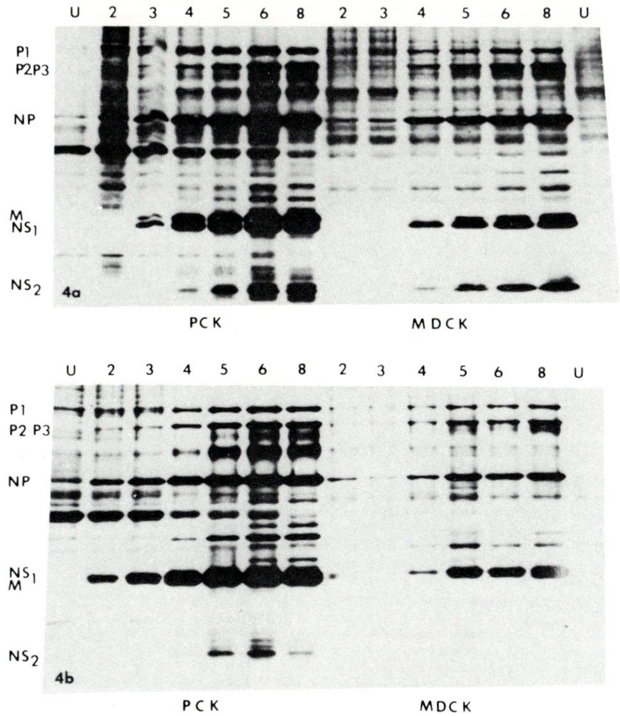

Figure 4. Time course of viral protein production. U = uninfected control. Numbers 2 through 8 represent the time pi when a 30 min pulse label was begun. a) = CR43-3 virus; b) = CR31-10 virus.

this bears upon the restricted host range phenotype is unknown.

SUMMARY

Growth curve analysis of the CR43-3 virus and its congenic virus CR31-10, showed that very little infectious virus or non-infectious, hemagglutinin-containing virus particles are released from MDCK cells infected with CR43-3 virus. Equilibrium sucrose gradient analysis of radioactively labeled pelleted material from 48 hr pi media from CR43-3 virus infected MDCK cells showed little virus particles produced. The viral proteins found in unfractionated pelleted material probably represent unassembled viral proteins released from lysed cells along with cellular membrane fragments. The M protein was conspicuous by its absence. Intracellular viral protein analysis also showed a decrease in the

M protein. In the nonpermissive system the amount of M protein produced in relation to the NS1 protein was two-thirds less than in permissive systems. At nonpermissive temperatures, viruses with a ts mutation in the NS gene also produced reduced quantities of M protein (11,12,13,14). The loss of M protein production may be symptomatic or causal in its relationship to the restricted host range phenotype.

ACKNOWLEDGEMENTS

We would like to thank Curt Smitka and Louise Allen for their expert technical assistance, and Dr. Mark Krystal for his helpful discussions.

REFERENCES

1. Maassab, H.F., Smitka, C.W., Donabedian, A.M., Monto, A. S., Cox, N.J., and Kendal, A.P. (1981). In "The Replication of Negative Strand Viruses" (D.H.L. Bishop and R.W. Compans, eds.), p. 395. Elsevier/North Holland, NY.
2. Maassab, H.F., and DeBorde, D.C. (1983). Virology, in press.
3. Buonagurio, D.A., Krystal, M., Palese, P., DeBorde, D.C., and Maassab, H.F. (1983) Submitted to J. Virol.
4. Cox, N.J., Maassab, H.F., and Kendal, A.P. (1979). Virology 97, 190.
5. Odagiri, T., DeBorde, D.C., and Maassab, H.F., (1982). Virology 119, 82.
6. Maassab, H.F. (1969). J. Immunol. 102, 728.
7. Reed, L.J., and Muench, H. (1938). Am. J. Hyg. 27,493.
8. Peacock, A.C., and Dingham, C.W. (1968). Biochemistry 7, 668.
9. Laemmli, U.K. (1970). Nature 227,600.
10. Smith, G.L., and Hay, R.J. (1982). Virology 118,96.
11. Almond, J.W., McGeoch, D., and Barry, R.D. (1977). Virology 81,62.
12. Almond, J.W., McGeoch, D., and Barry, R.D. (1979). Virology 92,416.
13. Wolstenholme, A.J., Barret, T., Nichol, S.T., and Mahy, B.W.J. (1980). J. Virol. 35,1.
14. Koennecke, I., Boschek, C.B., and Scholtissek, C. (1981). Virology 110,16.

SELF-INACTIVATION OF INFLUENZA A VIRUS AND AUTOLYSIS OF THE VIRUS-INFECTED CELLS

Kazufumi Shimizu,[1] Katsutoshi Aihara,[2] and Tsutomu Miyamoto

Department of Bacteriology
Nagasaki University School of Medicine
Nagasaki 852, Japan

SELF-INACTIVATION OF INFLUENZA A VIRUS AT LOW pH

To penetrate into cell, enveloped virus has two membrane barriers to break through, i.e., membrane of virus itself and membrane of cell. Fusion of virus membrane with cell membrane is a mechanism to bring topologically the viral nucleocapsid-core inside the cell membrane. Small and localized dissociation or perturbation on apposed membranes at a closely contacted area leads to fusion of the two membranes (1). The dissociation or perturbation would need to be confined to highly localized sites so that the integrity of the membranes would not be threatened. In this consequence, activity of virus to induce the initial membrane perturbation should be under restrictive control in its activation and inactivation also.

Membrane fusion and hemolytic activities of influenza virus have recently been recognized (2-5). The hemagglutinin (HA) glycoprotein has a key role in fusion and hemolysis (HL), and cleavage of the molecule and low pH environment are required for the activation (2-14).

Here we report self-inactivation mechanism of fusion and HL activities of influenza virus.

<u>pH Dependency of HL</u>. The profile of pH dependency of HL by influenza A/Udorn/72 (H3N2) virus was similar to those reported previously for other strains of Influenza A virus (3,4,6,14). The virus caused maximum HL at pH 5.3, but none at pH 7.2.

Present address: [1]Department of Microbiology Nihon University School of Medicine, Oyaguchi-Kami Machi, Itabashi, Tokyo 173, Japan; and [2]The Chemo-Sero-Therapeutic Research Institute, Kumamoto 860, Japan.

Copyright © 1984 by Academic Press, Inc.
All rights of reproduction in any form reserved.
ISBN 0-12-183501-4

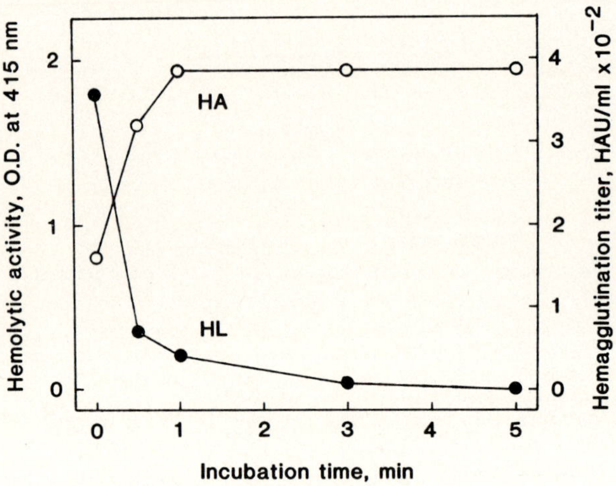

FIGURE 1. Decrease of HL activity of A/Udorn/72 virus during incubation at 37°C in ABS, pH 5.3.

Rapid decrease of HL Activity at Low pH. When virus was exposed to pH 5.3 at 0°C, massive aggregates of virions were observed in the electron microscope as described by others (15,16), and HA titer decreased to 1/10-1/100 probably due to the aggregation but HL activity to approximately 1/2. The remained HL activity rapidly diminished by subsequent incubation at 37°C in acetate buffered saline (ABS), pH 5.3, while HA titer slightly increased indicating that no further aggregation occurred during the incubation (Fig. 1). The inactivation at 37°C was irreversible, i.e., the change of pH from 5.3 to 7.2 did not restore the HL activity. No inactivation occurred by incubation at 37°C in phosphate buffered saline (PBS), pH 7.2.

The conclusion from the results is as follows. Decrease of the viral activities in low pH media at 0°C is only apparent. It is attributable to formation of virus-aggregates, whose HA molecule virtually retains HA and HL activities. In contrast, at 37°C HL activity of the molecule is inactivated substantially while HA activity is unchanged.

Prerequisite of HA Cleavage to the Low pH-induced Inactivation of Virus HL Activity. If virus having uncleaved HA (HA0) was incubated at 37°C in ABS, pH 5.3, no significant inactivation of HL activity occurred, i.e., the virus revealed full HL activity in the presence of trypsin which cleaves HA0 to HA1 and HA2 (TABLE 1). Such cleavage of HA has been

TABLE 1
PREREQUISITE OF CLEAVAGE OF HA TO THE LOW pH-INDUCED INACTIVATION OF HEMOLYTIC ACTIVITY

Virus[a] having	Preincubation (37°C 30 min) pH	Hemolytic activity (O.D. at 415 nm) Trypsin (25 μg/ml)	None
Uncleaved HA	5.3	0.895	0.167
	7.2	1.085	0.215
Cleaved HA	5.3	0.040	0.160
	7.2	0.720	1.045

[a]Virus grown in MDCK cells without trypsin was purified and used as virus having uncleaved HA (HA0). Virus having cleaved HA (HA1 + HA2) was prepared by treatment of the uncleaved HA-virus with trypsin.

TABLE 2
EFFECT OF LOW pH-INCUBATION ON VARIOUS VIRAL ACTIVITIES

Buffer	Incubation time (min)	Viral activities, %[a] HA	NA	HL	Inf.	RPase
ABS, pH 5.3	30	3	52	2	2	34
	0	3	67	41	85	82
PBS, pH 7.2	30	100	96	100	100	97
	0	100	100	100	100	100

[a]Percentage of each activity to the corresponding activity of virus incubated at 37°C for 0 min in PBS, pH 7.2.

shown to be essential for HL and membrane fusion (4,5,12,14). The present results demonstrated that the cleavage is also essential for low pH-induced inactivation of HL activity.

Concomitant Inactivation of HL Activity and Infectivity. Effects of the low pH-incubation on various viral activities were examined (TABLE 2). In this experiment, HA titer of virus suspended in ABS, pH 5.3, at 0°C was 3 % of that suspended in PBS, pH 7.2. After incubation at 37°C in ABS, pH 5.3, HL activity and infectivity greatly decreased, while HA, neuraminidase (NA), and RNA polymerase (RPase) activities were not changed as much. It was shown that incubation of virus at 37°C in low pH media causes selective inactivation of HL activity and infectivity, both of which are considered to be involved with membrane-fusion-inducing activity of HA.

As shown above, the requirements for inactiviation such as cleavage of HA, low pH, and high temperature were exactly the same as those for manifestation of membrane-fusion-inducing ability of HA except for no requirement of the presence of target membrane. This suggests that even in the absence of target membrane, if these three requirements are filled, HA abortively exerts its fusion-inducing mechanism to exhaust itself.

AUTO LYSIS OF VIRUS-INFECTED CELL

It is rational to assume that fusion-inducing mechanism of HA causes perturbation of virus membrane as well as that of cell membrane since both membranes should be ready for fusion. Influenza virus buds from the plasma membrane of the infected cell. Thus parts of the infected cell membrane consist of viral envelope. If such perturbation would take place at viral parts of the membrane, the infected cell-lysis might follow. Here we studied effects of activation of HA on plasma membrane of infected cells to demonstrate that HA causes perturbation on its own base-membrane.

HL Induced by Virus-Infected Cells. Expression of HA and NA activities on the surface of cells infected by A/Udorn/72 *ts* mutants (17,18) were determined at permissive (34°C) and nonpermissive (40°C) temperatures. As summarized in TABLE 3, ICRC282 was a NA-*ts* mutant, which expressed hemadsorption (HAD) but not NA activity at 40°C. On the contrary, NA233 was a HA-*ts* mutant, which expressed NA but not HAD activity at 40°C. The infected MDCK cells having HAD activity induced HL and cell fusion at 34°C or 40°C in ABS, pH 5.3, containing trypsin for HA cleavage. The presence or absence of NA activity did not give any significant effect on the extent of HL or cell fusion. This was considered to be another evidence for that HA has a key role in membrane fusion activity (2,4,5,9-14).

Lysis of Virus-infected Cells. As shown above, HA molecules on plasma membrane of the virus-infected cell were activated by cleavage and low pH and induced lysis of adsorb-

TABLE 3
EXPRESSION OF VIRAL ACTIVITIES ON THE SURFACE OF MDCK CELLS INFECTED BY A/UDORN/72 ts MUTANTS

Virus	Incubation temperature (°C)	Expression of viral activities[a]			
		NA	HAD	HL	Cell Fusion
Wild-type	34	+	+	+	+
	40	+	+	+	+
ICRC282	34	+	+	+	+
	40	-	+	+	+
NA233	34	+	+	+	+
	40	+	-	-	-

[a]+, expressed. -, not expressed.

ed chicken red blood cells (CRBC). In addition, it was indicated by the experiments with radioactive Cr-labeled cells that lysis of the infected cell itself also occurred at the same time (TABLE 4). Then the infected cell-lysis without adsorption of CRBC was examined. Lysis of the infected cell alone occurred by incubation at 40°C in ABS, pH 5.3, at the same extent as with CRBC adsorption (TABLE 4). NA activity was not required for the lysis. The lysis did not occur at pH 7.2 or if HA was not cleaved into HA1 and HA2. These results suggested that activated HA acts on its own base-membrane. To exclude the possibility that released virions from the infected cells might attach to the cells again and induce the lysis, effect of bacterial neuraminidase treatment of infected cells on the lysis was examined and the results were summarized in TABLE 5. The neuraminidase treatment was effective to destroy receptors for HA binding so that lysis of cells induced by virions from without was drastically inhibited by the treatment. The infected cell lysis still occurred after the identical neuraminidase treatment without any suppressive effect. This indicated that released virions did not contribute to the infected cell lysis and that HA induced the lysis without binding to the receptor containing sialic acid.

It was concluded that HA molecules on plasma membrane of the infected cell are activated by exposure to weakly acidic environment at 37°C after cleavage of the molecules and induce perturbation on their own base-membrane leading to cell lysis.

TABLE 4
LOW pH-INDUCED LYSIS OF MDCK CELLS INFECTED WITH A/UDORN/72 VIRUS

	Expression of		Cell lysis[a] (^{51}Cr release, cpm)	
Virus	HA	NA	With CRBC	Without CRBC
Wild-type	+	+	1582	1411
ICRC282	+	-	1633	1708
NA233	-	+	62	38

[a]MDCK cells prelabeled with radioactive Cr (10,000 cpm/dish, incorporated) were infected with each virus and incubated at 40°C for 5 hr in MEM containing trypsin. After adsorption of CRBC, ABS, pH 5.3, was added and cells were incubated at 40°C for 30 min. Then radioactive Cr, released in the fluid, was counted.

TABLE 5
EFFECT OF NEURAMINIDASE TREATMENT ON LOW pH-INDUCED LYSIS OF MDCK CELLS INFECTED WITH A/UDORN/72 VIRUS

		Cell lysis (^{51}Cr release, cpm)	
Lysis from	Virus	Pretreated with neuraminidase	None
Within[a]	Wild-type	2465	2388
	ICRC282	1940	2167
Without[b]	Wild-type	127	5038

[a]Infected cells were treated with Arthobacter ureafacines neuraminidase and autolysis of the cells were determined.

[b]None infected cells were treated with neuraminidase and lysis by adsorbed virions was determined.

A PLAUSIBLE FUNCTION OF HA IN MEMBRANE FUSION

The present studies demonstrated that an HA-function inducing perturbation on the membrane into which HA is

Cleaved HA

↓

Activation and instant inactivation
of a function
by exposure to weakly acidic environment at 37°C

↓

Perturbation on viral membrane

↓

Perturbation on apposed cellular membrane

↓

Membrane fusion or cell lysis

FIGURE 2. Schematic representation of the sequence of reaction mediated by influenza virus HA.

anchored via its C-terminus (19,20) is activated by exposure to low pH at 37°C after its cleavage into HA1 and HA2 and that after activation it is rapidly self-inactivated. The simultaneous occurrence of activation and inactivation suggested that the function only works once. The involvement of hydrohobic peptide at the N-terminus of HA2 in formation of an active site of HA for inducing membrane fusion has been suggested (15, 21). The C-terminus of HA2 is also required for membrane fusion (12,16), on which fatty acids are covalently linked (22). Considering those findings, we are tempted to assume that the activated HA has esterase-like activity to release the fatty acid from HA itself. The released fatty acid may be a membrane-perturbing molecule proposed by Lucy (23) and responsible for the autolysis of virus-infected cells. Membrane fusion or HL may be the secondary reaction leaded by the perturbation on viral membrane since HA free from viral membrane is not able to induce membrane fusion or HL (12,16) while the perturbation on viral membrane occurs without binding of HA to target membrane. Globular lipid micelles probably formed on viral membrane as a result of incorporation of the perturbing molecules might turn into a perturbing agent to the target membrane like lipid emulsions containing lysolecithin (24). The proposed sequence of the reaction mediated by HA is schematically represented in FIGURE 2.

REFERENCES

1. Hosaka, Y., and Shimizu, K. (1977). In "Virus Infection and the Cell Surface" (G. Poste and G. L. Nicolson, eds.), p. 129. Elsevier/North-Holland.
2. Huang, R. T. C., Wahn, K., Klenk, H. -D., and Rott. (1980). Virology 104, 294.
3. Maeda, T., and Ohnishi, S. (1980). FEBS Lett. 122, 283
4. Huang, R. T. C., Rott, R., and Klenk, H. -D. (1981). Virology 110, 243.
5. White, J., Matlin, K., and Helenius, A. (1981). J. Cell Biol. 89, 674.
6. Lenard, J., and Miller, D. K. (1981). Virology 110, 479.
7. Klenk, H. -D., Rott, R., Orlich, M., and Blodorn, J. (1975). Virology 68, 426.
8. Lazarowitz, S. G., and Choppin, P. W. (1975). Virology 68, 426.
9. Shibata, M., Maeno, K., Tsurumi, T., Aoki, H., Nishiyama, Y., Ito, Y., Isomura, S., and Suzuki, S. (1982). J. Gen. Virol. 59, 183.
10. Hosaka, Y., Seriburi, O., Moran, M. G., Yasuda, Fukai, K. and Nerome, K. (1982). Biken J. 25, 51.
11. Lenard, J., Bailey, C. A., and Miller, D. K. (1982). J. Gen. Virol. 62, 353.
12. White, J., Helenius, A., and Gething, M. -J. (1982). Nature 300, 658.
13. Hosaka, Y., Yasuda, Y., and Fukai, K. (1983). J. Virol. 46, 1014.
14. Maeda, T., Kawasaki, K., and Ohnish, S. (1981). Proc. Natl. Acad. Sci. USA 78, 4133.
15. Skehel, J. J., Bayley, P. M., Brown, E. B., Martin, S. R., Waterfield, M. D., White, J. M., Wilson, I. A., Wiley, D. C. (1982). Proc. Natl. Acad. Sci. USA 79, 968.
16. Sato, S. B., Kawasaki, K., and Ohnishi, S. (1983). Proc. Natl. Acad. Sci. USA 80 3153.
17. Shimizu, K., Mullinix, M. G., Chanock, R. M., and Murphy, B. R. (1982). Virology 117, 38.
18. Shimizu, K., Mullinix, M. G., Chanock, R. M., and Murphy, B. R. (1982). Virology 117, 45.
19. Skehel, J. J., and Waterfield, M. D. (1975). Proc. Natl. Acad. Sci. USA 72, 93.
20. Gething, M. -J., and Sambrook, J. (1982). Nature 300, 598.
21. Richardson, C. D., Scheid, A., and Choppin, P. W. (1980). Virology 105, 205.
22. Schmidt, M. F. G. (1982). Virology 116, 327.
23. Lucy, J. A. (1970). Nature 227, 815.
24. Ahkong, Q. F., Cramp, F. C., Fisher, D., Howell, J. I., and Lucy, J. A. (1972). J. Cell Sci. 10, 769.

USE OF AVIAN-HUMAN REASSORTANT INFLUENZA A VIRUSES AS LIVE VACCINE VIRUSES IN MAN

Brian R. Murphy[1], Alicia J. Buckler-White[1], Shu-fang Tian[1],
Mary Lou Clements[2], William T. London[3], Robert M. Chanock[1]

[1]Laboratory of Infectious Diseases
National Institute of Allergy and Infectious Diseases
National Institutes of Health
Bethesda, Maryland 20205

[2]Center for Vaccine Development
Department of Medicine,
University of Maryland, School of Medicine
Baltimore, Maryland.

[3]National Institute of Neurological and
Communicative Disorders and Stroke
National Institutes of Health
Bethesda, Maryland 20205

There is renewed interest in the development of a live attenuated influenza A virus vaccine since current inactivated influenza virus vaccines do not provide complete protection (1) and do not appear to retain their effectiveness when administered annually (2). Live attenuated influenza A viruses have been produced by transfer of genes from an attenuated donor virus to new epidemic influenza A viruses (3). Since resistance to influenza A virus is mediated by the development of an immune response to the hemagglutinin (HA) and neuraminidase (NA) glycoproteins (4, 5), live, attenuated reassortant vaccine strains were selected in which genes for these surface antigens were derived from the epidemic virus, while the non-surface protein ("internal") genes were derived from the attenuated parent. This process of gene exchange is readily achieved with influenza A viruses since they possess a segmented genome consisting of eight negative-stranded RNA segments that code for at least ten proteins (6-8). Although Although genes bearing temperature-sensitive mutations have been transferred to a series of new epidemic wild-type viruses and have rendered such viruses satisfactorily attenuated for man, the genetic instability of the attenuated phenotype represents an insoluble problem (3, 9, 10).

Copyright © 1984 by Academic Press, Inc.
All rights of reproduction in any form reserved.
ISBN 0-12-183501-4

There is a need for stable attenuated viruses that are unable to escape their attenuated phenotype. Many of the influenza A virus genes that have evolved over a long period in birds differ significantly in nucleotide sequence from corresponding genes of human influenza A viruses (11-15). Because of these marked differences, we would expect some avian influenza viruses to replicate inefficiently in human cells and thereby be attenuated. We would also expect such attenuated avian viruses to retain their attenuated characteristics after limited replication in man.

This concept was evaluated initially in squirrel monkeys because this primate develops illness after experimental infection with influenza A viruses (16). Ten avian influenza A viruses were evaluated in squirrel monkeys, and a spectrum of virulence was observed (17) (Table 1). Some avian influenza viruses replicated almost as well as the human influenza A virus. However, several avian influenza A viruses grow at least 1,000 times less efficiently than human influenza A viruses. We initiated studies with these latter avian influenza A viruses with the intent of producing avian-human reassortant viruses that are attenuated for man and that could be used as live virus vaccine strains.

TABLE 1

EFFICIENCY OF PLAQUE FORMATION OF AVIAN AND HUMAN INFLUENZA A VIRUSES IN TISSUE CULTURE AND THEIR LEVEL OF OF REPLICATION IN SQUIRREL MONKEYS

Influenza A Virus*	Antigenic Subtype	Reduction of Plaque Formation+ at 42°C Compared to 37°C (Log_{10})	Mean Log_{10} Titer of Virus ($TCID_{50}$/ml) in Tracheal Lavage Fluid
Avian			
MAL/573/78	H1N1	0.2	1.5
MAL/6750/78	H2N2	0.0	2.4
PIN/286/78	H4N8	0.0	2.4
PIN/358/79	H3N6	0.0	2.8
MAL/827/78	H8N4	0.6	2.9
PIN/119/79	H4N6	−0.1	3.0
TUR/5/79	H10N7	−0.1	3.1
MAL/88/76	H3N8	0.2	4.3
MAL/6874/78	H3N2	0.0	4.4
PIN/121/79	H7N8	0.0	5.0
Human			
Udorn/307/72	H3N2	>4.6	6.5
Wash/897/80	H3N2	>5.8	5.7

+ On MDCK or primary chicken kidney cell culture

* Each virus tested in at least four squirrel monkeys

Restriction of replication of the avian-human reassortant vaccine virus in the primate respiratory tract is effected by naturally occurring avian influenza virus genes rather than by mutant genes selected by limited passage of virus in an unnatural host.

One avian influenza A virus, A/Mallard/New York/6750/78 (H2N2), that was markedly restricted in replication in the trachea of squirrel monkeys, was evaluated as a donor of its non-surface protein genes for attenuation of virulent human influenza A viruses. Avian-human reassortant influenza viruses were produced by mating the avian influenza virus and a virulent human influenza virus in primary chick kidney culture at 37° and selecting progeny at 42° in the presence of antibodies to the surface antigens of the avian parent virus (18). As seen in Table 1, 42°C is restrictive for replication of human influenza A viruses. In matings involving three different virulent human influenza viruses (Figure 1), each reassortant virus isolated derived its surface antigen genes from its human influenza virus parent and its "internal" genes from its avian parent virus. Such reassortants will be referred to as "six-gene" reassortants. Like their avian influenza parent virus, each of the

Gene	Genotype of Avian-Human Influenza Reassortants from Mating of Avian^Δ Influenza Virus with Indicated Wild Type Human Influenza Virus^ϕ		
	A/Udorn/72(H3N2)	A/Washington/80(H3N2)	A/California/78(H1N1)
PB2	□	□	□
PB1	□	□	□
PA	□	□	□
HA	■	■	■
NA	■	■	■
NP	□	□	□
M	□	□	□
NS	□	□	□

Δ Avian virus was A/Mallard/New York/6750/78(H2N2)

■ = Gene derived from human wild type virus

□ = Gene derived from avian virus

ϕ Progeny from indicated mating selected in the presence of antisera to avian influenza virus surface antigens at 42°C, a temperature restrictive for human influenza virus.

Figure 1. Consistent transfer of 6 "internal" avian influenza virus genes to avian-human influenza virus reassortants with H1N1 or H3N2 surface antigens.

TABLE 2
RESTRICTION OF AVIAN-HUMAN INFLUENZA REASSORTANT VIRUSES IN SQUIRREL MONKEYS

Influenza A Virus	Number of Monkeys	Virus Replication in Trachea: Average Duration of Virus Shedding (Days)	Virus Replication in Trachea: Mean Peak Titer (Log_{10} $TCID_{50}$/ml) of Tracheal Lavage Fluid
Avian			
A/Mallard/6750/78	7	2.6^{ϕ}	2.7^{ϕ}
Reassortant‡			
A/Udorn/72	6	2.0^{ϕ}	2.2^{ϕ}
A/Washington/80	8	2.3^{ϕ}	2.7^{ϕ}
Human Wild Type			
A/Udorn/72	15	5.2	5.9
A/Washington/80	14	5.6	5.7

NOTE: Monkeys received $10^{7.0}$ $TCID_{50}$ of virus (0.5 ml intratracheally) and were infected in each instance.

ϕ Avian virus and reassortants significantly restricted in growth compared to wild type human influenza viruses

‡ Each reassortant contains 6 "internal" avian influenza virus genes.

avian-human reassortant influenza viruses produced plaques efficiently at 42°C indicating that one or more of the avian influenza genes that code for non-surface proteins specify growth at 42°C. The level of replication of two avian-human reassortant influenza viruses in the lower respiratory tract of the squirrel monkey was compared to that of their parental viruses (Table 2). The two avian-human reassortant influenza viruses were as restricted in growth in the monkey's trachea as their avian influenza parent viruses. In each instance the reassortant viruses were shed in lower titer and for a shorter duration than the human influenza virus parent. These findings indicate that restriction of replication of the avian influenza virus is a function of one or more of its "internal" genes. To investigate which of the avian genes was responsible for restricted repliction in primates, reassortant viruses were produced that contained human influenza virus surface antigens from the A/Udorn/72 (H3N2) virus and one or more of the internal genes derived from the avian influenza virus parent (Figure 2). Avian-human reassortant viruses that contained only an RNA 1, RNA 3, or

Influenza Virus$^{\Delta}$	Parental Origin of Genes in Avian-Human Influenza Reassortant Viruses								Virus Replication in Trachea ‡	
	RNA1	RNA2	RNA3	HA	NA	NP	M	NS	Average Duration of Virus Shedding (Days)	Mean Peak Titer (Log_{10} $TCID_{50}$/ml) of Tracheal Lavage Fluid
Human	□	□	□	□	□	□	□	□	5.3	5.1
Avian-Human Reassortant										
(a) 6 "internal" avian virus genes	■	■	■	□	□	■	■	■	0.3*	0.7*
(b) Single substitution of an avian virus gene	■	□	□	□	□	□	□	□	6.0	5.2
	□	□	■	□	□	□	□	□	5.0	5.4
	□	□	□	□	□	■	□	□	0.0*	0.5*
	□	□	□	□	□	□	■	□	3.3*, 2.0*φ	2.6*, 1.6*φ
	□	□	□	□	□	□	□	■	5.0	4.4

Δ Avian Virus was A/Mallard/New York/6750/78(H2N2). Human Virus was A/Udorn/307/72(H3N2).

‡ Each virus tested in at least four squirrel monkeys

* Statistically significant difference from wild type human influenza virus

φ Two independently derived reassortants with this gene constellation

□ = Gene derived from human virus

■ = Gene derived from avian virus

Figure 2. Effect of substitution of a single avian influenza virus gene on growth of human influenza A virus in monkeys.

Influenza Virus$^{\Delta}$	Parental Origin of Genes in Avian-Human Influenza Reassortant Viruses								Virus Replication in Trachea ‡	
	RNA1	RNA2	RNA3	HA	NA	NP	M	NS	Average Duration of Virus Shedding (Days)	Mean Peak Titer (Log_{10} $TCID_{50}$/ml) of Tracheal Lavage Fluid
Human	□	□	□	□	□	□	□	□	5.3	5.1
Avian-Human Reassortant										
(a) 6 "internal" avian virus genes	■	■	■	□	□	■	■	■	0.3*	0.7*
(b) 2 or 3 "internal" avian virus genes	■	□	□	□	□	□	□	■	3.5	3.2*
	□	□	■	□	□	□	□	■	6.0	5.1
	□	□	■	□	□	■	□	■	1.0*	0.9*
(c) 5 "internal" avian virus genes	■	■	□	□	□	■	■	■	2.0*	1.7*
	■	□	■	□	□	■	■	■	0.5*	0.8*
	□	■	■	□	□	■	■	■	2.0*	1.7*

Δ Avian Virus was A/Mallard/New York/6750/78(H2N2). Human Virus was A/Udorn/307/72(H3N2).

‡ Each was tested in at least four squirrel monkeys.

* Statistically significant difference from wild type human influenza virus

□ = Gene derived from human virus

■ = Gene derived from avian virus

Figure 3. Effect of various constellations of avian influenza virus genes on growth of avian-human influenza reassortant viruses in monkeys.

non-structural (NS) protein RNA segment of avian influenza origin did not exhibit restriction (i.e., they grew to the same level as their human influenza A/Udorn/72 parent). In contrast, avian-human reassortants that contained only the avian nucleoprotein (NP) or matrix (M) protein RNA segment were as restricted in their growth as their avian influenza parent. Avian-human reassortant influenza viruses containing two or more genes derived from their avian influenza parent virus were restricted in replication if they possessed an NP or M gene from the avian parent (Figure 3). In addition, the reassortant possessing the RNA 1 and NS genes from the avian parent manifested significant restriction of virus replication in the trachea. This suggests that specific genes, that by themselves do not restrict replication, can act together to effect a reduction in replication. However, the avian influenza NP and M genes appear to play the major roles in the host range restriction exhibited by the A/Mallard/78 (H2N2) virus and by reassortants containing its six "internal" genes. Since either of these genes is able to effect restriction by itself, restoration of virulence of an avian-human reassortant for man would require appropriate genetic changes in both genes or suppression of both genes. The extent of the nucleotide differences between the avian NP

TABLE 3
INFECTION WITH AVIAN-HUMAN INFLUENZA REASSORTANT VIRUS INDUCES RESISTANCE TO CHALLENGE WITH WILD TYPE HUMAN INFLUENZA VIRUS PARENT

Monkeys Administered 10^7 $TCID_{50}$ of Indicated Influenza Virus Intratracheally Four Weeks Pre-Challenge	Number of Monkeys	Virus Replication in Trachea after Intratracheal Administration $10^{7.0}$ $TCID_{50}$ of Wild Type Human Influenza Virus	
		Average Duration of Virus Shedding (Days)	Mean Peak Titer (Log_{10} $TCID_{50}$/ml) of Tracheal Lavage Fluid
Human Wild Type Virus^Δ	6	0	≤0.5
Avian-Human Reassortant	5	0.4	1.0
Avian Virus#	2	6.0	3.8
Placebo	6	5.3	5.8

Δ Human virus was A/Washington/897/80(H3N2)

Avian virus was A/Mallard/New York/6750/78(H2N2)

and M genes and the corresponding human genes are currently under investigation.

Significant resistance to challenge with wild type human influenza virus was observed in monkeys previously infected with the avian-human reassortant influenza viruses (Table 3). Thus, despite restricted replication of the avian-human reassortant influenza viruses, significant resistance was induced to virulent virus.

The avian-human A/Washington/80 reassortant influenza virus was examined for stability of phenotype, transmissibility, and safety in squirrel monkeys. This virus was passaged alternately 5 times in monkeys and tissue culture. Alternate passages in tissue were required because the amount of virus recovered from infected monkeys was not sufficient to infect other monkeys. Following this alternate passages series the reassortant retained its restriction of growth in the monkey's respiratory tract. Furthermore, the virus was not transmitted to susceptible monkeys housed in the same cage.

Since avian influenza A viruses can cause enteric and systemic infections in their natural hosts, it was important

Gene	Genotype of Reassortant Virus Produced by Mating the Indicated Pair of Viruses$^{\phi}$	
	A/Mallard/Alberta/573/78 × A/Udorn/307/72	A/Pintail/Alberta/121/79 × A/Udorn/307/72
PB2	□	□
PB1	□	□
PA	□	□
HA	■	■
NA	■	■
NP	□	□
M	□	□
NS	□	□

□ = Gene derived from avian influenza virus; i.e., from the Mallard or Pintail virus

■ = Gene derived from human (A/Udorn/72) influenza virus

ϕ Progeny from indicated mating selected in the presence of antisera to avian influenza virus surface antigens at 42°C, a temperature restrictive for human influenza virus.

Figure 4. Genotype of avian-human influenza reassortant viruses.

that we define the pathogenesis of an avian-human reassortant virus in primates. Monkeys were sacrificed on days 2, 4, and 8 after administration of the avian-human A/Washington/80 influenza reassortant, and twenty-six tissues were removed and assayed for virus. Virus was recovered only from the respiratory tract of the monkeys, indicating that the transfer of the six "internal" RNA segments did not lead to systemic spread or enterotropism. Instead, these "internal" avian influenza genes imposed a reduction in the level of virus replication in primate respiratory tissue.

Two additional avian-human reassortant viruses were produced by mating the A/Mallard/Alberta/573/78 (H1N1) or A/Pintail/Alberta/121/79 (H7N8) avian virus with the human influenza A/Udorn/72 (H3N2) virus. Reassortant viruses containing the HA and NA genes from the human virus and the other six RNA segments from the avian parent (Figure 4) were evaluated for their level of replication in the upper and lower respiratory tract of squirrel monkeys. Each reassortant virus was as restricted in the lower respiratory tract as its avian influenza parent (Table 4). Thus, with three successive avian influenza viruses tested, the "internal" genes of the avian parent virus constituted the primary determinants of restriction of replication in squirrel monkey respiratory epithelium. By evaluating a

TABLE 4
DECREASED REPLICATION OF AVIAN-HUMAN INFLUENZA REASSORTANT VIRUSES IN SQUIRREL MONKEYS

Influenza Virus	Virus Replication in Trachea*	
	Average Duration of of Virus Shedding (Days)	Mean Peak Titer (Log_{10} $TCID_{50}$/ml) of Tracheal Lavage Fluid
Avian A/Mallard/78	2.7^{ϕ}	2.3^{ϕ}
A/Mallard/78 × A/Udorn/72 Reassortant	3.3^{ϕ}	3.4^{ϕ}
Avian A/Pintail/78	4.7^{ϕ}	5.1^{ϕ}
A/Pintail/78 × A/Udorn/72 Reassortant	4.0^{ϕ}	4.8^{ϕ}
Human A/Udorn/72	6.0	6.3

NOTE: Avian viruses were A/Mallard/Alberta/573/78(H1N1) and A/Pintail/Alberta/121/79(H7N8)

* Each virus tested in six squirrel monkeys

ϕ Significantly reduced compared to human influenza A/Udorn/72(H3N2) virus

TABLE 5
RESPONSE OF SERONEGATIVE ADULTS TO INFLUENZA A/WASHINGTON/80 (H3N2) AVIAN-HUMAN REASSORTANT OR WILD-TYPE VIRUS

Influenza A Virus Administered	Dose of Virus ($TCID_{50}$)	Number In Group	Percent Infected	Virus Shedding (Nasal Wash)		Percent with Immunological Response		Percent with Indicated Illness	
						Serum	Nasal Wash		
				Percent Shedding	Peak Mean* log_{10} Titer ($TCID_{50/ml}$)	HI, NI, and/or IgG ELISA Antibody	ELISA IgA Hemagglutinin Antibody	Febrile, Systemic or Both	Any Illness+
Avian-Human Reassortant	$10^{5.0}$	12	17	17	1.0	8	8	0	0
	$10^{6.0}$	13	67	8	0.6	46	25	0	0
	$10^{7.0}$	19	84	11	0.6	74	37	0	0
	$10^{7.5}$	20	80	10	0.5	79	47	0	0
	$10^{8.0}$	19	100	49	0.7	89	74	11▲	11
Wild Type	$10^{6.0}$	24	96	88	3.6	91	93	38	46

* Data from each infected volunteer were used for calculations.

\+ This category also includes upper respiratory tract illnesses.

▲ Afebrile illness of less than 24 hours duration.

NOTES: Transmission of reassortant virus to six contact controls was not observed. Virus was not recovered from blood or rectal swabs.

HI = hemagglutination-inhibiting; NI = neuraminidase inhibiting; and ELISA = enzyme linked immunosorbant assay.

large series of avian influenza viruses that differ in their level of replication in primates, it should be possible to identify one whose "internal" genes reproducibly confer an acceptable balance of attenuation and immunogenicity on avian-human reassortants.

Considered together these findings have implications for the production of influenza A vaccine viruses that are attenuated for man. A preliminary evaluation of the "six-gene" avian-human A/Washington/80 influenza reassortant virus in man indicates that it is attenuated and immunogenic but not transmissible (Table 5). The reassortant virus grows poorly in the human respiratory tract which is consistent with its satisfactory level of attenuation. The human infectious dose 50 of the avian-human reassortant is similar to that of an influenza A/Washington/80 cold-adapted reassortant virus. Neither viremia nor growth of reassortant virus in the intestinal tract was detected.

These preliminary trials are encouraging because they demonstrate that avian-human reassortant viruses can be safely given to man and that these reassortants are attenuated by virtue of their reduced replication in the respiratory tract. We are currently evaluating a series of avian-human reassortant influenza viruses derived from different avian influenza A donor viruses. By evaluating these reassortants in primates and man, it should be possible to identify an avian donor influenza virus whose "internal" genes specify an acceptable balance between attenuation and immunogenicity. By evaluating a series of single gene reassortants (ie, one avian-seven human genes) in man, it should be possible to identify avian influenza genes that confer a satisfactory balance between attenuation and immunogenicity.

REFERENCES

1. Meyer, H. M. Jr., Hopps, H. E., Parkman, P. D., Ennis, F. A. (1978). Am. J. Clin. Pathol. 70, 146.
2. Hoskins, T. W., Davies, J. R., Smith, A. J., Miller, C. L., Allchin, A. (1979). Lancet i, 33.
3. Chanock R. M., and Murphy, B. R. (1980). Rev. Infect. Dis. 2, 421.
4. Murphy, B. R., Kasel, J. A., Chanock, R. M. (1972). N. Engl. J. Med. 286, 1329.
5. Virelizier, J.-L. (1975). Immunol. 115, 434.
6. Palese, P., Sculman, J. L., Ritchey, M. B. (1978). Perspect. Virology 10, 57.
7. Lamb, R. A., Etkind, P. R., Choppin, P. W. (1978). Virology 91, 60.

8. Lamb R. A., and Choppin, P. W. (1981). Virology 112, 729.
9. Tolpin, M. D., Clements, M. L., Levine, M. M., Black, R. E., Saah, A. J., Anthony, W. C., Cisneros, L., Chanock, R. M., and Murphy, B. R. (1982). Infect. Immun. 36, 645.
10. Tolpin, M. D., Massicot, J. G., Mullinix, M. G., Kim, H. W., Parrott, R. H., Chanock, R. M., and Murphy, B. R. (1981). Virology 112, 505.
11. Baez, M., Zazra, J. J., Elliott, R. M., Young, J. F., Palese, P. (1981). Virology 113, 397.
12. Rohde, W., Harms, E., Scholtissek, C. (1977). Virology 79, 393.
13. Scholtissek, C., Harms, E., Rohde, W., Orlich, M., Rott, R. (1976). Virology 74, 332.
14. Scholtissek, C., Rohde, W., von Hoyningen, V., Rott, R., (1978). Virology 87, 13.
15. Scholtissek C., and Von Hoyningen-Huene, V. (1980). Virology 102, 13.
16. Murphy, B. R., Sly, D. L., Hosier, N. T., London, W. T., Chanock, R. M. (1980). Infect. Immun. 28, 688.
17. Murphy, B. R., Hinshaw, V. S., Sly, D. L., London, W. T., Hosier, N. T., Wood, F. T., Webster, R. G., Chanock, R. M. (1982). Infect. Immun. 37, 1119.
18. Murphy, B. R., Sly, D. L., Tierney, E. L., Hosier, N. T., Massicot, J. G., London, W. T., Chanock, R. M., Webster, R. G. and Hinshaw, V. S. (1982). Science 218, 1330.

Index

J

L